Technology, Adaptation,
and Exports

Technology, Adaptation, and Exports

HOW SOME DEVELOPING COUNTRIES GOT IT RIGHT

Vandana Chandra, Editor

THE WORLD BANK
Washington, DC

ISBN-10: 0-8213-6507-X
ISBN-13: 978-0-8213-6507-6
eISBN-10: 0-8213-6508-8
eISBN-13: 978-0-8213-6508-3
DOI: 10.1596/978-0-8213-6507-6

Library of Congress cataloging-in-publication data has been applied for.

Contents

BOXES

FIGURES

TABLES

Foreword

IN MOST DEVELOPING COUNTRIES, TECHNOLOGICAL ADAPTATION is indispensable for rapid economic growth, especially export-led growth. When developing-country producers, small and large, successfully adopt superior production technologies (adapting them as necessary), the economy can grow, with improvements in income generation and poverty reduction.

As global markets become more competitive, poor countries are finding it increasingly difficult to free themselves from dependence on primary products and to diversify into manufactured exports, whose added value translates into wealth. The difficulty may stem from a small market or poor business climate, but often it is due to products that fall short of the sophisticated standards required to do business in world markets. It is technology that allows producers to close that gap by closely following the lead of innovative competitors in the industrial countries.

But acquiring and adapting technology is not a simple matter. Individual firms often lack the incentive, the expertise, and the resources to set out in pursuit of the innovators. Often they have trouble gathering the information they need to identify opportunities. In such cases, governments in developing countries, working with their development partners, can make a significant contribution to growth by encouraging or enabling the private sector to seek, acquire, adapt, and deploy modern technologies of production while investing in the skilled workforce that is vital for technological adaptation and building the dedicated institutions that will foster the process of modernization.

This book studies 10 cases in which developing countries as diverse as Chile, India, and Uganda leveraged technological adaptation to achieve technological parity with their competitors in world markets. We hope that the cases will become beacons for other countries. It is time for governments and their development partners to join together to use technology to underpin the growth necessary for successful poverty reduction strategies.

Danny M. Leipziger
Vice President and Head of Network
Poverty Reduction and Economic Management
The World Bank

Acknowledgments

THE WORLD BANK'S WORK PROGRAM ON how countries have used technological adaptation to achieve faster export growth was championed by Gobind Nankani, then vice president of the Bank's Poverty Reduction and Economic Management (PREM) Network, and Yaw Ansu, then director of PREM's Economic Policy Unit. The program was carried out under Yaw Ansu's guidance in 2004 and 2005. At the time, lackluster growth in many low-income countries, especially those of Sub-Saharan Africa, was reminding economists that although macroeconomic stabilization, liberalization, and privatization are necessary, they are not sufficient to propel growth in developing countries. I am indebted to Mr. Ansu for his intellectual leadership in helping me to initiate a study on a subject that is well known to be central to growth but is, even today, only reluctantly recognized by mainstream economists and policy analysts in the World Bank.

Many people have my gratitude for their contributions to this study. External reviewers, Professors Richard Nelson (Columbia University), Vernon Ruttan (University of Minnesota), and David Mowery (University of California, Berkeley), provided detailed comments and invaluable suggestions on the drafts. The late Professor Sanjaya Lall's comments helped me to focus the study on issues relevant to low-income Africa. Peer reviewers Carl Dahlman, William Maloney, Alfred Watkins, and Shahid Yusuf offered many constructive comments throughout the process and helped to transform the study into a forum for lively discussion inside and outside the World Bank.

Within the Bank, Danny Leipziger has supported the study since becoming PREM vice president. Vikram Nehru, now director, and Dana Weist, sector manager of PREM's Economic Policy Unit, supported the study and have been instrumental in disseminating its findings.

Their support brought the study to the world. Neil Macpherson, Derek Byerlee, Eija Pehu, and Willem Jansen of the Bank's Agriculture and Rural Development anchor graciously provided useful suggestions throughout the process. Neil Macpherson also organized financial support, provided in part by the U.K. Department for International Development.

Yaw Ansu, Gobind Nankani, and Alan Gelb are among the scholars who, in public discussions, have emphasized the importance of technological adaptation in promoting economic growth. It is a deep belief that I share with them. The authors of the case studies came together in several workshops in 2004 and 2005 to discuss with outside speakers and other participants how governments in developing countries have leveraged technological adaptation to propel faster growth.

Three workshops and the 2005 PREM conference sessions provided me with many constructive ideas and lessons on the subject of technological catch-up in developing countries. I thank the speakers, Phillip Griffith (Millenium Science Initiative, Institute for Advanced Study, Princeton) and Alice Amsden (Massachusetts Institute of Technology), for sharing their views at the workshops on technological catch-up in developing countries. This study also benefited from the contributions of Jose Luis Guasch, Ijaz Nabi, John Page, Roberto Zagha, Warrick Smith, Sushma Ganguly, Ian Johnson, Jim Adams, Benno Ndulu, Eric Talroth, and the participants across the various networks of the World Bank. Their insights, comments, and suggestions proved valuable in making the workshops a success and in enriching this study.

Finally, I thank Steven Kennedy for editing the book.

Contributors

José Miguel Benavente is a professor at the Department of Economics, University of Chile. He has been working on research and development and innovation at the Senate Finance Commission since 2004 and helped to prepare the New Chilean Strategy Plan for Innovation during March 2006 for the Innovation Commission for Competitiveness. He was adviser to the governments of Chile and El Salvador, and he was a consultant at the Central Bank of Chile, the World Bank, the Inter-American Development Bank, and the United Nations Commission for Latin America. He gratefully acknowledges J. Katz, C. Sanhueza, J. Olivari, E. Bordeu, K. Schroder, F. Almeda, and R. Echeverría for useful comments.

Subhash Bhatnagar is an adjunct professor at the Indian Institute of Management (IIM), Ahmedabad, in India. He held several positions at IIM in the past including chaired professor, board member, and dean. He also is a consultant adviser to mainstream e-government at the World Bank in Washington, DC. He has published eight books and several papers on information communication technologies and development. He thanks V. Gupta and A. Rathore of IIM, Ahmedabad, for research support.

Vandana Chandra is an economist in the World Bank's Economic Policy and Debt Department of the Poverty Reduction and Economic Management Network. She has worked on the sources of growth and employment in several African countries. Her current areas of research include technological change, export diversification, competitiveness, and tertiary education in the context of economic growth.

Jorge Katz is acting director of the Division of Production, Productivity, and Management, ECLAC, Santiago, Chile. He teaches technology and innovation at the University of Chile and has published extensively on technology and industrial restructuring in Latin America and on the structure and behavior of the health sector. He would like to acknowledge valuable comments from J. M. Benavente, L. Vidal, and S. Radic.

Shashi Kolavalli, formerly a development consultant, is now a senior research fellow at the International Food Policy Research Institute in Accra, Ghana. He previously worked at the Indian Institute of Management, Ahmedabad, India, and the Internal Crop Research Institute for the Semi-Arid Tropics, Hyderabad, India. His current research interests are development strategies and governance.

Rose Kiggundu is a researcher working on innovation in livestock agriculture at the International Livestock Research Institute, Ethiopia. She appreciates the support of the Department of Fisheries Resources in Uganda and of fish-processing and exporting firms in the country for making it possible to begin this research in 2002 and for providing additional information in 2004.

John A. Mathews holds the chair of strategic management at the Macquarie Graduate School of Management, Macquarie University, Sydney, Australia. His research covers global management and development, with a focus on East Asia. His recent books include *Tiger Technology* (2000); *Dragon Multinational* (2002), and in 2006, *Strategizing, Disequilibrium and Profit*. He gratefully acknowledges the assistance of Dr. Mei-Chi Hu and E. Allrot, and dedicates his paper to the late Professor Sanjaya Lall, who was its inspiration.

Gopal Naik is a professor of economics and social sciences at the Indian Institute of Management, Bangalore, in India. He is associated closely with the Center for Public Policy and works in the areas of agricultural policy and agribusiness. He would like to thank S. Deshpande, C. Reddy, and P. Babu for their assistance.

Rajah Rasiah is a professor of technology and innovation policy at the University of Malaya. He specializes in foreign direct investment, learning and innovation, and competitiveness. He has field experience in Latin America and the United States; South and East Asia; the Netherlands and Germany; and Southern and East Africa. His palm oil paper benefited considerably from discussions with J. Gopal and data from R. Felix.

Meri Whitaker, formerly a development consultant, is now project codirector of the Virgin Islands Experimental Program to Stimulate Competitive Research in the U.S. Virgin Islands. She worked previously at the International Crops Research Institute for the Semi-Arid Tropics, Hyderabad, India, and the American University in Cairo, Egypt.

CHAPTER 1

Technology, Adaptation, and Exports—How Some Developing Countries Got It Right

Vandana Chandra and Shashi Kolavalli

THE HISTORY OF ECONOMIC DEVELOPMENT IS replete with cases in which governments have tried unsuccessfully to promote the use of new technologies to accelerate growth.[1] The many reasons for failure include bad luck, poor governance, flawed policy design, and weak implementation. That said, developing-country policy makers must continue to look for innovative ways to increase economic growth, especially through exports. The challenges they face are daunting, not least because global competition is rising and taking new forms, driven by rapid technological change and the growth of global production networks in which sophisticated skills and capabilities are required at the entry level, even for manufactured goods as simple as T-shirts.

At the same time, it is easier than ever before to gain access to technology. Knowledge of new production techniques—often bundled with financial, management, and technical skills—is carried across national boundaries to be paired with natural resources and local factor markets in new global production centers. For many developing countries, especially low-income ones, easier access to technology offers renewed hope for improving competitiveness and growth. To realize that hope, however, countries must become able to apply the new technologies. In this task, the developing countries are not alone. The government of almost every country that has experienced significant, sustained growth has been involved in a major way in promoting technological development and in supporting the private sector in the quest to improve productivity.

This volume studies 10 cases of developing countries that successfully adapted new technologies to catch up with developed countries in a particular industry. They demonstrate that latecomers can leapfrog across several stages of technological development. In every case,

Table 1.1 Ten Cases of Technological Adaptation for Export Competitiveness

Sector	Industry	Country
Agriculture	Maize (corn)	India
	Grapes	India
	Floriculture	Kenya
Agro-industry	Palm oil	Malaysia
	Wine	Chile
Fisheries	Nile perch	Uganda
	Salmon	Chile
Manufacturing	Sophisticated consumer electronics, computers, and communications equipment	Taiwan (China)
	Components, assembly, and testing	Malaysia
Services	Software development	India

Source: Authors.

economic activity in the industry was nonexistent or, at best, nascent until concerted public intervention and other factors enabled private firms and farmers to harness new technologies and grow fast.

Our aim is to understand why and how these countries got it right. What did technological adaptation actually mean in each case, and what drove it? What difference did it make? What role did government play? Do the cases hold lessons that may be useful for other developing countries?

We picked for our study a range of industries where learning how to use new technologies and upgrade them was the driver of development. Among the criteria for selection of the cases were the contribution of the industry to national production and exports, the acquisition of a visible share of the global export market, and entry into a quality-conscious developed country market (table 1.1). Of the 10 cases, nine involve export industries; in only one case (maize in India) is production oriented predominantly to the domestic market. Likewise, only one of the cases, that of electronics in Taiwan (China), deals with the much-studied club of newly industrialized countries of East Asia that are recognized as icons of miraculous technological adaptation.

Our purpose is to highlight the institutional structures and policies that allowed countries to acquire, adapt, and disseminate technologies to hone their export competitiveness. Although we emphasize technological adaptation, other factors essential for catch-up are discussed as well. Chief among these is a stable macroeconomic, political, and social environment supported by a legal system that enforces contracts and protects property rights. Such an environment is a necessary condition for sustained export growth, but it is not sufficient to account for the assimilation of knowledge that is always necessary to put a sector on a high-growth trajectory.

This chapter is presented in five sections. The first discusses the nature of technological adaptation. The second explores the various forms of technology transfer. In the third section we summarize the outcomes of the 10 cases that make up our study. The fourth section attempts to account for performance by asking why technological adaptation occurred where it did. In the last section we draw out the lessons of the study.

The lessons of the cases are most useful for low-income countries, especially those of Sub-Saharan Africa. They illustrate different stages of economic development in a range of countries. One case is taken from a high-income country (Taiwan), four from middle-income countries (Chile and Malaysia), three from low-income India, and two from low-income Sub-Saharan Africa. Until fairly recently, the economies reviewed in this volume were at stages of economic development quite similar to those of Sub-Saharan Africa today. They produced primary goods made mostly by less-skilled labor. By adroitly adapting technologies innovated in developed countries, however, they quickly diversified away from primary products toward manufactured goods and services produced mostly by skilled labor. In some industries their products eclipsed those of the developed countries. The latecomers that matured into industrialized developing countries mastered, in a period of one or two decades, innovations that had evolved for more than a century in the West and that required sizeable fixed costs in human and financial resources. Spectacular rates of export growth were a testimony to the latecomers' technological capabilities. Most remarkably, after catching up, these latecomers retained their competitive edge.

Several aspects of the latecomers' success made them attractive candidates for our study. Some industries made substantial contributions to the domestic economy and captured a visible share of global exports in relatively short order. In recent years, annual growth rates of more than 30 percent helped the Indian software industry to acquire a 3.3 percent global share of outsourced information technology (IT) services. Its sales increased to $15.6 billion in 2003/4 from $1.86 billion in 1996/97; it now services nearly one-half of the Fortune 500 companies. Remarkable growth in electronics led to a 10-fold increase in Taiwan's per capita income over the last four decades. Between 1989 and 2001, Taiwan achieved a share of nearly 13 percent of global electronics exports; today it produces 5.5 percent of the world's semi-conductors. Malaysia's electronics industry is one of the world's largest exporters of semiconductor devices, electrical goods, and appliances.

Agro-processing industries performed equally well. In the last 30 years, Malaysia has moved from being an exporter of crude palm oil processed in Europe to a global leader in processed oils and fats for household and industrial use. Its share of global production of processed palm oil increased to about 50 percent from just 10 percent in 1995, and its plantations, processors, and manufacturers are generally regarded as operating at the industry's technological frontier. Within two decades, Chile emerged as the third-largest producer of farmed salmon in the world. Its share in world production of salmon and trout ballooned from 1.5 percent in 1987 to 35 percent in 2002. It is also the world's tenth-largest wine producer and fifth-largest wine exporter, having increased its share of global exports from less than 0.5 percent in 1988 to nearly 5 percent in 2002. Kenya is the largest producer of flowers in Africa, the leading supplier to Europe, and the world's third-largest exporter of cut flowers. From small beginnings half a century ago, the floricultural industry has grown to dominate Kenya's horticultural exports. The value of floriculture exports grew by more than 300 percent between 1995 and 2003, while the rest of the economy, including the agricultural sector, stagnated.

The three other industries in this volume are noteworthy because of their importance in national production and exports. Production of maize in India has grown following the liberalization of imported seed technology. As the primary ingredient in chicken feed, maize is the driver of India's fast-growing poultry industry. It is also important in pharmaceuticals and for other industrial uses. Maize production in India provides an interesting contrast to its production in Sub-Saharan Africa, where many countries have liberalized trade but not benefited similarly from the adaptation of modern technology. Uganda's fishery exports to the European Union are another example of an industry that has bestowed significant benefits on the national economy. The industry has grown to become, after coffee, Uganda's largest export commodity, accounting for a significant share of the European market. The recent emergence of grape exports from India and the beginnings of a wine industry are an interesting case of technological upgrading to gain entry into the quality-conscious markets of the European Union and of value addition. India is the world's largest fruit producer, accounting for some 10 percent of world production. Even though grapes account for just 3 percent of India's fruit production, the industry has a large export potential.

The 10 cases may also be classified by the nature of the competitive advantage they offer—natural or acquired. Floriculture in Kenya, Nile perch fishing in Uganda, and salmon farming and wine making in Chile are examples of natural resource–based industries in which the resource provides a significant advantage over others, even limiting potential competition in some cases. Maize and grapes in India and oil palm in Malaysia are examples of natural resource–based industries in which the resource itself does not provide a significant advantage. Electronics in Malaysia and Taiwan and software in India illustrate an acquired competitive advantage based on the development of physical and human capital.

Understanding Technological Adaptation

In the 1960s, several economists tried to explain how technologies could be diffused among countries. Gerschenkron (1962) believed that differences in nations' ability to innovate technology and adapt it to their particular circumstances were the primary cause of differences in per capita income between countries, and that the ability to appropriate what others had innovated was the essence of the latecomer's advantage.[2] Arrow (1962) posited that technological progress was endogenous because superior technology was embodied in new capital goods and could be acquired through learning by doing. That concept, however, could not explain the improvements in technology that arose from individual firms' investments in R&D. By assuming the presence of an explicit technology-producing sector, Uzawa (1965) came close to explaining how diffusion could help transfer better technologies from the well-developed to the less-developed countries.

While agreeing that the primary cause of differences in per capita income between countries is the technology gap, some writers in this literature proposed an alternative definition of technology: "Technology or the know-how of how to do things is embedded in organizational structures (firms, networks and institutions, etc.), and is more often than not difficult and costly to transfer from one setting to another" (Fagerberg 1994, citing Nelson 1981). The cases in this volume provide strong support for this school of thought.

The newly industrialized countries of East Asia grew by seeking out and learning to use technologies that were new to them but in use elsewhere (Pack and Westphal, quoted in Kim and Nelson 2000). So-called technological learning was critical in that process because, although machinery and equipment could be purchased, the knowledge needed to use them could not. To develop it, a special public effort was required.

The differences in the economic performance of countries that have access to the same set of technologies and similar levels of investment in physical and human capital suggest that technological learning is important in its own right and requires special policy attention (Lall 2000; Kim and Nelson 2000). The newly industrialized countries of East Asia illustrate the point—they achieved high levels of growth despite investments and policies that were comparable to those of less successful countries (Pack 2000). The difference was technological learning.

Technological learning is a complex process because several of its elements are "tacit"—or embodied deeply in people and organizations. Making good use of new technologies requires the deliberate building of such tacit capabilities (information, skills, interactions, and routines) to handle technology (Lall and Urata 2003). Upgrading technology often requires even more sophisticated capabilities that may need to be developed. The entrepreneurial and managerial skills needed to develop technological capabilities are themselves tacit rather than explicit—they are a part of the firm's invisible assets (Amsden 2001). The tacit element of learning makes the adaptation of technology more difficult as its complexity increases.

The sort of learning that leads to capability development occurs in two stages (Lall and Urata 2003). In the first, that of *technological mastery*, firms learn to use simple production technologies. In the second, that of *technological deepening*, they tackle more complex tasks involving greater value added. Technological deepening is required to progress from technological adaptation to innovation.

Degrees of Technological Complexity in Production and Processes

There is considerable variation in the complexity of the technologies used by producers in the 10 case studies, and those differences have direct implications for catching up and attaining competitiveness.

The most complicated of the technologies studied here are those used in the design and manufacture of complex electronic components, their assembly into consumer and industrial products, and their testing. The sophistication of the leading-edge technologies needed to produce electronics hardware and assemble it in a continually shifting product cycle requires a very high level of technological mastery. Edging into technological deepening in electronics requires the capability to manufacture integrated circuits, capability in design and fabrication, use of submicron technology, and, eventually, world-class innovations. While both Malaysia and Taiwan have mastered the technologies needed to produce today's electronics hardware and software, only Taiwan has become capable of innovation, a testament to the challenges posed by the tacit elements of electronics technology.

Palm oil refining and wine making involve process technologies that are easier to absorb than the manufacturing technologies needed to produce electronics. Those process technologies may be embodied in capital goods, intermediate inputs such as chemicals, and knowledge needed to control production processes. In palm oil, the complexity of the processes increases as the product is upgraded from crude to processed oil and into high-grade oleochemicals for industrial use. The subtler technologies involved in wine making depend on firm-specific processes of fermentation, blending, and aging to meet international consumer tastes. Viticultural technologies also play an important role in the quality of wine. Because of fine differences among varieties and in geographic microclimates, standardization is not possible. The tacit knowledge and individual skills required to produce a taste-sensitive product are far greater in wine making than in palm oil refining, which can employ more standardized industrial processes.

The complexity of the technology used in our resource-based cases varies widely. In natural resource–based industries, new technology often must be substantially altered and adapted to take advantage of local raw materials and labor (Ruttan 2001). The technologies used in the production of maize and grapes in India and palm oil are examples of such technologies. Differences in the technologies used arise from variation in the sensitivity of production processes to environmental conditions, management practices, and product perishability. Examples of the technologies used in the production of Chilean salmon and wine, Ugandan fish, and Kenyan flowers fall in this category.

Technologies for high-value primary production include genetic material; hardware; fertilizer, feed, pesticides, and other treatments; and, finally, the knowledge of how to combine and manipulate various inputs to obtain outputs with desirable attributes.

The hardware used in fisheries and floriculture includes capital goods such as production tanks and feeding systems for salmon farming and greenhouses for floriculture, both used to create a controlled environment. It also includes the systems needed to administer or deliver other inputs.

The genetic material used (for Chilean salmon or Kenyan roses, for example) is important because it determines potential yields and quality. Developing or gaining access to suitable genetic material—that embodies desirable product characteristics, is suitable for production under various environmental conditions, and readily accommodates other inputs used in production—is a critical concern in all natural resource–based production. The case of Indian maize, Kenyan flowers, and Chilean salmon are good examples of initially imported genetic material (seeds, flower plantings, and fish eggs, respectively) adapted to local conditions. The bulk of the knowledge to adapt and apply the inputs may be generalizable and readily available to researchers (if not to most producers) as a public good.

Other inputs embody technologies of different complexities. In salmon farming, for example, these include nutrients (fertilizers, salmon food, and growth hormones) and other products that affect growth; in floriculture they are organic or inorganic products that provide protection against pests and diseases (vaccines, pesticides).

The last element of production technology is the knowledge required to combine and modify inputs so as to produce outputs with certain attributes (taste, appearance, shelf-life, disease- and drought-resistance), in greater quantities, or with greater efficiency. This knowledge has to be developed at each location, as the complex set of conditions that influence production is likely to vary. Each salmon operation, for example, is unique. And varieties of maize may be imported, but the best practices for exploiting each variety need to be developed locally.

Technological learning takes place when engines of diffusion spread knowledge to producers. Suppliers of intermediate inputs pass on some knowledge, but often the inputs, once acquired, must be further adapted to the local environment. (The uniqueness of each salmon-farming operation is an example.) Individuals with technical training or a scientific understanding of particular production processes are in a better position than laypersons to extrapolate from generalized knowledge to locally suitable applications. Tacit knowledge of production processes can also be developed by observing practices—such as pruning of grape vines or plant-protection practices in a greenhouse in floriculture—in technologically superior operations.

Products based on natural resources stand or fall on appropriate postharvest technologies. Salmon in Chile and Nile perch in Uganda are processed after harvest, requiring appropriate facilities and tools, skills, and the understanding to design and control procedures that will ensure compliance with international phytosanitary standards. The objective of these postproduction processes is to convert the products into forms preferred by consumers and suitable for transport to distant markets. Flowers and grapes involve less postharvest transformation, apart from packaging, but like other perishable exports, they require cooling. Additional postharvest care is needed where value added is greater, as in the case of Kenyan exports of table-ready flower arrangements to British supermarkets.

The technology of software development is distinct from production or process technologies in manufacturing and agro-processing. Software production requires technical capabilities of individuals and both technical and managerial capabilities of organizations. Those technical capabilities include the capacity to diagnose problems or visualize applications and to design suitable responses. Managerial capabilities include the means of producing those solutions using software engineers.

What Does It Mean to Upgrade Technology?

The common understanding of upgrading, largely associated with the East Asian experience, is that of moving from a low-skill task, such as assembling components, to production of complex components. While assembly can be learned easily, the production of components requires greater capabilities. There are analogous transitions for the technologies involved in our cases. They involve making new products of higher value, moving up the value chain, and eventually producing inputs.

The technologies involved in the production of the genetic material, hardware, and other inputs are far more complex than the production of the outputs to which they contribute. In other words, the biological and chemical technology and processes needed to create them are more fundamental than the processes used to apply them. It is one thing for an industry to master the application of sophisticated technologies; it is another thing to move upstream to develop or even adapt those technologies (Ruttan 2001). Indicators of upgrading, or deepening, in each of our cases are listed in table 1.2.

Technological deepening in these cases may be indicated by a movement to higher-value products (to roses from carnations, for example), to products of higher quality (chilled rather than frozen fish fillets and products that meet safety standards), and to products with greater value added (refining crude palm oil and producing oleochemicals). Upgrading technologies in resource-based industries may also be associated with the development and production of inputs that embody the production technologies.

Moving up the value chain, upgrading information technologies in software development may have two facets: (i) providing total solutions (including diagnosis, design, code writing, and testing) in addition to writing and testing code; and (ii) developing products that represent solutions to widespread needs or problems, as opposed to developing customized software for a specific situation or a client. A shift from variable to fixed-fee contracts, which enables software developers to benefit from efficiency gains, is often an indicator of a firm's maturity and capabilities. Another indicator, in cases of

Table 1.2 The Indicators of Technological Adaptation

Case	Technology	Indicators of upgrading toward the technological frontier
Low-end electronics in Malaysia	Assembly and testing of components to produce consumer and industrial goods	Transition from assembly to the production of complex components; activities that involve higher level of skills
High-end electronics in Taiwan (China)	Assembly, testing, and production of complex components to produce sophisticated consumer and industrial goods	Transition from assembly to the production of complex components; activities that involve higher level of skills Innovation-grade technological capabilities evident in the growing number of electronics patents held by Taiwanese Integrated-circuit fabrication and design and use of submicron technology processes
Software in India	Assessment/diagnosis of application/problem Design Code writing Testing	Transition from data services and labor-intensive services to projects and software products Transition from code writing and testing to total solutions Movement from process-oriented IT exports to exports of finished products
Maize in India	Yield potential of seed varieties Package of practices and seeds Dissemination of technologies	Development of plant-breeding capabilities evident in scientific R&D to adapt foreign technologies to local conditions
Grapes in India	Practices to produce grapes meeting EU quality standards Postharvest care technologies	Improving quality and yields Adherence to global phytosanitary standards (HACCP and Euro Gap compliance)

Oil palm in Malaysia	Oil palm production technologies, including new varieties Crude and processed palm oil refining technologies Oleochemical technologies	Techno-diversification and domestic value addition Refining of crude oil and production of upstream inputs such as machinery and downstream products for the oleochemical industry Global leader in oleochemicals
Salmon farming in Chile	Production tanks, feeding mechanisms, and infrastructure for processing Eggs, varieties, salmon food, vaccines, etc. Practices, processes-related knowledge	Decreasing production costs Development of local knowledge Production of capital goods and inputs, especially feed Ability to comply with global phytosanitary standards
Wine making in Chile	Fermentation practices and equipment Internationally preferred varieties Grape production practices Expertise in wine making	Quality premium in global markets
Nile perch in Uganda	Infrastructure to transport fish from capture to processing units Fillet processing facilities with acceptable layout to minimize costs Adoption of standard procedures for food and safety (HACCP)	Processing that retains freshness and nutritional value Meeting quality requirements Developing new value-added products Ability to meet global phytosanitary standards
Flowers in Kenya	Greenhouses and postharvest care facilities Chemicals and other inputs Production knowledge	Flower quality Environmental impact of production

Source: Authors.
Note: HACCP = Hazard Analysis and Critical Control Point.

outsourcing of IT services, is the transfer of research and development (R&D) capabilities to the host from the client country.

As will become clear in the section on technology transfer, production technologies—hardware, plant-ware, software—are traded because it is more lucrative for innovators to trade them than to exploit them themselves, often because they do not enjoy favorable natural conditions (such as year-round growing conditions in the case of flowers in Kenya and good soil and climate for wine grapes in Chile) or low production costs (as with IT services in India and electronics in East Asia).

Underproduction of Technology and the Logic of Government Intervention

The tacit aspects of new technologies need to be acquired and developed locally. This takes time and involves significant risk (Nelson and Nelson 2002; Lall and Urata 2003). Government can play a critical role in mitigating risks, thus smoothing the way for the diffusion of new and better technologies to increase the productivity and export competitiveness of all firms in an industry (Rodrik 2000). For example, Kremer (1998) argues that governments should acquire patent rights so as to be able to provide incentives to innovate and diffuse technology. "With the benefit of hindsight, it is clear that leading industrial clusters in the OECD economies, Singapore, and Taiwan (China) owe their existence to government actions" (Yusuf 2003).

Few disagree over the rationale for public interventions, which are designed to counter circumstances in which new technologies will be "underproduced." The first such circumstance, often called the *traditional spillover,* arises when government intervention is necessary to attract enough foreign firms to invest in an activity whose collective benefits exceed those of individual firms. For example, it is hard for individual firms to invest in and capture the benefits of biological and chemical research in maize production, salmon farming, or palm oil processing (Ruttan 2001). The second, called an *informational spillover,* arises when market prices do not reveal the profitability of new investments in new technologies, thus deterring firms from discovering them. An example is Taiwan's electronics industry in its early stage of development, when private firms were unwilling to explore and adapt new technologies because they were unaware of payoffs in the industry. The third, called a *coordination spillover,* is related to scale economies. Coordination spillovers arise when large-scale complementary investments in inputs must be made before individual firms will find it profitable to invest in new technologies (Rodrik 2004).

There is less agreement on the aptness and extent of government involvement in shaping the context and institutional structure surrounding firms and in creating an environment that stimulates the effort to adapt (Porter 1990).

Institutions, understood as a set of rules augmented by a system of incentives and penalties and enforced by organizations, play a central role in technological catch-up by translating the social, industrial, and technology policies that governments adopt to support technological adaptation (Freeman 2002). Critical institutions may be weak or missing altogether in many developing countries, which some say explains the divergence between rates of technological adaptation in the developed and developing worlds (Abramovitz 1986) because technology, or knowing how to do things, "is embedded in organizational structures," such as firms, networks, universities, and government agencies (Nelson 1981).[3]

There is no blueprint for the institutional arrangement that best delivers technological adaptation. For each function—market creation, regulation, stabilization, and legitimization—that institutions perform, there is an array of possible arrangements and policy choices. Inevitably, institutions must be context-specific, that is, shaped by initial conditions, history, socioeconomic setting, and political regime, among other factors. Institutional innovations, such as combinations of unorthodox and orthodox policies to perform a function, have been successful in spurring technological adaptation in several countries, but attempts to transplant them from one country to another have generally failed (Rodrik and Subramanian 2003).

National efforts to provide public support for technological adaptation, while simultaneously encouraging institutional development, typically take some form of an industry-specific policy (box 1.1).

Institutional Frameworks for Technological Adaptation: Common Elements

The institutional arrangements that supported technological learning in our 10 cases were shaped by history, individual and institutional capabilities, and the socioeconomic importance of the industry to the government. Getting it right in each case can be attributed largely to appropriate synchronization of the elements of industry-specific policies with the institutions necessary to motivate learning among exporters. A good example is the Malaysian palm oil industry, for which government provided a comprehensive package of public support. It funded industry-specific R&D and skills development, provided financing, built physical infrastructure, and offered tax incentives to

Box 1.1 Institutional Support for Industry-Specific Government Policies

Although techniques and styles differed from case to case in our sample, an industry-specific government policy was evident in each. With one exception (that of maize in India), governments fostered the use of new technologies in our sample industries to promote specific nontraditional exports. They did so, sometimes weakly, through a mix of private sector–led and government-facilitated institutions and policies. The specific combination of public and private capabilities determined the flavor and potency of the approach. An industry-specific policy was most vigorous and explicit in the case of electronics, palm oil, and salmon, where governments were well endowed with technical and management capabilities that created a web of institutions to support learning. Its weakest reflection was in Uganda's fisheries, where weak institutions could, at best, regulate technological standards in a preferred sector to promote technology-enhanced exports. In the remaining cases, one or another approach was taken to support a favored industry—the unmistakable signature of an industry-specific government policy.

Latecomer status helped most of the governments to avoid the perils of (ineffectual) industry-specific policies. By tapping new technologies innovated elsewhere, the latecomers avoided the risks of innovation. Those risks arise from the experimental nature of technological innovation, which requires large fixed investments of financial resources, human capital, and time but may not be successful. Adapting known and tried technologies reduced substantially the risks of industry-specific policies compared to their potential benefits. The East Asian economies moved up the technological ladder by choosing from a large inventory of proven technologies rather than developing new ones, as the developed countries had to do (World Bank 1993).

Failure in implementing industry-specific policies was also avoided because governments were committed to specific export *industries,* not firms; industry performance, easily measured by export growth rates, is not subject to misrepresentation. Aware of the potency of new technologies in improving competitiveness, several governments concertedly developed industry-specific learning and research institutes in which to adapt existing technologies and, in some cases, develop new ones. However, this was possible only where domestic research capabilities were strong.

Several governments tried to ensure that small producers would benefit as export industries grew. Examples include Kenya's efforts to enable small producers to participate in horticulture and floriculture exports, and India's public research on segments of the maize seed market that might not be served by private seed companies. Malaysia's resettlement of poor peasants enabled smallholders to participate in the production of a crop that had been confined to large estates.

encourage firms to transition from exports of crude to processed palm oil and oleochemicals.

The institutional arrangements for technological adaptation in the 10 cases showed some broad similarities, but the policy instruments used to implement them varied considerably from one country to another, and sometimes even from industry to industry in the same country. The common features of the institutional frameworks in the 10 cases were as follows:

- *Common goal-targeting.* Governments valued and supported technological learning in the favored industries for different reasons. In some cases, public support for technology development was provided because of the government's commitment to export-led growth in a specific nontraditional industry: electronics, palm oil, salmon, or wine. In others, technological learning received support because the industry was a valuable source of foreign exchange. Here, Ugandan fisheries, Kenyan floriculture, and Indian software exports in their early stage of development are examples. Later the government supported them because they were an important source of export growth. Palm oil and grapes are examples of government's joint support for an industry that generated foreign-exchange revenues while fulfilling social objectives.

- *Political vision.* Specific industries enjoyed strong political commitment to technological adaptation. Taiwan presents the extreme example. There the government managed the electronics export industry. Low economic inequality and a highly skilled workforce nurtured social cohesion, which, in turn, gave the state the mandate to pursue policies that would put electronics exports on the global map. The cases of salmon in Chile and of oil palm and electronics in Malaysia are other examples where political commitment to the industry approached the level of a guiding vision. In the remaining cases, where the industry was not part of a national vision and a closed economic environment prevailed, politicians valued the contribution of technological learning and did what they could to facilitate it. Commitment grew with the economic visibility of the industry, or when social concerns were raised.

- *Rewarding winners, abandoning losers.* Except for electronics in Taiwan, salmon, and palm oil in the early stage of development, governments did not pick the industries that later blossomed. Instead, they nurtured those that they found to be growing faster than others. In the wine industry, the Chilean government extended general support for horticulture but targeted wine exports when it saw them pick up. Firms in the wine industry were given direct or indirect public assistance to adapt new technologies and achieve exportability. Generally, governments offered the same level of public support to all exporters in the preferred industry and let the discipline of the market prevail. Within an industry, governments bestowed rewards on firms that performed well and penalized poor performers by letting them fall out of export competition. In the Indian software industry for example, the privilege to import computer hardware was extended to all firms that could export. The rewards varied significantly. The practice of providing public support to favorite firms within an industry was not followed.

- *Private sector–led export development.* The private sector was the driver of exports and the beneficiary of technological learning, while government played a facilitating role, especially in promoting adaptation and the diffusion of new technologies. Even in cases where the first firms were born in government incubators, as in the case of Taiwanese electronics and Chilean salmon, they were privatized as soon as they achieved commercial viability. Government then turned its attention to facilitation, coordination, and regulation. Instead of fostering conglomerates, Taiwan's government developed clusters of new firms each time an industry transitioned to independent technological development.

- *Competition.* Other than a few limited episodes of protectionist policies while Taiwanese electronics firms were mastering new technologies, governments in the case studies valued competition among domestic and foreign firms. Entry into and exit from the industry were open. Multinational corporations (MNCs) were given special privileges (often as incentives) in most of the industries. Technologically advanced foreign firms participated in Chile's salmon exports industry as early as the 1970s, while government was incubating domestic firms; in the 1990s, the industry turned into a foreign-dominated oligopoly. Foreign firms that brought in new technologies were welcomed in Uganda and Kenya. Where domestic capabilities were stronger, MNCs played a lesser role.

- *Property rights and the rule of law.* Without exception, property rights were preserved and played an important role in attracting MNCs with new technologies, as well as domestic firms that sought to adapt those technologies to produce export-quality goods. Kenya's floriculture industry, for example, was exempted from the wave of Kenyanization (nationalization) that stifled nearly all other productive industries.

The institutional arrangements for technological adaptation in the 10 cases may be understood in terms of the relative weight of the public and private sectors. That weight, in turn, reflects differences between private and public capabilities to design incentives for learning, differences that affected the quality of technological adaptation in the cases under study. For simplicity, we classify the cases into two broad types of institutional settings—one led by the private sector, the other by the public sector.

Privately Initiated, but Supported by Government   Under this type of institutional arrangement, private firms took the initiative to seek access to new technologies, and the government facilitated that access. Often a policy reform, such as deregulation or liberalization, triggered the adoption of new technologies in nascent sectors. Where private capabilities for learning were strong, private institutions compensated partially for weak public institutions and achieved fairly sophisticated technological upgrading. But, where both private and public capabilities were weak, public-private partnerships, too, were weak—with attendant implications for technological learning.

The three Indian cases reveal different institutional settings, reflecting industry specificity.

- In the case of software exports, early public investments in tertiary technical education could be construed as a government policy to develop a cadre of high-tech capabilities for sophisticated industries in the future. The crucial distinction was that in the 1960s, the Indian government had not picked the software industry as the winner; a few decades later, the market did. In the highly protected economy of the 1980s, as private firms initiated software exports, the government lent support through special privileges to import hardware, a critical input in sharpening the sector's technological edge.
- In the case of grapes, private initiatives on the part of small growers caught the government's eye because of the sector's export potential. Generous government support for technological upgrading and market development, mostly through public–private partnerships, was critical in scaling up grape exports.
- In maize, the government triggered technological upgrading by lifting a ban on imported maize seed technology, unleashing healthy competition between public and private laboratories, both domestic and foreign, to adapt imported technology to local conditions. Public investments in maize research and access to publicly developed breeding material enabled domestic

companies to enter the seed sector and compete effectively with multinationals. Past public investment in scientific skills enabled local scientists to participate in technological adaptation, an advantage missing in the African cases.

The institutional setting for floriculture and horticulture in Kenya was initially based on reciprocity between government and private growers. In the early stage, in spite of foreign domination, floriculture received favored treatment, probably because it generated income for local small growers, as well as foreign exchange, which was in short supply. Later, as private technological learning helped to scale up exports, the government demonstrated increased support. When Kenya became a global leader in cut flowers, political commitment intensified, but weak public and private scientific capabilities continue to constrain technological learning and research in Kenya.

In the extreme case of Uganda, weak private and public capabilities constrained the development of public–private institutions needed to enhance technological learning. Initiatives by foreign firms successfully stimulated and spurred an export industry, but their weak technological capabilities prevented them from moving into higher value products. Weak public capabilities could, at best, oversee compliance with the European Union's phytosanitary standards.

Strong Public Commitment   The second type of institutional arrangement is found where governments have a strong political commitment to upgrade technology in chosen industries or sectors. Various interventions, often in combination, are used in such cases. In our cases, government and private firms benefited from efficient institutions, and public–private partnerships were common.

The leading characters in the case of electronics in Taiwan were a committed government and efficient public and private institutions that worked in tandem to score technological advances. In the cases of salmon in Chile and electronics in Malaysia, private and public capabilities were not as strong, but the public institutions were dynamic. As the industry matured and scaled up, government's role changed from one of seeding firms to one of supporting and then regulating firms.

The ability to envision, implement, and continually modify policies to keep abreast of a highly dynamic industry was a manifestation of strong public management and monitoring capabilities; these were matched by firms' technical and managerial capabilities. A high savings rate (when translated into a large pool of domestic capital) gave government the autonomy to execute its policies without any foreign pressure to abandon technological catch-up.

Technology Transfer and the Game of Catch-Up

Latecomers attempt to catch up with technology leaders through transfers of technology. Transfers occur in various ways, including internal transfers in the context of foreign direct investment (FDI), and external transfers through licenses, contracts, and sales of equipment or know-how and so on. However, simply attracting FDI may not be effective due to the tacit elements of effective technology transfer.

Transfers Through MNCs: Tricky

Most developing countries use FDI to acquire technologies, enabling them to achieve technological mastery, but not necessarily deepening (Lall and Urata 2003). MNCs move technologies from one country to another to take advantage of differences in natural, physical, and human capital. They may invest in upgrading local capabilities, but only if the skills base in the host country is already expanding and local supplies are improving. Meanwhile, technology transfer through FDI may actually conflict with local capability development. Without host-country policies to develop local capabilities, MNC-led exports are likely to remain technologically stagnant, leaving developing countries unable to progress beyond the assembly of imported components. Because FDI is attracted to local factors, countries need to develop technological capabilities that can draw FDI into the more complex activities that characterize every strong industrial economy.

Because innovation is expensive and uncertain, MNCs usually centralize their R&D in countries with a good skills base and scientific and technological infrastructure. They transfer only the results of their R&D, as opposed to the research processes, to less well-endowed sites. As a result, those sites may never develop the capability for local technological development if they rely solely on MNC investment for their technological capabilities.

One solution to this limitation is variously referred to as the national innovation system in the developed-country context and the national learning system in less-developed countries. The first is a system that improves the capability of a developed country to innovate by pushing back the technological frontier. The second enhances the capability of a developing country to absorb and improve on existing technologies. Both imply an institutional environment in which actors demand and supply capabilities that place an industry, sector, region, or country on the path of technological deepening.

Channels of Technology Transfer in the Ten Cases

Our cases reveal a variety of channels through which technologies may be transferred.[4] A rough indication of their importance is illustrated in table 1.3. The table highlights two points: first, a single magical channel for the successful transfer of new technologies does not seem to exist. Countries followed a variety of channels to achieve the same goal. Second, some channels were used by all countries, although to varying degrees. Here we elaborate on the channels and identify the cases to which they are most relevant.

Internal transfers through FDI. When a parent firm transfers technologies to subsidiaries, some of the capabilities for making use of those technologies also are transferred. Because the transfer is internal, even the best, most closely guarded technologies may be transferred, but the parent firm typically takes measures to ensure that proprietary technology and know-how remain within the firm (Lall and Urata 2003). Internal transfers through FDI initiated the electronics sectors in Malaysia and Taiwan, floriculture in Kenya, and Nile perch fisheries in Uganda, although in the last case the technology was rudimentary and had to be upgraded to sustain exports. In addition to investments in production itself, in which investors supplied the necessary technologies, direct investments often were made in the production and supply of inputs. These included investments by multinational companies in the production of maize seeds in India, rose-breeding operations in Kenya, wine production in Chile, and salmon-food production in Chile.

External transfers through licensing agreements. External transfers of technology occur through licensing, subcontracting, or contracts for original equipment manufacturing (OEM). A limitation of this channel is that buyers must develop their own marketing capabilities (Lall and Urata 2003). Salmon farming in Chile and Nile perch fisheries in Uganda offer contrasts in this regard. The international firms that invested in salmon farming in Chile now sell directly to global retailers such as Wal-Mart, while the regional investors in Ugandan fisheries depend on intermediaries in Europe for transport, marketing, and product development. These importers did little to upgrade Uganda's fish-processing technologies when processors faced a ban on their exports to the European Union after 1997.

The best manufacturing technologies are unlikely to be transferred through external, arm's-length transactions between organizations, because innovators would then be unable to capture the full benefits of the exploitation of those technologies. However, the same does not apply in process-oriented and natural resource–based industries, where benefits can be maximized by selling the technology to numerous

Table 1.3 The Importance of Various Technology Transfer Mechanisms in the 10 Cases

Sectors	Foreign direct investments	Licensing agreements	Imports of capital goods/inputs	Local industry development and participation	Contracts/ consultants	Local R&D	Diaspora, tech parks
Software in India	Low	n.a.	High	High	Medium to high	Low	High
Electronics in Taiwan	High	High	High	High	Medium to high	High	High
Electronics in Malaysia	High	Low	High	Low	Medium	Low	Medium
Oil palm in Malaysia[a]	Low	Low	Medium to high[d]	High	Low	High	Medium
Salmon farming in Chile	High[b,c]	n.a.	Medium to high[d]	High	Low	Medium to low	n.a.
Wine in Chile	Medium	n.a.	High	High	High[e]	Medium to low	n.a.
Grapes in India	n.a.	n.a.	High	High	Medium to high	High	n.a.
Maize in India	Low	High[f]	Low	High[g]	n.a.	Medium to high	n.a.
Fisheries in Uganda	High	n.a.	Medium	Low	High[h]	Low	n.a.
Floriculture in Kenya	Medium to high[b]	High	High	Medium to high[i]	High	Low	n.a.

Source: Authors.

Note: n.a. = Not applicable.

a. Includes technologies for producing crude palm oil and for processing the oil.

b. Investments in the supply of inputs: varieties of maize, roses, and salmon food.

c. FDI became important only in the 1990s. Few production technologies were transferred through FDI, although the industry is still based on imported capital goods and other inputs.

d. Production tanks, initially imported, are now imitated locally.

e. Independent grape producers who produce on contract with wineries upgrade their production technologies to meet the requirements of buyers. *Proyectos de fomento* and Chilean Wine Corporation finance consultants to assist small firms.

f. Payments for hybrids that need to be purchased annually.

g. Capability of local firms to make use of genetic material that becomes available from national and international research programs.

h. Product development assistance comes from buyers and consultants abroad.

i. Greenhouse managers are largely expatriates, although the number of African managers is increasing.

producers. Here, external transfers in various forms may be the norm and often will include the best technologies. The only issue in such cases is whether policies and institutions permit the innovator to capture enough benefits; that is, whether international property rights are respected, as they were in our cases of salmon, maize, and flowers.

Licensing appears to have been a less common means of technology transfer, except in the case of electronics in Taiwan, where it was widely used. If royalty payments for plant varieties are considered as a form of licensing, licensing could be said to apply in the maize and floriculture cases as well. Licensing can enable rapid acquisition of product and process know-how, while preserving local control over adaptations and modification. However, it does not encourage local firms to keep up with advances in technology. A significant level of local technological capability is required to put licensed technology to work.

Capital goods imports allow local firms to acquire new technology without incurring the transactional baggage of licenses and FDI. But imports are useful only if the importing country has the technological capability to adapt and use the imported material (Dahlman and others 1985). The role of imported capital goods in electronics in Malaysia and Taiwan is well known. Imports into India of computer hardware, during a period of strict controls on most imported inputs, were critical in enabling India's software export industry to achieve technological parity, although the computers did not directly enhance software-writing capabilities. Imported inputs are also essential tools of the trade in precooling equipment for grape exports in India, palm oil refineries in Malaysia, fish tanks and other equipment for salmon farming, tanks and barrels for wine making in Chile, and greenhouses for floriculture in Kenya.

Technology is transferred through sales in three ways. Materials transfers involve imports of materials such as seeds, machines, and turnkey factories. While they do not require further adaptation, transfers of materials also have tacit elements that can complicate adaptation. Design transfers, or transfers of blueprints, manuals, and formulae, must be adapted, thus requiring engineering capabilities in the host country. Capacity transfers—transfers of scientific knowledge and technical capabilities—require substantial investments by the host country in education and training, as well as the development of scientific and technical capacity to adapt new technologies for local use (Ruttan 2001).

Local adaptation or development. The domestic capabilities needed to benefit from design and capacity transfers can be built in several ways. One way is through technical interactions among firms and between firms and professionals hired to adapt new technologies for local use. A second is by sending nationals to developed countries to learn about new technologies. Since the 1950s, many thousands of developing-country students have traveled abroad for higher education in scientific and technical fields. Benefiting from study abroad or study tours requires a solid foundation of secondary education with good grounding in engineering, math, and the natural sciences (Dahlman 1994). Local industry development refers to the acquisition of tacit knowledge through trial and error and assimilating practices in technologically superior operations within and outside the country. Entrepreneurship based on experience gained while working in U.S. firms was particularly significant in the Indian software industry. In the case of maize and grapes, strong domestic research capabilities facilitated public–private partnerships to acquire, absorb, and adapt new technologies. The tacit aspects of some process technologies in primary production can also be absorbed through observation. Indian grape growers, as well as Chilean salmon and wine producers, have traveled abroad to observe and learn from more advanced operations.

Contracts and consultants. Technology transfer through contracts occurs when foreign buyers impart to local producers the knowledge they need to meet product or service specifications. Such transfers from clients to producers are common in electronics and software production. But small-scale grape producers also learn through contracts about the preferences of wine producers in Chile. The importers of Nile perch assist processing firms in Uganda with product development. Consultants are used to various degrees by Indian grape producers, Ugandan fish processors, and Kenyan flower producers. The extent of their involvement depends on local capabilities to absorb and build on the information obtained from them. Wine producers in Chile benefit from regular exchanges with visiting oenologists from around the world.

Formal research and development (R&D). Investments in firms and public institutions are another mechanism used to identify, acquire, and adapt foreign technologies. Although public research systems performed relevant research in many of the case countries, the extent of their contribution to the development of the 10 industries is unclear. Taiwan has been fairly successful in promoting research by local firms and in its public research institutions. Research in Malaysia allowed the palm oil industry to develop without recourse to external production technologies. Generous financial incentives were available for R&D, including genetic research and training. Advanced scientific skills in the sector have enabled Malaysia to set the world technological frontier in palm oil cultivation and processing. In most of our cases, however, research done in established research institutions was not specific or advanced enough to be helpful in upgrading technologies, making it necessary to seek other forms of technology transfer. Examples include industry-specific research on grapes in India and on wine and salmon farming in Chile.

Harnessing the diaspora. Returning students and expatriates played an important role in disseminating technological knowledge to domestic firms in many of our cases, either as CEOs of firms set up with government support or at high ranks in publicly funded technology-promotion institutes. Indian software and Taiwanese electronics benefited from this channel. Governments have tried to induce graduates to return through lucrative financial packages, which few poor governments can afford for long. And, as the Malaysian experience attests, technically skilled professionals require more than a good salary; they want to be assured of the presence of networks of similarly qualified scientists and professionals. We do not know why Malaysian professionals did not return; the author of the case cites cumbersome bureaucracy as a primary cause. Beyond immediate working conditions, the social and policy environment in the home country may also be a deterrent to returning home.

Other dissemination channels, such as clusters and technology parks. Technology parks, icons of productive public–private partnerships in Taiwan, evolved into organisms in which high-tech firms seeking and adapting cutting-edge technologies flourished in close proximity, benefiting from the synergies created by technological progress.

Government's Role in Technology Transfers: The Myriad Forms of Industry-Specific Government Policies

The tacit quality of technological learning motivated some governments to design policies and interventions to spur technology transfers through the mechanisms discussed in the previous section. Interventions to promote the use by firms or farmers of a particular transfer mechanism were industry specific.

Some of the measures that governments took to support technology transfer were conventional (generic) policies that would be considered appropriate anywhere for private sector development. These included maintaining a stable macroeconomic environment, enabling liberal trading polices that allowed foreign firms free entry and exit, protecting property rights, and enforcing the rule of law. Less conventional interventions and policies were customized to the needs of specific industries. These complemented the industry-specific institutional arrangements discussed earlier.

Some governments intervened in the early stages of a nascent industry's development to build a critical mass of firms or farmers engaged in technological adaptation. However, where the industry's evolution was privately initiated, most government interventions were fairly moderate, even if clearly industry-specific.

The industry-specific public interventions observed in the 10 cases fall into several categories, each of which is described below.

- Negotiating with MNCs
- Spinning off domestic firms
- Facilitating acquisition and dissemination of technologies
- Promoting exports
- Developing industrial clusters
- Providing regulatory services
- Supporting industry organizations and coalitions
- Meeting technical manpower requirements

Negotiating with MNCs. In Taiwan, the government sought not only to attract FDI to manufacture electronics, but also to ensure that the technological capabilities of local firms would rise to match those of foreign firms. By negotiating contracts and founding joint ventures, Taiwan ensured that technology was transferred smoothly and seamlessly from foreign to local firms, giving rise over time to a strong, domestically owned electronics industry. In Malaysia, successive rounds of negotiations, generous fiscal incentives, and other privileges lured MNCs to invest in the electronics sector, even after Malaysia began to lose its cheap-labor advantage. However, except for a small number of Malaysian entrepreneurs who were able to learn from the MNCs and start their own firms, the transfer of technologies remained limited, largely because of the shortage of technical manpower.

Spinning off domestic firms. To help domestic firms emerge, governments may provide financial support, demonstrate feasibility, or make available technologies developed in domestic research organizations. Examples include Taiwan's Industrial Technology Research Institute, which retained shares in the firms it spun off. Fundación Chile and the Corporación de Fomento (CORFO) started the first commercial-scale salmon-farming operation in Chile. Domestic firms were reluctant to invest in salmon farming or semiconductors until public entrepreneurship demonstrated the potential returns to investment in the new activity—an example of informational spillover.

Facilitating acquisition and dissemination of technologies. The 10 cases show that governments facilitated access to technology in several ways.

- The Chilean and Indian governments helped producers of salmon, wine, and grapes to visit technologically superior operations abroad to learn about new developments and apply them to their operations.
- Governments supported research devoted exclusively to a specific industry or sector. Such research took place in institutes, universities, and firms. National research centers exist for grapes in India, palm oil in Malaysia, and electronics in both Malaysia and Taiwan. Collaborative R&D ventures among firms, public research institutes, and trade associations often received financial assistance from government. With technological learning and industry growth as their objective, such ventures were seen in Taiwan's electronics sector. In Chile, collaboration between a public–private institute, institutions of higher learning, and local firms led to the local production of salmon eggs. The Chilean and Malaysian governments supported private research in firms and academic organizations. In India, public and private bodies cooperated on maize research.
- Taiwan's China Productivity Council and India's wine parks spread know-how among small firms and producers.
- Taiwan and Malaysia imposed local-content regulations in the early stages of development of their electronics industries to ensure the transfer of technologies from MNCs to local parts and components firms. Until 1984, Taiwan assessed tariffs on color TVs and banned imports from Japan to protect the island's color TV industry.

Promoting exports. Export-promotion measures enabled industries to develop new markets for products and helped smaller organizations benefit from exports. Indian organizations assisted grape producers in finding a market for their produce. Chilean development organizations helped two industry associations of small and medium-size firms to organize technological diffusion and marketing for their members. As individual firms were not in a position to establish a national brand identity, Chile's government helped with the building and promotion of a national identity for the country's wine. Malaysia gave MNCs and domestic exporters generous tax incentives to locate in free trade zones from which they could export electronics. The government offered incentives to trigger exports of palm oil and to induce firms to move into higher value products such as processed palm oil, processed palm kernel oil, and palm kernel cake. In Kenya, government made large investments in refrigerated facilities at the airport to enable local floriculturists to start exporting.

Developing industrial clusters. Where clustering could enhance efficient transmission of information, governments, often in partnership with firms, invested in cluster-conducive infrastructure. Public investments in land and utilities were needed to persuade firms to relocate to Taiwan's pioneering Hsinchu Science Park, which was designed to create an "industrial ecology" in which high-technology industries could flourish. Other examples include the Penang cluster in Malaysia, wine parks to promote forward linkages from grapes in India, and the software technology parks of India (STPI), with international gateways at 39 locations equipped with IT and telecom infrastructure and enjoying single-window permitting.

Providing regulatory services. Regulatory services are important where exports are fresh and must meet stringent phytosanitary, environmental, process quality, or social performance standards. Two cases of contrasting effectiveness are Uganda's fisheries and Chile's salmon farming. Chile was quick to introduce the legal basis for regulations for its fish industry, to upgrade them to meet the international requirements, and to set up capable organizations for their enforcement. In Uganda, on the other hand, poorly defined regulations and poor enforcement because of inadequate resources led to a ban after exports repeatedly failed to meet the European Union's health and sanitation requirements.

Chile appears to have been able to develop and enforce regulations on its own, whereas in Uganda it was necessary to build government capabilities with help from multilateral organizations. In Kenya, the floriculture industry took the lead in developing codes of practices and lobbying the government to establish the Kenya Plant Health Inspectorate Service to ensure that exports were not threatened by consumer concerns about the impact of the industry on the environment.[5]

Some governments are proactive in helping industries develop and adhere to standards as a means of enhancing the quality of the country's exports. India's support to the software industry and grape exporters is an example. The Indian government has, with the help of leading certification agencies in the world, initiated training in the country to ensure that it is up to date in software testing and to set standards for Indian software firms to aspire to meet. The grape producers' export development agency has started a program to help small producers monitor pesticide residues at reduced costs. To help small fishermen meet EU certification, the government of Uganda has established state organizations that test and certify fish for producers at subsidized costs.

Supporting industry organizations and coalitions. Industry organizations in export sectors are particularly effective in ensuring that members meet the quality standards demanded in global markets.

Governments and bilateral agencies have assisted in establishing such organizations. The U.S. Agency for International Development supported the formation of the Fresh Produce Exporters Association of Kenya, which was one of the first to develop a code of practices and lobby the government for policies that have made horticulture a thriving sector in the country. Chile's CORFO and Fundación Chile have played similar roles in salmon farming, working with producers to meet phytosanitary certification requirements. In Taiwan, Industrial Technology Research Institute (ITRI) took the lead in putting together R&D consortia that enabled small firms to benefit from public R&D and gain access to government research infrastructure. The Indian states of Maharashtra and Karnataka helped to organize cooperatives that have enabled small producers to participate in grape exports. The Malaysian palm oil licensing agency works with producers to link local palm oil processing firms, oil palm genetics labs, and agro-industrial engineering education. It set up the Palm Oil Research Institute of Malaysia.

Meeting technical manpower requirements. Meeting the technical manpower requirements of growing industries is a challenge in developing countries. Those requirements are particularly stringent for industries that work with complex technologies, such as electronics. A common short-term solution is to attract qualified nationals working abroad. The governments of Taiwan and Malaysia offered attractive packages and investment opportunities to induce their nationals working abroad, especially in the United States, to return home. Taiwan appears to have done this more effectively than Malaysia. Its Science and Technology Congress in 1978 and the "Scheme to Strengthen the Cultivation and Recruitment of High-Technology Experts" are clear examples of a government committed to assembling a critical mass of technical skills. The government of India permits software companies to grant stock options to attract qualified staff from abroad, although the dynamic software industry has been able to attract the Indian diaspora without resorting to special lures. Kenya has permitted qualified foreign nationals to work in the country, although it is planning restrictions so that greater investments will be made in developing the capabilities of locals. Many countries offer tax benefits to companies that train their employees. Kenyan floricultural operations, for example, depend a great deal on in-house training to meet their requirements.

Increasing technical education is a long-term strategy. Malaysia has expanded its support for technical education but has not been able to match its competitors in electronics, Taiwan and the Republic of Korea. Large-scale investments in the past in technical and managerial education in India played a key role in propelling the software sector when it took off in the late 1980s; they also paid off in adapting foreign technologies to local conditions in maize and grapes. Long-term investments in education and training involve more than what engineers generally mean when they talk about technology. While important aspects of technological learning are indeed structured or embodied in machinery or other physical artifacts, they also involve modes of organizing, coordinating, and managing activities. These latter capabilities often are much more difficult to develop than the needed engineering know-how.

Outcomes of the Ten Cases

We evaluate the outcomes of the 10 cases in terms of global competitiveness, technological parity with global leaders, control over global supply chains, domestic control of the industry, and domestic capabilities (table 1.4).

Global competitiveness here refers to the ability to attract FDI and export products of a technological sophistication on par with those of world leaders. Where FDI is not relevant, global competitiveness is defined as the competitiveness of products and services in the domestic market. Technological parity is somewhat difficult to measure, as not all of the cases have benchmarked technologies. It may be closely related to competitiveness in many cases, however, because it influences competitiveness through quality and costs. The extent of control over global supply chains refers to the ability to manipulate supply chains to one's advantage and to survive global marketing challenges. Domestic control refers to the participation of domestic firms in the sector and has implications for both technological deepening and value addition. Domestic capabilities relate primarily to human capital, skills, and the capabilities of workers engaged in the sector to absorb and implement technologies. The idea of domestic capabilities also extends to the capabilities of the government to design appropriate incentives for firms.

All the industries considered here are globally competitive, except Indian maize, which is produced for a vast domestic market, and grapes that are exported to the European Union in season. But the cases show important differences in domestic control and domestic capabilities, the two aspects that are relevant for technological deepening. The dissimilarity between the electronics sectors in Taiwan and Malaysia are well known. Malaysian electronics is dominated by MNCs that depend on parent companies for new technologies. Despite rapid growth, value addition in the Malaysian electronics sector is much lower than in neighboring countries that are close

Table 1.4 Outcomes of the 10 Cases

Sectors	Criterion of success				
	Global competitiveness	*Global parity—level of technology/ability to overcome technological barriers*	*Global dominance in marketing*	*Domestic control of industry*	*Domestic capabilities*
Software in India	Growing exports; global competitiveness in software services	Moved up from code writing to total solutions; inching into product development	High marketing capabilities for services; weak for products	Strong domestic control	Fairly high levels of capability; capability-based competitiveness
Electronics in Taiwan	Global dominance in electronics	Close to technological frontier; successful upgrading into higher-value-added products; innovation-grade technologies	Both domestic firms and MNCs exercise substantial control over supply chains	Considerable control by domestic firms	High domestic capabilities
Electronics in Malaysia	Global presence in electronics	Globally competitive in some of the technologies; failure to upgrade to higher-value-added products; threatened by competition from lower-cost production centers	Substantial control over global supply chains, with the participation of major producers	Sector largely controlled by MNCs	Limited domestic capabilities; shortage of engineers and skilled individuals
Oil palm in Malaysia	Global dominance in production and marketing	Acquired process technologies; substantial local development	Substantial control over global supply chains	Substantial domestic control over production and processing	Adequate domestic capabilities

Salmon farming in Chile	A dominant producer; globally competitive	Advanced technologies; timely assimilation of developments elsewhere	Global producers with links to international retailers	Substantial domestic presence in the industry	Growing domestic capabilities, but still an imitator
Wine in Chile	Leading exporter; globally competitive in quality and cost	Adoption of global best practices	Global marketing capabilities weaker than those of competitors	Considerable domestic control	Growing domestic capabilities
Grapes in India	Narrow window of opportunity for exports; not competitive globally	Technologies adequate only to meet minimum standards	Weak marketing capabilities; no brand image	Total domestic control; small producers participate through cooperative societies	Limited but improving domestic capabilities
Maize in India	Modest growth in the production and dissemination of technologies; questionable global competitiveness	Technologies, particularly varieties, not on par with what is available globally	No exports; industry serves large domestic poultry market	Substantial domestic presence in seed supply	Fairly high domestic capabilities in the seed industry
Fisheries in Uganda	Growing exports; fish supply is a limitation	Just adequate for exports; successful learning of processing technologies	Weak control over supply chains or participation in product development	Control by regional MNCs; weak domestic control	Weak domestic capabilities; largely manned by unskilled staff
Floriculture in Kenya	Growing exports; competitive; global dominance	Access to most advanced technologies	High levels of control over global supply chains	High level of domestic control; substantial participation of native Africans	Weak but growing domestic capabilities (of Kenyan Africans)

Source: Compiled by authors from case studies.

competitors. Malaysia's sector is threatened by the emergence of low-cost production centers in China and Indonesia. By contrast, Malaysia has far greater control over its palm oil industry.

Some resource-based industries have remained as production centers rather than technology centers, particularly the Chilean salmon farming, Nile perch fisheries, and Kenyan floriculture. In Chile and in Kenya, support industries that produce some inputs have developed domestically, but the bulk of the research and development that supports salmon farming occurs in other countries. Although rose breeders have operations in Kenya, these operations appear to be largely field trials. Both Kenya and Uganda depend on external consultants. Local research stations and organizations do not appear to have become engaged in the development of the sectors. In the case of Indian maize, on the other hand, technical education and substantial public investments in research are providing technical skills to the private seed industry, which is now developing. In the case of grapes, although technology was initially acquired through imported equipment and consultants, local research organizations and private companies are now building on foreign technologies.

Why Did Technological Adaptation Occur Where It Did?

Chen and Dahlman (2004) attribute the success of the East Asian countries to the presence of a "knowledge economy" in which technical and managerial skills were combined with dynamic information and innovations systems that permitted firms and research centers to tap into global knowledge networks. Noting the role of strong university–industry linkages and high-quality research in the development of the so-called miracle economies, Yusuf (2002, 2003) suggests that governments in other developing countries aspiring to catch up can facilitate such linkages. But dynamic information and innovations systems and university–industry linkages were relatively underdeveloped in the cases examined here (except for Taiwan), even where the industries were sophisticated and enjoyed significant global visibility (such as electronics and palm oil in Malaysia and salmon in Chile). Higher skills, by contrast, were a critical factor in spurring technological learning in every case.

The ten cases in this volume suggest that many of the factors that allow countries to attain technological parity with global leaders in a given industry are the same as those that produce overall economic development. However, a few special factors are vital for technological catch-up—notably an environment that nurtures the capabilities to learn and apply new technologies and the visible and facilitating hand of government. In varying degrees, such factors became the turning points in the development of the 10 industries.

In each case, several factors combined to develop the industry (see table 1.4). The relative importance of each is not always clear. Different factors may have contributed to technological adaptation at different stages in the evolution of the industries studied. Some factors (or policies) used to good effect several decades ago are now proscribed by the World Trade Organization or other bodies. But others remain pertinent options for developing countries today.

Success Factors

In all 10 cases, government maintained a *reasonably conducive macroeconomic* environment, although, in each case, the environment also included special incentives and institutions. During the period from 1980 to 2004 (the period for which we have data), the real exchange rate, a fundamental determinant of export growth, depreciated (except in Kenya, where it appreciated after 1995). Chile, India, and Uganda underwent fairly sharp real depreciations in the 1980s. Inflation, another key macroeconomic variable, displayed generally declining trends over the 1990s. In Chile, Kenya, and Uganda, inflation was significantly above 10 percent in the early 1990s. By the end of the 1990s, it was substantially below 5 percent in all countries except Kenya, where it was well within 10 percent.

Attracting FDI has been an important strategy in technological adaptation in all cases except those of palm oil in Malaysia and grapes and software in India. The definition of FDI used here is broad, in several cases covering foreign investments directed toward the production of inputs rather than the output under discussion. Input production in some cases led to the development of backward linkages that contributed to the scaling up of the sector. Chilean salmon is an example. The technologies involved in the production of the genetic material, an input into salmon food, greatly improved the efficiency of production, contributing to the viability of salmon exports. In Taiwan, a competitive industrial base in electronics was built and continually upgraded by local enterprises in partnership with MNCs, largely with support from public organizations. In countries where entrepreneurial skills were lacking, MNCs substituted for it to varying degrees.

The measures that countries have taken to attract FDI include the creation of export processing zones, tolerance of foreign ownership in an economy otherwise closed to foreign ownership, the building of infrastructure, and reductions in the costs of establishing and doing business (including tax incentives). In Taiwan and Malaysia, the electronics sectors successfully attracted FDI and continue to do so. Liberalization of the economy enabled technological upgrading in the Chilean wine and salmon industries, initiated large-scale floriculture in Kenya, and developed the Nile perch processing industry in Uganda. Endowments of natural resources (soil, climate, and water) created significant comparative advantages that drew foreign investors. Industry-targeted policies that permitted the import of capital goods, inputs, and expertise were critical to the development of maize, grapes, and software in India and floriculture in Kenya. Seeds for maize, postharvest machinery for grapes, computers for software development, and plantings for flowers—all were brought in under targeted policies.

In two of the three cases in which FDI did not play a leading role, the influence of foreign technologies was important at a later stage of the industry's development. In the case of Indian maize, foreign technology came in the form of imported seeds that were adapted by domestic and some foreign firms to local conditions. India's software exports, like its grape exports, were initiated by domestic firms, but, in the case of software, the firms benefited from prior exposure to the U.S. software industry. Later, as exports picked up, investments from the Indian diaspora in the Silicon Valley helped to fuel expansion. *The palm oil industry in Malaysia was seeded by the government and showcased as a leading export industry without any FDI participation.*

Investments in *human capital* have been critical to the development of all industries studied here, making the difference between being able to replicate rather than buy foreign technologies (Amsden 2001). In the service and manufacturing sectors, in particular, success depended on having a supply of well-trained engineers and other technical and managerial skills. Software development in India is the obvious case—knowledge embodied in software engineers and accumulated over a long period. Although controversial in the 1950s and for a time thereafter, heavy investments in engineering and management education were critical in creating the mass of technical professionals that drive India's IT industry today. Investments in technical education at home and postgraduate education abroad sped the development of a critical mass of technical skills that supported technological deepening in Taiwan. In the 1950s, Taiwan had four universities and four colleges; by 1989, the numbers had risen to 42 universities and 75 polytechnics. Taiwan is one of the few economies in East Asia that has developed innovation-grade technological capabilities. Malaysia, on the other hand, was successful initially in attracting investments into labor-intensive electronics operations by offering literate and disciplined labor, but it has been less successful in technological deepening because of inadequate technical skills. Thus, it still must buy from abroad most of the electronics technology it needs. By contrast, Malaysia became the world's leading producers of oleochemicals by channeling rubber cultivation and processing skills into the oil palm sector, with accompanying investments in domestic R&D, genetics education, and training.

Investments in the *life sciences and biotechnology* played a critical role in the development of agro-processing and agriculture. The Malaysian palm oil case displays a clever transfer of plantation-specific forestry skills away from the country's rubber plantations and toward palm oil, a forest crop that required similar skills. Public funding helped to promote research and training in oil palm genetics and agro-industrial engineering education. In India, the scaling up of an existing maize industry and the emergence of export-oriented grape production provide good examples of how a domestic pool of scientific skills underpins the transfer of technology to domestic firms to sustain technological adaptation. In collaboration with the private sector, India's government finances a variety of national and state-level research organizations to promote R&D. The lack of similar skills in Uganda has constrained the sector from moving toward technological deepening and higher-value fish products. In floriculture in Kenya, the paucity of indigenous biotechnology-related skills has thwarted the development of domestic capability to adapt foreign technologies.

Between these extremes lie the cases of salmon farming and wine production in Chile. In both cases, Chilean traditions and technical skills have facilitated mastery of foreign production technologies, but the scarcity of high-level scientific skills has stalled progress toward the development of domestic capabilities.

In addition to engineering and scientific skills, a critical mass of *entrepreneurial talent* capable of engaging with foreign firms, participating in technology transfer, and doing business with global buyers played a critical role in technological deepening and adaptation across cases. In some cases the pool of business skills was enriched by a tradition of entrepreneurship, but in several others investments in management or business education were critical. The business skills that emerged from the rich tradition of wine making and fishing in Chile were sufficient to enable domestic producers to engage foreign firms in successful technology transfer. In India, strong entrepreneurial skills were reinforced by discrete long-term public investments in advanced management education that produced a strong cadre of software entrepreneurs. The concentration of a large pool of young entrepreneurial talent in Bangalore drew the attention of the Indian diaspora, further enlarging the pool of entrepreneurs in the industry. Even without the diaspora (or foreign involvement) a large pool of small-scale farmers with experience in agriculture was sufficient for the scaling up of maize and grape production in India. In Taiwan, and to some extent in Chile, a different type of investment was made to create new entrepreneurs. As private firms were not willing to start and nurture businesses in a high-risk sector with new technologies, government agencies underwrote the risks by incubating the first few firms and nurturing them until they became economically viable, at which point they were sold to private buyers. The Taiwanese government offered lucrative financial packages to woo back its diaspora and so deepen the pool of entrepreneurial skills. The shortage of entrepreneurial skills has inhibited the emergence of large domestic firms in Ugandan fishery. In Kenyan floriculture, there are large domestic firms with managerial talent, but the paucity of scientific skills has stalled local research and upgrading.

Several countries have used *restrictive policies* to shape technological adaptation to their advantage. The major policy instrument used by the Taiwanese government until the mid-1980s was the local-content requirement imposed on foreign producers, which forced them to upgrade the capacity of local suppliers of parts and components. The local-content policy was a direct intervention by the government aimed at fostering a local parts and components industry. Another example from Taiwan is tariffs and, later, a ban on imports of color TVs from Japan to give Taiwanese firms time to benefit from domestic sales while building up export capability. Malaysia's government levied taxes on exporters of crude palm oil to encourage them to move into processed palm oil. In Uganda, a ban was levied on the export of unprocessed fish to neighboring countries to support domestic processing.

The governments of case-study countries made other industry-specific interventions, sometimes with the support of aid donors. Examples include the seeding of firms with new technologies in Taiwan's electronics sector and Chile's salmon farming industry, financial support for foreign–local R&D collaboration in Taiwan, publicly funded research to complement private R&D in maize in India, export processing zones in Malaysia, clusters and science and technology parks in Taiwan and Malaysia, lucrative packages to induce expatriates to return to Taiwan and Malaysia, technology transfer agreements and joint ventures to develop the Taiwanese electronics industry, support to develop markets and acquire technologies (grapes in India, wine and salmon farming in Chile), support for industry organizations (Chilean wine, Kenyan floriculture), and comprehensive crop and processing industry development (Malaysian oil palm). For fast-growing sectors that generated foreign exchange, there were special rewards—such as the privilege of importing inputs for India's software industry at a time of import restrictions, exemption from government regulations, the construction of special airport infrastructure for Kenya's flower exporters, and tax exemptions for Malaysia's electronics industry.

Technological Catch-Up in the Industrialized World

Achieving competitiveness through technological change has been an important policy concern for industrial nations. East Asia's impressive economic performance, driven by technological adaptation, is documented in numerous case studies that explore how firms, industries, markets (with their imperfections), and countries grow and compete. Examples include Amsden (1989), Amsden and Chu (2003), Lall (1992, 1996), Mathews and Cho (2000), and Nelson and Pack (1999). Opinions differ on the factors that were critical to the success of the miracle economies. Amsden (1989) and Amsden and Chu (2003) put the weight on the role of organizations like ITRI in Taiwan that intervened heavily to promote the adaptation of new technologies; others like Noland and Pack (2003) have argued that the return of the diaspora, in-house learning capabilities, and so on, may have been the driving force behind much of Taiwan's success. In either case, the contribution of the social capabilities of governments in East Asia to support export-led growth through technological change is well acknowledged.

However, the East Asian countries were not the first to catch up with the developed countries through technological adaptation. In this section, we briefly review a few examples of how some developed countries achieved parity with a technology innovator.

In the United States, the federal government played an important role in the technological development of almost every industry that is now competitive on a global scale (Ruttan 2001). The technological lead that the United States took over European countries after World War II was based on a resource-rich, capital-rich economic base and scale-intensive technologies, but also on high-tech industries that emerged from large public (and private) investments in education (Fagerberg 1994). In the same period, the National Institutes of Health solidified U.S. leadership in pharmaceutical and biomedical research, and the Department of Defense funded research that led to U.S. leadership in semiconductors, computers, software, biotechnology, and the Internet. The government's antitrust policy is believed to have played a role in the development of the software industry, largely independent of computer hardware manufacturing (Gordon 2002).

The damage done to the European economies during the war allowed the United States to take a large lead in productivity and innovation.

That gap narrowed as the industrial economies reached greater "technological congruence" through changes in the international economic environment (especially in the area of trade), improvements in "social capabilities," and the development of new institutions to promote technological change (Fagerberg 1994).[6]

Germany's productivity growth in the postwar period depended largely on government support that encouraged firms to adopt and adapt new technologies as they became available, rather than to concentrate on domestic inventions. Examples include the rapid adoption of computer-integrated manufacturing in the Federal Republic of Germany in the 1980s, and the technological upgrading of old firms in the former German Democratic Republic in the 1990s. Knowledge spillovers have been internalized with the help of Germany's public system of research and technology transfer, which has a mandate to support innovation in small and medium-size firms that cannot afford in-house R&D departments. The Arbeitsgemeinschaft Industrieller Forschungsvereinigungen (AiF) is a network of industrial associations that supports and performs cooperative research by small and medium-size enterprises. Many German universities and polytechnics engage in collaborative R&D under contract with private firms. The funding sources for research institutes, too, are both public and private (Siebert and Stolpe 2002).

In Japan, too, mastery of technology preceded the emergence of the capacity to innovate. Japan's emphasis on continuous improvement of production processes in publicly identified strategic sectors began with the importation of key technologies from western countries and the establishment of ambitious industrial standards by the Japanese government and industry. Early industrial policy encouraged sectors that faced income-elastic demands in international markets, had increasing returns, and facilitated learning through interaction with foreign firms. In later years, Japanese industrial policy focused narrowly on technological promotion, that is, on setting up a large-scale technological base for very-large-scale integrated circuits, other electronics, and semiconductors. The Ministry of International Trade and Industry (MITI) provided foreign-exchange privileges to favored firms and ensured that they were treated fairly in acquiring new technologies from foreign firms (World Bank 1993). Other factors that encouraged innovation were long-term development of high-quality human capital, teamwork between manufacturing employees and management, consistently high levels of R&D investment (funded mostly by the private sector), adequate property rights, and specialized worker training in production (Posen 2002).

Conclusions and Lessons

The cases in this volume deal with a wide variety of industries, technologies, economies, and societies lying for the most part outside the high-tech club of East Asia's miracle economies. The fine details of institutional arrangements, their sociopolitical environment, why particular interventions were made or not made, and whether other choices might have been better in certain cases are just a few of the issues not covered in this study. These form the agenda for future research.

The lessons of the cases are especially pertinent for the countries of Sub-Saharan Africa, because they suggest that it is possible for a developing country to exploit its latecomer advantage to catch up relatively rapidly with the developed countries in a particular industry by acquiring and adapting technologies innovated in the industrialized world.

Technological adaptation and upgrading was not easy in any of the cases we studied. In addition to machines and equipment, adaptation also meant mastering the aspects of technology that are buried in people and organizations (Lall and Urata 2003). Mastering this "tacit" side of technology required technical and managerial capabilities, and therefore time and resources. There were no shortcuts to technological learning and capability building, not even in the sophisticated case of Taiwanese electronics. The risks associated with the tacit element of technology reduced the incentives for firms to invest in technological adaptation, even when all players would benefit from such investments. To prevent spillovers from deterring firms from adapting new technologies, governments played a critical role in facilitating the transfer of technology from the developed countries. The nature of the interventions varied between one industry and another and often within similar industries across countries.

In encouraging specific nontraditional export industries to adapt new technologies, the case-study governments used a blend of conventional and industry-specific policies. The role of government was most effective when its support for *specific* industries was embedded in institutions and policies that were internally consistent, had an explicit purpose, and were blessed with political commitment (box 1.2). Chile's salmon-farming industry, for example, benefited from a strong industry-specific policy. In Kenya's floriculture industry, by contrast, the government's policy toward floriculture was initially weak but became stronger as the industry became a clear success and an indispensable source of foreign exchange.

Latecomers can avoid failures in industrial policy if the discipline of the global marketplace is left to be the judge of good performance. In the cases studied here, governments' commitment to support specific export *industries,* not firms, preserved overall efficiency and high rates of export growth. Within a favored industry, a firm's eligibility for reward—public support for adapting superior technologies—was tied to its export growth rate. Government interventions to promote the use of known and tried technologies minimized enormously the risks of failure associated with the costs of innovation, especially compared with the likely benefits.

A variety of familiar factors led firms to acquire, learn, and upgrade technologies to reach and retain global competitiveness. Two seem to

---

Box 1.2 A Common Set of Industry-Specific Government Measures

Our case-study governments adopted a common set of industry-specific measures to promote technological adaptation:

- They set goals to support a nontraditional or nascent industry because it was a valuable source of export-led growth (electronics, salmon, and wine) or foreign exchange (software in the early stage of development, fishery, and flowers) or jointly satisfied social objectives and earned foreign exchange (grapes, palm oil, fishery, and flowers).
- Political commitment in some cases took the form of a guiding national vision; in others, politicians valued the industry and did what they could to facilitate its development, sometimes in an otherwise closed economy.
- Governments rewarded winners and abandoned losers by conditioning public support on exports, thus ensuring that only the good performers reaped rewards for learning to use new technologies.
- With a few important exceptions (electronics, palm oil, and salmon), the government did not pick winning industries. Instead, it nurtured those it found growing faster than others. It pursued private sector–driven, export-led growth by facilitating the private sector's ability to adapt and master new technologies and sharpen its export competitiveness. In the few cases where it incubated the initial firms, it privatized them when they became commercially viable. As the industry grew, the government modified its role.
- Competition among domestic firms, and between domestic and foreign firms, was generally unrestricted.
- And the rule of law and property rights, including observance of international property rights, applied in all cases.

have mattered most: underlying institutions and government policies and interventions. Industry specificity was important in both.

Rarely can a single institution or government policy, acting alone, produce successful technological adaptation. However, as shown in the 10 cases examined here, a set of distinct government policies and complementary institutions that together accomplish certain vital tasks can do the job. Strong and effective institutions require "good organizational structures and modes of management in the public sector" (Nelson 2003). The *quality* of public and private capabilities jointly determines the extent to which institutions affect technological learning.

In the cases we studied, institutional arrangements were distinguished by where the strongest capabilities lay. Where *private* technological capabilities were strong, private firms initiated technological learning, while the government facilitated the process. Even where government capabilities were weak, fairly high quality technological adaptation occurred. The Indian cases are good examples. But where public and private capabilities both were weak, technological learning, especially upgrading, was weak. The cases of fishery and flowers illustrate this.

The other cases reflect institutional arrangements in which *government capabilities were strong enough* to design efficient and dynamic organizations that could promote technological learning in private firms and accelerate industrial exports and global visibility. In these cases—electronics, palm oil, and salmon—the extent of the learning was limited by private capabilities.

In every case, getting it right depended upon the degree of synchronization between institutions and government policies to motivate learning among exporters. The hallmark of these policies was industry specificity, which offers some useful lessons for other developing countries.

The success of *efforts to spin off new firms, seed new technologies, and encourage firms to invest in R&D* depends critically on public and private capabilities and can be quite risky. Taiwan was successful in building an indigenous electronics sector and Chile in initiating commercial salmon-farming operations. Successful governments gained a thorough understanding of the industry they sought to influence, remained flexible when implementing strategies, and exhibited the integrity and political commitment to preempt rent-seeking from interventions that gave administrators discretionary powers. Taiwan's government developed its strategies with the help of some of the leading industrialists and academics in the world, and its efforts were guided by professionals who had worked in industries that were at the technological frontier. However, these interventions required good governance and extraordinary capabilities in the public sector that are in short supply in many developing countries.

Openness to FDI and trade in inputs can be an important means of obtaining access to the best technologies. Some technologies were transferred through direct investments in the production of primary products and in the development and sale of technological inputs, including varieties (of fish, flowers, maize, grapes, oil palm) with greater production potential. Manufacturing technologies were transferred through FDI and through purchases or licensing of inputs and capital goods. In some cases, technology was acquired through government research initiatives. In software services, foreign technologies were acquired by local firms through direct, on-site learning and through outsourcing contracts. The recognition and protection of the property rights of foreign firms, notably breeders, has been a critical element in obtaining access to advanced proprietary technology.

Governments are more willing to adopt open economic policies where there is a *potential for exports*. But here, too, differences can be seen between the cases. Chile, Malaysia, and Taiwan were more open to investments and trade than were India, Kenya, and Uganda, though all of them began with restrictive import-substitution policies. Governments' facilitative policies were designed to promote exports from particular industries, usually in the form of exemptions to economy-wide regulations or special privileges to enhance the ability of firms to adapt new technologies. Facilitative policies usually emerged after lobbying by industry associations or recognition by the government of export potential.

Technological deepening, defined as an increase in the complexity of technologies in use, requires *public investments in tertiary technical and scientific education and research*. In the 10 cases, the countries with superior technical education and substantial public investments in research had higher levels of participation in technically more complex operations. A distinction can be made between Malaysia and India, on the one hand, and Uganda and Kenya, on the other. In the former, dissemination of technologies to other industries was higher, whereas in the latter, superior technology remained restricted to certain island industries, preventing technological deepening.

Public research may be an essential ingredient of an adaptation system, but it contributes significantly to technological upgrading only under certain conditions. It is usually most effective when it is pursued in tandem with a growing industry that is exposed to competitive pressure and demands excellence. Unfortunately, public research in developing countries often falls short of excellence because of limitations in manpower and other resources.[7] Even in Taiwan, where ITRI is believed to have played a significant role in the development of the electronics sector, the licensing of new technologies, the return of graduates, and in-house learning in local multinational companies may have been more important than domestic research capabilities (Pack 2000).

In the absence of links with an industry, public research may not be focused and effective. In such cases, public universities become ivory towers, as in Uganda, Kenya, and, to a lesser extent, Chile. Some useful models that emerge from the cases include the research consortia established by Taiwan, the new arrangement that the Chilean wine industry is developing with research institutions, and the linking of public grape and maize research in India with producer organizations. With funding from the sectors, and greater accountability to and guidance from a sector that is exposed to market forces, public research can make more relevant contributions, particularly in promoting the absorption and dissemination of technologies.

Firms adapt new technologies in response to competitive forces in domestic and international markets, not because they have access to government R&D facilities and support programs. The incentive for profit making and growth is the trigger, and in this sense technological adaptation is demand-driven. Some examples of well-established industries in which technologies were not upgraded until firms entered export markets are grapes in India, wine in Chile, and fish processing in Uganda. Exports and competition, especially in the global marketplace, guide technological upgrading and act as an efficient disciplining device.

Other forms of public action can be useful in specific cases. The governments of Chile and India provided *regulatory services* to enable exporters to comply with the phytosanitary standards of the developed countries. The governments of India, Malaysia, and Taiwan developed *industrial parks* to leverage synergies between MNCs, domestic firms, and public research centers (electronics, software). Support for *industry coalitions and organizations* helped to communicate the industry's needs to government in some cases.

Timing, location and initial conditions are important for technological adaptation. Government interventions in several cases were important in protecting an industry at a critical stage of its development but could not be repeated today. For example, the tariff bans used by Taiwan to protect the development of its color TV industry would not be permissible today under WTO rules and regulations. Similarly, the role played by Fundación Chile and CORFO in incubating firms for the Chilean salmon farming sector may have been critical at the time but might not be necessary for technological adaptation today.

However, Taiwan's strategies of setting up a council of foreign advisors, facilitating the building of technological capability, negotiating contracts with foreign investors to the benefit of local firms, and luring back its diaspora are still relevant for latecomers.

*   *   *

The crux of sustainable technological progress in a developing country rests on the quality of its technical and managerial capabilities in the private and public sectors. Building those capabilities requires large investments over time. Contrasting the cases of Indian maize and grapes with that of Ugandan fisheries, and the case of Taiwan's electronics industry with that of Malaysia, we found in the more successful cases deeper private and public capabilities to master new technologies and retain competitiveness—and a much clearer record of public investment in developing technical and managerial skills.

The cases illustrate interventions which, in the initial stages of an industry's development, can substitute for the shortage of technical and managerial skills in developing countries. Such interventions make heavy demands on the government's management and technical capabilities, which are weak in most developing countries. Public–private organizations, often operating in collaboration with an international counterpart, can compensate for some deficiencies in capability. Precisely how these practices can be harnessed by governments for local application will depend on country-specific (and possibly even sector- or industry-specific) institutional arrangements. For this reason, rote replication of another country's interventions can be an unfortunate choice, but paying close attention to the common elements of the institutions and policies that nurtured technological learning in successful cases can provide useful pointers.

Endnotes

1. Notable among them are the Brazilian informatics industry, the Finnish television industry, and the Korean fashion industry.
2. Earlier, Solow (1956) conceived technology as an exogenously produced public good that is freely available to all countries. Because all countries have free access to the same technology, their per capita incomes should grow at the exogenously determined rate of growth of technological progress. These assumptions could not explain the wide divergence in per capita income growth among countries.
3. Developing countries may lack the "social capabilities" needed to design effective institutions. Ohkawa and Rosovsky (1973), quoted in Fagerberg

(1994), coined the term social capability to describe factors that enhance a country's ability to import or engage in technological or organizational progress. The social capabilities of the newly industrialized countries were important in shaping efficient institutions.

4. The broad categories adduced here are adopted from Dahlman (1994).

5. Collective action by the industry appears to have been feasible in this case because of the presence of a large firm that organized the industry.

6. Abramovitz (1993, cited in Fagerberg 1994) explains that the other industrial countries were slow to catch up with the United States in the first decades after the war because of "technological incongruence"—the difficulty in applying new U.S. technologies in foreign settings.

7. Examples of sector-oriented public research that were unable to deliver the technologies initially needed for exports are horticultural research in the *initial* stages of the grape sector's development in India, fishery research in Uganda, and viticultural research in Chile. In India's case, private initiatives triggered the demand for technology, after which government moved to fulfill that demand by seeking foreign technologies. Now, publicly funded R&D is being targeted at the sector. In Uganda and Chile, weak domestic research capacity has led the sector to depend on imported technologies. To raise domestic research capabilities in the short run, for example, governments may consider public–private initiatives involving foreign firms, as in the case of Taiwan, until domestic capabilities are developed. In the long run, however, global competitiveness depends on strong public-sector technical training, higher education, and research institutions that support industry (Ruttan 2001).

References

Abramovitz, M. A. 1986. "Catching Up, Forging Ahead, and Falling Behind." *Journal of Economic History* 46: 385–406.

Amsden, A. A. 1989. *Asia's Next Giant: South Korea and Late Industrialization*. New York: Oxford University Press.

———. 2001. *The Rise of the Rest—Challenges to the West from Late-Industrializing Countries*. New York: Oxford University Press.

Amsden, A. A., and W. W. Chu. 2003. *Beyond Late Development: Taiwan's Upgrading Policies*. Cambridge, MA: MIT Press.

Arrow, K. J. 1962. "The Economic Implications of Learning by Doing." *Review of Economic Studies* 29: 155–177.

Chen, D. H. C., and C. J. Dahlman. 2004. "Knowledge and Development: A Cross-Section Approach." Policy Research Working Paper 3366, Washington, D.C. August.

Dahlman, C. J. 1994. "Technology Strategy in East Asian Developing Economies." *Journal of Asian Economies* 15 (4): 514–572.

Dahlman, C. J., B. Ross-Larson, and L. E. Westphal. 1985. "Managing Technological Development: Lessons from the Newly Industrialized Countries." World Bank Staff Working Paper 717, Washington, D.C.

Fagerberg, J. 1994. "Technology and International Differences in Growth Rates." *Journal of Economic Literature* 32 (3): 1147–1175.

Freeman, C. 2002. "Continental, National, and Sub-National Innovation Systems: Complementarity and Economic Growth." *Research Policy* 31: 191–211.

Gerschenkron, A. 1962. *Economic Backwardness in Historical Perspective.* Cambridge, MA: Belknap Press.

Gordon, R. J. 2002. "The United States." In B. Steil, David G. Victor, and Richard R. Nelson, eds., *Technological Innovation and Economic Performance.* Princeton, NJ: Princeton University Press.

Kim, L., and R. R. Nelson, 2000. *Technology, Learning, and Innovation.* Cambridge, England: Cambridge University Press.

Kremer, M. 1998. "Patent Buyouts: A Mechanism for Encouraging Innovation." *Quarterly Journal of Economics* 113 (4): 1137–1167.

Lall, S. 1992. "Technological Capabilities and Industrialization." *World Development* 20 (2): 165–186.

———. 1996. *Learning from the Asian Tigers: Studies in Technology and Industrial Policy.* London: Macmillan.

———. 2000. "Technological Change and Industrialization in the Asian Newly Industrializing Economies: Achievements and Challenges." In L. Kim and R. R. Nelson, eds., *Technology, Learning and Innovation.* Cambridge, England: Cambridge University Press.

Lall, S., and S. Urata, 2003. *Competitiveness, FDI, and Technological Activity in East Asia.* Cheltenham, England: Edward Elgar.

Mathews, J. A., and D. S. Cho. 2000. *Tiger Technology: The Creation of a Semiconductor Industry in East Asia.* Cambridge, England: Cambridge University Press.

Nelson, R. R. 1981. "Research on Productivity Growth and Productivity Differences: Dead Ends and New Departures." *Journal of Economic Literature* 19 (September): 1029–1064.

———. 2003. "A Program of Study Involved in Technological and Economic Catch Up." Unpublished draft.

Nelson, R. R., and K. Nelson. 2002. "Technology, Institutions, and Innovation Systems." *Research Policy* 31: 265–272.

Nelson, R. R., and H. Pack. 1999. "The Asian Miracle and Modern Growth Theory." *Economic Journal* 109 (457): 416–436.

Noland, M., and H. Pack. 2003. *Industrial Policy in an Era of Industrialization: Lessons from Asia.* Washington, DC: Institute for International Economics.

Pack, H. 2000. "Research and Development in the Industrial Development Process." In L. Kim and R. R. Nelson, eds., *Technology, Learning and Innovation*. Cambridge, England: Cambridge University Press.

Porter, M. E. 1990. "The Competitive Advantage of Nations." Glencoe, IL: Free Press.

Posen, A. S. 2002. "Japan." In B. Steil, David G. Victor, and Richard R. Nelson, eds., *Technological Innovation and Economic Performance*. Princeton, NJ: Princeton University Press.

Rodrik, D. 2000. "Institutions for High-Quality Growth: What They Are and How to Acquire Them." NBER Working Paper 7540, National Bureau of Economic Research, Cambridge, MA.

———. 2004. "Industrial Policy for the Twenty-First Century." Unpublished paper prepared for UNIDO. September. Accessed in March 2006 at http://ksghome.harvard.edu/~drodrik/papers.html.

Rodrik, D., and Arvind Subramanian. 2003. "The Primacy of Institutions— And What This Does or Does Not Mean." *Finance and Development*, International Monetary Fund, Washington, D.C. June.

Ruttan, V. W. 2001. *Technology, Growth, and Development—An Induced Innovation Perspective*. Oxford and New York: Oxford University Press.

Siebert, H., and Michael Stolpe. 2002. "Germany." In B. Steil, David G. Victor, and Richard R. Nelson, eds., *Technological Innovation and Economic Performance*. Princeton, NJ: Princeton University Press.

Solow, R. M. 1956. "A Contribution to the Theory of Economic Growth." *Quarterly Journal of Economics* 32: 165–194.

Uzawa, H. 1965. "Optimum Technical Change in an Aggregative Model of Economic Growth." *International Economic Review* 6 (1): 18–31.

World Bank. 1993. *The East Asian Economic Miracle—Economic Growth and Public Policy*. New York: Oxford University Press.

Yusuf, S. 2002. "Remodeling East Asian Development." *ASEAN Economic Bulletin* 19 (1): 6–27.

———. 2003. *Innovative East Asia: The Future of Growth*. Oxford and New York: Oxford University Press.

CHAPTER 2

India's Software Industry

Subhash Bhatnagar

Introduction

THE OBJECTIVE OF THIS STUDY IS to understand how the Indian software industry has been able to catch up to the technological standards of global leaders—successfully access, learn, and develop—while others in the developing world lag behind. The focus is on explaining factors that contributed to the phenomenal growth of software exports from India, including the role of institutions and government policy.

The Indian software industry has been a remarkable success story. It has grown more than 30 percent annually for 20 years, with 2008 exports projected at close to $60 billion. India exports software services to more than 60 countries, with two-thirds of the total going to the United States, including half of all Fortune 500 companies.

Economic policy has undergone substantial revision driven by this industry since India started opening up. Foreign exchange reserves are high, markets greatly influence policy, and a string of coalition governments have not deviated from economic liberalization. However, the benefits are uneven, as the very poor have been little affected. Unemployment is still high, and huge bureaucracies still yield to corruption. While problems remain, India is an emerging economy fueled by techno-savvy manpower and a world-class information technology (IT) industry.

Indian software firms quickly moved up the value chain, from performing low-cost programming abroad to providing comprehensive software-development services for overseas clients. Early government investment in technical education created a series of elite technical and management institutes that responded to a severe global shortage of technical manpower. Drawing on this abundant pool of Indian technical manpower, Indian firms sent English-speaking, trainable, and wage-hungry staff to on-site client facilities in the United States.

Indian professionals in Silicon Valley built networks based on their personal reputations and used their growing influence within U.S.

companies to help Indian companies take advantage of expanding opportunities for outsourced IT work. Once the potential of software exports was demonstrated, the Indian government helped build a high-speed data-communications infrastructure that allowed overseas Indians to return home and set up offshore sites for U.S. clients. The Indian "brand" image for affordable speed and quality grew.

Intense quality and productivity improvements built client value, and today these Indian companies deliver a wider range of software-development tasks, as well as benefits in new service segments, such as product design and information-science (IS) outsourcing. Many firms have met top certification requirements for quality standards in demand around the world. New data-protection practices are moving quickly to the top of the agenda.

A Remarkable Success Story

Even two decades ago, the image of India in the world was one of a country beset with poverty, unrestrained population growth, and substandard competitiveness. Today, due largely to the dramatic growth in the Indian software industry, India is an emerging economy with techno-savvy manpower building an impressive IT industry edifice. Parts of India continue to be beset with problems of poor infrastructure and poverty, but such an image has clearly receded into the background.

The industry has grown more than 30 percent annually throughout the last two decades. From about $50 million in exports in the late 1980s, the industry grew around 30 percent a year to more than $200 million exported by 1993 (table 2.1). In the boom years of the mid- and late 1990s, software exports grew 50–60 percent annually, reaching $6 billion by 2001. Even during the infamous "dot com" bust, software exports continued to grow by about 25 percent annually, which significantly outpaced growth in the software industry anywhere else in the world. Today India's software industry is back to a healthy 33 percent growth rate, with export projections for 2008 close to $60 billion.

In the last seven years, output has grown dramatically, from $1.86 billion in 1996/97 to $16.5 billion in 2004/5. The bulk of the growth came from exports: by 2003/4, nearly 85 percent of sales were in the form of exported software services (table 2.2).

Software exports in the last few years have accounted for 15 to 20 percent of all exports from India. This share has grown from less than 2 percent in 1995 and is expected to grow to nearly 26 percent of all

Table 2.1 Growth of Indian Software Exports

Year	Exports of software ($ millions)	Growth over previous year (percent)
1987/88	52	n.a.
1988/89	67	29
1989/90	100	49
1990/91	128	28
1991/92	164	28
1992/93	225	37
1993/94	330	47
1994/95	450	36
1995/96	734	63
1996/97	1,100	49
1997/98	1,759	60
1998/99	2,600	48
1999/2000	3,400	31
2000/2001	5,300	56
2001/2	6,200	17
2002/3	7,100	15
2003/4	9,200	30
2004/5	12,200	33

Sources: Compiled from Bhatnagar and Madon (1997) and NASSCOM (2001, 2002, 2003, 2004, 2005).
Note: n.a. = data not available.

Table 2.2 Annual Sales of the Indian Software Industry *($ billions)*

Year	Total	Export sales	Domestic sales
1996/97	1.9	1.1	0.8
1997/98	2.9	1.8	1.2
1998/99	4.0	2.6	1.4
1999/2000	5.3	3.4	1.9
2000/2001	7.8	5.3	2.5
2001/2	8.7	6.2	2.5
2002/3	9.9	7.1	2.8
2003/4	12.8	9.2	3.6
2004/5[e]	16.5	12.2	4.3

Sources: Compiled from NASSCOM (2004, 2005).
Note: Figures may vary because of changes in the conversion rate of the U.S. dollar from one fiscal year to another. e = estimate.

exports by 2008. This growth is particularly impressive in a period when growth in exports from other sectors, such as jewelry, garments, and manufacturing, has also been high (NASSCOM 2002, 2004).

The software industry's share of GDP has grown correspondingly. IT's share of overall GDP in 2004/5 was 4.1 percent. It is likely to grow to 7 percent by 2008 (with software and services alone accounting for 2 percentage points).

Software exports are primarily information and software services rather than products. While firms export to more than 60 countries, the United States represents nearly half the global market and is the primary destination.[1] The United States (69.4 percent), United Kingdom (14.5 percent), Japan (3.0 percent), Germany (2.8 percent), and Singapore (1.8 percent) account for nearly 91.5 percent of India's software exports.

In 2005 India's share of the global market for outsourced IT services stood at 3.3 percent. It served nearly half of all the Fortune 500 companies. India's market share has grown rapidly, and in terms of absolute share its position is second only to the United States.

The report begins by briefly characterizing the organization of the industry today. A historical overview of the software industry explains how India's software sector has evolved since the early 1970s. Important explanatory factors are discussed in the following sections, including the role of government investment in technical education, the role of the Indian diaspora in promoting and facilitating early growth of the industry, the facilitating role of government policies, and the role of the industry association in promoting the Indian software "brand" abroad and in lobbying for favorable policies. Final sections discuss the impact of the industry on the Indian economy, sector strategies for moving up the value chain, and the role of research and development (R&D) in the industry. A short discussion of the future concludes the chapter.

Industry Organization

The Indian software industry has a pyramidal structure, with a few large indigenous firms dominating the sector (table 2.3). Among the 3,000 firms exporting software from India, the three largest firms each have more than $1 billion in annual sales.[2] At the other extreme, the smallest 2,900 firms have annual sales of less than $10 million, with most less than $2 million. The largest five firms account for 32 percent of software exports, while the smallest firms own a 14 percent share.

Table 2.3 Structure of Indian IT Services and Software Exports Industry, 2002–4

Annual sales ($ millions)	Number of companies		
	2001/2	2002/3	2003/4
>210	5	7	9
105–210	5	5	8
52.5–105	15	15	24
21–52.5	27	41	53
10.5–21	55	71	56
2.1–10.5	220	244	367
<2.1	2,483	2,644	2,653
Total	2,810	3,027	3,170

Source: Compiled from NASSCOM (2005: 74).

Smaller firms play a more significant role in the domestic market, where they supply software services to small and medium domestic firms in different sectors. For their export contracts, the smaller firms have primarily entered into a variety of linkages with individuals and small companies abroad for marketing access. There are few horizontal partnerships between small and large vendors. Some early efforts by established firms to subcontract to smaller ones failed, because the smaller partner tried to make direct contact with overseas clients. In recent years, a few small firms have developed products for the domestic market with some success, but exports have been very limited.

Multinational companies arrived in India relatively late and account for only about a quarter of exports. Multinational firms use their Indian operations primarily as export platforms. Increasingly multinationals are setting up shop in India to conduct sophisticated software-development activities and as a captive source of R&D, utilizing India's pool of highly trained engineers.

Many Indian firms have been started by entrepreneurs who acquired some wealth and experience working in larger established firms and then set up new companies. Many of the corporate leaders did their graduate studies in the United States and/or worked in Silicon Valley, so they have a keen understanding of the software-development process. The high profitability and relatively low risk of the industry has attracted a large number of professionals. Also, entry costs are relatively low. To start a software company does not require huge investments in land, plants, or machinery. Most of the assets can be acquired on lease without a high up-front investment. The lead time for generating revenues is also much shorter than in many other industries.

Industry Growth

Beginnings in Bodyshopping

The birth of the software industry in India began in 1970 with the entry of Tata Consulting Services (TCS) into the domain of outsourced application-migration work. In the late 1960s, the Tatas, a large conglomerate, created TCS as a central service center for the Tata Group of companies. A few young MIT-trained Indian professionals were recruited, and a large computer system was imported. With IBM having been thrown out of India, the concept of outsourcing application-development work had become a necessity for Indian companies. Utilizing its excess computer capacity, TCS began doing outsourced application work for organizations such as Central Bank of India and Bombay Telephones. Within a few years, TCS began sending young Indian engineers to a joint-venture partner in the United States, Burroughs, for training. The trainee engineers excelled at doing platform conversions, and TCS started earning conversion assignments for its engineers in Germany and elsewhere.[3]

Later, a new company named Tata-Burroughs was formed. Tata was keen to exploit the personnel placement, or "bodyshopping," opportunities, whereas Burroughs was interested in selling hardware to the Indian market. After a few successful years, the partnership was broken at the behest of Unisys, which had by then acquired Burroughs in the United States, and the company was rechristened as Tata Information System Limited. A U.S.-trained Indian electrical engineer took over management of TCS in 1969. He used his influence in the Institute of Electrical and Electronics Engineers to further promote TCS and founded the Computer Society of India with fellow scientists and professionals from the Tata Institute of Fundamental Research. Many of these professionals later moved to government and became very influential policy makers. These early networks played a very useful role in overcoming severe administrative and procedural constraints in India's otherwise closed economy during the 1970s and 1980s. Following the success of TCS, many other companies were set up in India.

Beginning in the 1970s, a growing shortage of engineers for the expanding computer industry in the United States and Europe, an oversupply of Indian engineers relative to domestic demand, and a growing international reputation for the skills of Indian engineers provided an opportunity for bodyshopping, in which Indian firms such as TCS sent Indian engineers overseas to do software programming on-site, mostly in American firms for limited, billable projects.

During the first phase (1968–84) of exports, four types of companies interlinked in direct and indirect ways to facilitate bodyshopping (Biao 2002). There were established companies in India, such as TCS and Infosys Technologies, which supplied programmers to large multinationals in IT and non-IT sectors primarily in the United States. These multinationals also recruited programmers through local U.S. companies, such as Mastech (now iGate) and Information Management Resource, established by Indians living in the United States. Such companies, in turn, recruited manpower through local search agents (small companies run by Indians in the United States). These agents, from several states in the United States, would contact local agents in India from a multitude of small companies and operators. The responsibility of collecting résumés, forwarding them to U.S. placement agents, and preparing visa and contract finalization with the programmers was done by the agents in India. The programmers were paid low wages. Commissions were charged by different members of the supply chain. Sometimes, there were subagents dispersed in different towns and cities in India. There was an interesting network among revolving players. Programmers who returned to India after a stint overseas would join the pool of software engineers who could be hired by the established companies in India. Often, programmers sent on-site by large Indian companies would move laterally to another assignment in the United States through a local U.S. agent to prolong their U.S. experience. Later, they would return to India and be in the market for local Indian agents to hire them. The Indian diaspora played a key role in the bodyshopping exports. Arora and others (2001) also report several instances where Indian immigrants in the United States helped U.S. buyers to locate Indian suppliers. Field interviews with U.S. customers reported that the impetus for outsourcing to India came from employees of Indian origin.

The development of bodyshopping links between firms in the United States and India was due mainly to the large Indian diaspora in the United States, many of whom worked as professionals in the American IT industry. They promoted and facilitated connections between U.S. firms and firms or agents in India who could supply programmers for on-site work in the United States. The successful growth of bodyshopping was due to the skills of Indian entrepreneurs and the steady supply of low-cost and trainable Indian engineers. Bodyshopping was, and continues to be, an attractive strategy for new entrants into the industry, requiring nothing more than knowledge and established relations with a few potential clients.

The severe shortages in skilled technical labor for the growing IT industry in the West and the liberal immigration policies of the

United States fueled the emergence of bodyshopping. For example, in the 1990s, annual growth of IT expenditures on equipment in the United States was 24 percent, and in Germany and Britain, just under 20 percent. At the end of the 1990s, the shortage of programmers, systems analysts, and computer engineers was estimated at about 346,000 in the United States and 30,000 in Canada.

The Era of Outsourcing was Facilitated by the Government

While the initial development of India's software industry was based primarily on on-site bodyshopping work in U.S. firms, in recent years, the trend has been increasingly for Indian firms to conduct software development for U.S. clients offshore in India. This shift was the result of a maturing of India's software industry, its international reputation in the last 15 years, and the development of necessary infrastructure and communications technologies in India that has made offshore work possible.

As the Indian software industry matured, increasing client confidence in Indian capabilities and quality standards enabled Indian firms to move their work offshore. With maturity has come a goal to move up the value chain. Many new companies were set up in the 1980s by entrepreneurs with ambitions of creating world-class software-development centers. Firms that had started primarily as subcontractors for technical manpower gradually shifted to managing complete parts or phases of projects, and then to delivering complete solutions from India. During this phase, most companies made significant efforts to assimilate good practices in project management and quality and to acquire internationally recognized quality-standards certification. The National Association of Service and Software Companies (NASSCOM) played an aggressive role in promoting the India brand abroad. In some ways, during this period, India was building a launching pad for the eventual take-off of its software-service industry.

During this period, the Indian government played a facilitating role in advancing the industry and enabling offshore work in India. Recognizing the growth potential of the software industry, the government in the 1980s took key policy actions to open up the sector. Further policy reforms enacted since the late 1990s have facilitated the development of telecommunications and other infrastructure required for offshore work. A policy change in 1998 effectively ended a monopoly on Internet-service-provider (ISP) gateways that allowed India's private sector to offer needed bandwidth to the growing industry. Two years ahead of the World Trade Organization (WTO) commitment,

Table 2.4 Comparison of Indian Software Export Revenue by Delivery Location, 1988–2005 *(Percent)*

	Type	
	Delivered at overseas client site	*Delivered in India*
1988	90	10
1995	66	33
1998/99	54	44
1999/2000	57	44
2000/2001	56	44
2001/2	45	55
2002/3	43	57
2003/4	36	64
2004/5	29	71

Sources: Compiled from Kumar (2001: 4,280) and NASSCOM (2005: 58).

India liberalized international long-distance in 2002. In 1990, the government created software-technology parks (STPs) in 39 locations across India to provide software companies with access to high-speed data communications and single-window clearance for regulatory compliance. While few of the larger firms have made use of the STPs, they have provided opportunities for new firms to launch, and smaller firms to grow, with little investment.

Table 2.4 shows the shift in the last 15 years from client sites overseas to "offshore" business in India. The revenue from services provided in India increased from only 10 percent in 1988, to 33 percent in 1995, to more than 60 percent by 2003/4.

The Indian software industry is now in its third phase: take-off. Today, most leading companies are operating in the high-end software-services business and are also making efforts to enter the products segment. A new breed, led by second-generation software entrepreneurs, are setting up product-oriented companies. The industry has weathered ups and downs in the global market, maintaining a high rate of growth. It moved to center stage in the domestic media because of its visibility in the United States, high-market capitalization, and wealth creation for its employees. It is a source of national pride and, as a consequence, continues to attract disproportionate government attention. The government set ambitious software-export targets and has provided the policies to enable the industry to achieve those targets. Indian companies have fine-tuned the "offshore model" and project their brands as service companies that are recognized in the media for their leadership in adopting the best management practices. Companies have moved further up the value chain, improving productivity and targeting new geographies, vertical domains, and businesses.

Investment in Technical Education

Beginning in the 1960s, India's public investments in technical education provided the foundation for the growth of the IT industry. A lack of adequate opportunities for Indian engineers in the domestic economy ensured an abundant supply of high-quality and cost-effective workers for India's export-software industry.

In the 1960s, the government created a series of elite institutes for higher education in engineering and management, in collaboration with leading universities in the United States. Five Indian institutes of technology (IIT) and two Indian institutes of management (IIM) were set up in the 1960s. The IITs were set up through technical collaboration with the most industrialized countries of the time. At current prices, the cost of setting up an IIT was perhaps 10 billion Indian rupees (US$200 million). The IIMs were set up with active collaboration from two leading business schools in the United States. The cost of setting up an IIM at current prices would be approximately 1.5 billion rupees (US$30 million). The annual cost of undergraduate education at IIT Delhi is 150,000 rupees per student (US$3,000, excluding the capital investment and depreciation)[4] and that of graduate education at IIM Ahmedabad is 280,000 rupees (US$5,600) per student.[5] Both types of institutions attracted a large number of U.S.-trained Indian faculty. Subsequently, most state governments set up regional engineering colleges (REC) that attracted students from all parts of the country.

Professional education in India attracts large numbers of applicants. Most of the IIT, REC, and IIM admissions programs are able to choose one out of 100 or more applicants. This selectivity and reasonably good training produced the high-caliber engineers that formed the backbone of the software industry in its early years. A unique feature of these high-caliber engineers was their willingness to work as programmers, partly due to the shortage of lucrative jobs in a closed economy. IIT engineers who moved to the United States and those who worked for companies in India in the 1970s and 1980s built a reputation that helped Indian companies procure software-development contracts in the initial years.

Subsequent expansion of technical and management education in India helped fuel the numbers needed for the high growth. India has a large higher education system with about 253 universities and nearly 13,000 colleges producing 2.5 million graduates every year. Nearly

300,000 engineering degree and other graduates enter the workforce every year (table 2.5). Since engineers from any specialty are willing to work in the software industry, so far, supply has kept pace with the demand. Many initiatives by the Ministry of Human Resource Development (HRD) (discussed later) helped to multiply technical institutions and technically qualified graduates.

Recently, privatization of technical education produced an ever-increasing technical labor supply to meet demand (table 2.6). By the end of the last decade, there were an estimated 660 engineering colleges in the country.

Given the estimated demand, manpower will not be a bottleneck. Manpower projections for the software sector in 2008, when India expects to export $60 billion worth of software, also indicate sufficient labor supply. Estimates indicate that there would also be newer activities, such as products and technology services, which would employ 140,000 professionals.

Much of this manpower supply is due to the privatization of technical education. In 1999, output from private institutions outstripped the state output (table 2.7). Even though the number of engineers has increased, quality has not deteriorated significantly. In 1969, the IITs produced around 1,350 engineers.

The private cost of starting an engineering college that produces 500 students per year is currently about 500 million Indian rupees. The private institutions receive no government funding.

One reason for the concentration of software companies in the south is the proximity of the locations to a very large number of engineering colleges.

The HRD ministry played a facilitating role in ensuring adequate supply and quality of the technical labor force. Ministry policies encouraged the creation of private engineering colleges and industry IT-training institutions. With the proliferation of new private colleges and IT-training institutions, the HRD ministry developed mechanisms to ensure quality control, including the establishment of an All India Council for Technical Education to regulate technical education and an accreditation system run by professional societies, such as the Computer Society of India, to monitor private training institutions.

Additionally, the introduction of a master of computer applications (MCA) degree in many universities in the late 1980s was aimed at producing graduates with the combination of technical and management skills required for the expanding IT industry. While the pool of MCA graduates became a primary source of recruitment, the programs tended to be stronger in technical, rather than management, skills. Recognizing that raw technical recruits are generally unprepared to

Table 2.5 Labor Supply in the Indian Information-Technology Sector
(Thousands of persons)

Labor supply	2002/3	2003/4	2004/5	2005/6	2006/7
Engineering graduates	259	215	284	348	382
Degree (four-year)	129	112	155	210	235
Diploma (three-year)	130	103	129	138	147
IT (computer science, electronics, telecom) professionals	126	141	165	181	193
Engineering IT graduates (degree)	81	95	100	111	117
Engineering IT graduates (diploma)	45	46	65	70	76
IT professionals entering workforce	72	80	94	103	109
Engineering IT graduates (degree)	47	55	58	64	68
Engineering IT graduates (diploma)	25	25	36	39	41
Non-IT engineers entering IT workforce	40	40	40	40	40
Graduates in other disciplines entering IT workforce	35	30	30	30	30
Total fresh IT labor supply	147	150	164	173	180

Source: NASSCOM (2005: 158).

Table 2.6 The Indian IT and IT-Enabled Services (ITES) Sectors
(Thousands of professionals employed)

	1999/ 2000	2000/ 2001	2001/2	2002/3	2003/4	2004/5ᵉ
Software exports sector	110	162	170	205	270	345
Software domestic sector	17	20	22	25	28	30
Software, captive in user organizations	115	178	224	260	290	322
ITES, BPO	42	70	106	180	253	348
Total	284	430	522	670	841	1,045

Source: NASSCOM (2005: 156).
Note: e = estimate; BPO = business processing outsourcing.

Table 2.7 Engineering Colleges in India, 1998–99

Region	Number of colleges	Enrollment capacity	Enrollment in self-financed colleges (% total enrollment)
Central	50	9,470	52
East	25	4,812	26
North	140	25,449	42
West	140	34,165	74
South	308	82,597	79
Total	663	156,493	69

Source: Arora and others 2001.

work immediately as software engineers, most large companies rely on extensive training divisions.

Since engineers were willing to work as programmers in a domestic environment with few job opportunities, growth was also driven by larger salaries in the IT industry abroad.

Tackling the Manpower Issue: Firm-Level Efforts

Human-resource development (HRD) is critical in software companies where 95 percent have formal training divisions and learning-needs analysis programs. Minimum training per employee is 40 hours. This covers both technical and behavioral training, and the proportions vary between managerial and technical positions.

A 2003 survey by Hewitt Associates and NASSCOM provides insights into recruitment and training practices (NASSCOM 2004:

189–191). A large proportion of companies spend about 7 percent of total employee costs on recruitment. Nearly 26 percent of the total manpower requirement is met through campus recruitment of fresh graduates. In fact, nearly 44 percent of such companies source campus recruits from engineering institutions only. Nearly 17 percent of the companies also recruit from management campuses. Most companies run large lateral-hire programs that are based on written aptitude and technical tests followed by interviews.

Statistics reveal that among innovations in the software industry, the most significant interventions have taken place in human resources. For example, nearly 60 percent of companies have formal employee suggestion systems from which 28 percent of suggestions are actually implemented. Another study (Bhatnagar and Dixit 2004), of two large organizations, reports how special attention is paid to organizational innovations that meet the challenges of external and internal imbalances. They suggest that current software-service activity has built-in incentives to innovate up the value chain toward more complex services, software products, and hardware-software integrated products.

In terms of rewards and recognition, a majority of companies use market data to determine basic pay. Employment and wages in the software sector have increased over the last decade but not enough to erode India's competitive advantage. The differential between client countries and India remains very high (table 2.8). In terms of competition, countries comparable to India in overall cost/quality/delivery metrics have significantly higher wages than India. Moreover, expanded capacity of Indian engineering colleges will ensure that the supply is adequate for the likely demand in the next five years.

With the entry of many multinationals in the Indian market, there is competition for the best talent. The top 10 companies reportedly have retention rates over 90 percent, indicating a fairly stable environment. It is interesting that Indian companies are neck and neck with multinationals in these surveys (Dataquest 2003), indicating the highly professional nature of the HRD function in the industry.

Nearly 87 percent of the companies reported a routine performance-management process in which managers and employees together set performance goals, and 30 percent of the organizations had a 360 degree feedback system. A key measure of employee satisfaction is the attrition rate. Most large companies have created work environments that contain attrition to low levels. This is a significant benefit to assure clients that disgruntled employees are not distributing confidential information. To date, hardly any cases of this crime have been reported.

Table 2.8 India Compared with Competing Countries, 2003/4

Parameter	Canada	China	India	Ireland	Israel	Philippines	Russian Federation
Export industry size (US$ millions)	3,780	1,040	9,500	1,920	900	640	165
Export-focused professionals	45,000	26,000	195,000	21,000	15,000	20,000	5,500
IT employee costs (US$ per year)	36,000	9,600	5,000–12,000	25,000–35,000	25,000	7,000	7,000
Number of CMM-5-certified companies	NA	2	60	0	0	n.a.	3
Quality of IT labor force	High	Low quality	High	High to moderate	High	Moderate	High quality
Infrastructure	Good	Average	Average	Good	Good	Good	Poor
Unique positives	Near shore, highly compatible culture with the United States and United Kingdom	Large IT workforce	Abundant and skilled (English, highly qualified, exposed to clients) workforce, robust project management experience	Large MNC presence, early start	Large product development (shrink wrapped) experience	Good English skills and cultural compatibility	High quality engineers
Main negatives	High costs	Lacks project managers	Ordinary infrastructure	High costs	Regional unrest	Lacks project managers	Unstable economy

Source: NASSCOM 2004.
Note: n.a. = data not available.

The Role of the Indian Diaspora

The success of Indian IT professionals in the United States was a significant factor in the development of the software sector in India. The stream of U.S.-educated Indian professionals who joined the IT industry in Silicon Valley and met technical, managerial, and entrepreneurial success created a positive image of the capabilities of Indian professionals. By the year 2000, Indians headed 972 Silicon Valley technology companies, accounting for $50 billion in sales and nearly 26,000 jobs. Indians headed about 3 percent of technology companies started between 1980 and 1985; 10 years later, they headed up to 10 percent of the companies.

The Indian diaspora also expedited bodyshopping by showcasing the value of Indian programmers and fostering connections between software firms in the United States and India. Some have returned to work for multinationals that have established Indian subsidiaries, while others have launched firms in India. A few straddle both countries, able to speed the transfer of know-how about emerging markets and technologies and willing to nurture long-term relationships across borders.

In spite of the large English-speaking, technically educated Indian population, the number of doctorates awarded in the United States to scholars from India has been far fewer than in China and Taiwan during the 15-year period from 1985 to 2000. Whereas the number for India varied between 500 and 1,000 per year, the number for China has varied between 2,000 and 3,000 in the last 10 years. Interestingly, Taiwan, as a small country, has had more U.S. doctorates awarded than India.

Government's Facilitating Role

Although the story of the Indian software industry is a story of private initiative, the government played a supporting role with public funding of a large, well-trained pool of engineers and management personnel who could forge the Indian IT industry into a world-class treasure in a short time. Early government support came from a few visionary civil servants who championed the cause and helped the industry find its way through a labyrinth of regulations, making exemptions wherever possible. Later, policies that encouraged local firms and direct foreign investments were introduced.

The government targeted software exports once the market identified the industry's potential and created the necessary institutions. As early as 1972, the Department of Electronics introduced a policy to permit duty-free imports of computer systems if importers would promise to export software and services worth twice the value of the imported computers within a specified time. This policy helped a number of leading companies in their inception stage. In the 1980s, the department gave software developers a further boost by initiating policies friendly to software exports. It formed a software export–promotion council and liberalized import rules for materials needed for the industry. Software was explicitly targeted as a key sector for export promotion. In the late 1990s, the government created four major task forces comprising chief executives of leading software companies to study the sector and recommend actions and then acted on most of the recommendations.[6] At that time, the Department of Electronics became the Ministry of Communication and Information Technology. This was followed by the IT Act to address a large number of issues. In addition to these federal interventions, many states promoted the local software industry by improving infrastructure and IT education, and by providing more facilitating environments.

With the beginning of economic reforms in the early 1990s, efforts were made to attract foreign as well as domestic investment. Foreign companies were permitted to establish fully owned subsidiaries in the electronics-export processing zones. Within the Ministry of Finance, there was greater recognition of India's comparative advantage in the sector, as it abolished entry barriers for foreign companies, made available fast, low-cost data-connection facilities, and reduced and rationalized duties, taxes, and tariffs.

The Reserve Bank of India adopted several measures to support the IT industry. These included the simplification of the filing of Software Export Declaration Form (SOFTEX) and the acquisition of overseas parent-company shares by employees of the Indian company. Also, companies whose software sales were over 80 percent could grant stock options to nonresident and permanent-resident employees, foreign exchange could be freely remitted for buying services, and companies that executed contracts in "computer software" abroad could use income up to 70 percent of contract value to meet contract-related expenses abroad.

Tax holidays were given on company profits, although the government is progressively phasing out these deductions. Tax breaks from corporate income and tax on profits was available to units in any free-trade zone, any software-technology park, or any special economic zone to the extent of 100 percent of the profits derived from the business. These deductions will not be available from fiscal 2009/10 onward.

Indian direct investment in joint-venture (JV)/wholly owned subsidiaries abroad was simplified and a fast-track window is available for large investments. IT software and services companies in India can acquire companies overseas through American Depository Receipt/ Global Depository Receipt stock swaps without prior approval for up to $100 million or 10 times the export earnings of the previous year.

While the government has enacted significant reforms in the area of intellectual property rights (IPRs), and has joined the WTO and trade-related aspects of IPRs, the reforms have, so far, not led to a surge in patents in the Indian software industry, nor have IPRs been perceived as effective in protecting innovations in the Indian software industry (Gupta 2004).

Several policy reforms in the telecom sector helped accelerate the domestic and export industry. In 1998, a national telecom policy was announced to clarify the role of the regulator, transition from license fee to a revenue-sharing model, and open domestic long-distance to private operators. The ISP gateway monopoly ended in 2000 and permitted private companies to set up international gateways. In 2002, international long-distance was liberalized two years ahead of WTO commitments and competition increased in cellular markets. As a result, India's teledensity, the number of phones per 100 people, increased to five, and cellular penetration overtook the landline penetration.

Recognizing the growing need for manpower in the software industry, the Ministry of Human Resources Development took the following actions:

- Helped create and expand computer-science departments in existing engineering colleges
- Eased policies in order to enable private sectors to open educational institutions without public funding. A large number of engineering colleges were opened in the private sector
- Introduced quality-control systems for engineering colleges and other IT-training institutions, such as the All India Council for Technical Education and an accreditation system run by professional bodies, such as the Computer Society of India, to monitor private training institutions
- Encouraged the private sector to open training institutions. At its peak, nearly 1 million Indians were being trained in a year, with the IT-training industry earning some 10 billion Indian rupees in 1998 with no government subsidy

Software Technology Parks

The creation of NASSCOM in 1988 and later establishment of software technology parks (STPs) in 1990 represented a fundamental approach to policy making for the software industry. An important institutional intervention was the establishment of STPs to provide infrastructure for private companies to export software. Established in 39 locations, including most major towns, they provided ready-to-plug IT and telecom infrastructure. STPs also allowed single-window clearance for all regulatory matters. The benefits and approvals for STPs are similar to those of export-oriented units. Incentives provided in the export-import policy are also applicable to STP members.

The companies registered with these parks account for about 68 percent of software exporters. Many of these companies have not benefited from the actual STP infrastructure in any significant way. Perhaps the major contribution of these STPs was to enable new enterprises to launch and small and medium enterprises (SMEs) to grow. Already-established companies merely registered with these parks but did not use the infrastructure that was created.

The performance of STPs has been variable. Where the environment was right, the STPs enabled SMEs to set up and grow. On the other hand, in Gujarat, total sales from 60 to 70 SMEs was 1,000 million Indian rupees (US$22 million), miniscule in comparison with industry norms. The Gandhinagar STP had a membership of 300 companies, many of which may have been attracted because of the incentives. However, only 60 to 70 are active. Out of the 5 Mbps (megabits per second) bandwidth available for use, hardly 2 Mbps is being utilized.

One of the STPs' key contributions is providing high-speed data-communication services to the industry. The software technology parks of India (STPI) had international gateways at 39 locations (2003). For the last mile, users can connect through point-to-point and point-to-multipoint microwave links, and terrestrial fiber/copper cables were used (where feasible). The up time of STPI connections is 99.9 percent. STPI works with major international telecom operators, such as AT&T, Sprint, MCI, Intelsat, and British Telecom. STPI offers two main services: Softpoint service, secure and exclusive digital circuits for data and voice transmission; and SoftLink, Internet access on a shared basis.

Cluster Development

The software industry in India has been concentrated in Bangalore, Hyderabad, Chennai, Mumbai, Delhi, and Pune. Well-researched reasons to explain why these locations have become fertile centers have not been propounded. Many centers do not necessarily have the best infrastructure. The one reason often suggested is the availability of a large pool of locally trained manpower as the distribution of engineering colleges closely mirrors the distribution of the software industry. The other significant reason may be the attractiveness of these locations for young and upwardly mobile professionals (van Dijk 2002). Most have a strong cosmopolitan character. Other authors (Srinavas 1998) have reported the importance of a lower cost of living and favorable climate as reasons for choosing a location lending support to this argument. For example, Bangalore perhaps boasts of the best education system in India and, therefore, is very attractive as a place for professionals.

The presence of progressive chief ministers and special state government benefits to attract firms may explain the growth of the Hyderabad center, but other locations thrived without such political support.

Because of the high degree of professionalism in most exporting companies, there is consistency in the quality of experienced manpower that sometimes moves laterally from one company to the other. Other than this movement, there is not enough evidence of horizontal linkages between IT firms located in the clusters. Even though five to six centers account for more than 90 percent of software exports, the typical clustering effect associated with Silicon Valley does not seem to exist in these centers. Perhaps a lack of informal knowledge exchange exists because many companies view other companies as close competitors. Most Indian companies operate in a narrow market space, such as in the U.S. market, in two to three verticals, which essentially account for just 5 percent of the total global outsourced market. Some researchers have corroborated the absence of these linkages in the domestic market (Basant and Chandra 2004) and found very little evidence of horizontal interaction in a case where there were four Indian subcontractors in Bangalore working for the same multinational.

The Role of the Industry Organization

NASSCOM, India's software-industry association, was founded in 1988 and has been a vocal and potent force in lobbying for policy reforms, including rules that limit access to capital markets, issuance of stock options, easing rules on foreign-currency transactions, and improving telecom infrastructure.

NASSCOM played a significant role in establishing a brand image for India in the global software-services markets by participating in global trade fairs and events and organizing learning events in India that feature prominent experts from major markets. Through its annual reports, NASSCOM has become the most reliable source of data and information about the Indian software industry. NASSCOM activities were influenced by the dominant software players, who share a great commonality of interest in terms of policy recommendations and the Indian brand. NASSCOM also had a very dynamic leader (in Dewang Mehta) whose contribution was widely acknowledged by the Indian media.

NASSCOM's membership grew from 38 members in 1988 to more than 1,000 firms in 2005. It was most effective in policy concerns and brand promotion abroad. NASSCOM was less effective in representing small and medium enterprises, or domestic, rather than export, firms.

Impact of the Industry on the Indian Economy

The success of the Indian software industry has had wide-ranging effects across the Indian economy. Policy changes to enhance exports are facilitating the rapid development of a domestic IT market, offering efficiency gains through adoption of information technologies. In sharp contrast to even a decade ago, Indian business, government, and consumers have ready access to the newest software products and imported hardware.

The very high standards of management practiced in Indian IT firms and the tremendous employment opportunities offered by the industry have had significant effects on the confidence, aspirations, and work ethic of young professionals in India. The leading software firms have pioneered a movement to modernize Indian management practices, adopting practices of creative organizations with less hierarchical structures and strong work ethics. In order to comply with international norms to participate in international capital markets, IT firms have set new standards in accounting and corporate governance. They have offered unprecedented high-paying employment opportunities for the young and educated labor force, particularly for women professionals.

Moving Up the Value Chain

The leading firms have moved up the value chain in software services, developing organizational and managerial capabilities that enable them to offer more comprehensive services than merely low-cost programming. One sign of maturity is that the industry increasingly procures fixed-price contracts, rather than the time-and-materials contracts of earlier years. With the greater risk of fixed-price contracts comes flexibility in organizing work, greater management control, and an opportunity to earn higher returns as efficiency improves.

Revenue per worker is increased, indicating a move up the value chain—from an average of $9,000 in fiscal 1995/96 to $20,500 in 2000/2001—but revenues are still lower than what they are in product-based companies.

In order to build client value, companies have expanded their capacity to service a wider range of software-development tasks, as well as to move into new services, such as product design and information-services outsourcing. Software development includes analysis and specification of requirements, software design, writing and testing of software, and delivery and installation. Indian companies are trying to move beyond only writing and testing, which require the least skill and account for only a small portion of the overall project costs, to higher skill levels that require deeper business knowledge of the industry for which software solutions are being developed.

In their quest to climb the value chain, India's software firms ensured product quality and reliability by adopting internationally recognized standardized work processes. An increasing number of firms have met international certification requirements for key quality standards. For many, this was an exercise in brand building, but the processes and procedures put in place left their hallmark on the quality of software products and services.

Firms seek certification from various sources, beginning with quality-management practices that meet ISO 9000 standards to ensure consistent and orderly execution of orders. The next stage focuses on software engineering and certification under the People Capability Maturity Model (CMM) framework of the Software Engineering Institute (SEI) at increasing levels of process maturity. Another stage focuses on aligning internal practices with the CMM, which is a framework to guide attracting, motivating, and retaining a talented technical staff. The Six Sigma methodology ensures end-to-end quality across all company operations and focuses on improved customer satisfaction by reducing defects, with a target of virtually defect-free processes and products. As of December 2003, India had 65 companies at SEI CMM Maturity Level 5. In October 2002, the SEI of Carnegie Mellon University published a list of high-maturity organizations as part of its Survey of High Maturity Organizations and High Maturity Workshop research.[7] The full set of 146 high-maturity organizations includes 72 Level 4 organizations and 74 Level 5 organizations. Of the 87 high-maturity organizations assessed outside the United States, 77 are in India.

Because most Indian software firms are export-oriented and serve clients around the world, meeting globally acceptable frameworks and standards has been critical to validating their credentials to new clients, who often demand that vendors adopt ISO and CMM standards.

The reasons for the success of the quality improvements can be grouped in three categories: people based, business related, and management related (Jalote 2001). The Indian software industry primarily delivers services, which have embraced globally software-process improvement (SPI) more than those who deliver products. As Indian companies serve worldwide clients who demand that their vendors adopt standards such as ISO and CMM, companies were motivated to certify their credentials and used these frameworks to also deliver real software-process improvement. As companies moved to an offshore model, SPI became a necessity to succeed. Managing subcontracted work typically requires monitoring structures to contain risk. This imposes a degree of formality at the interface between the users and developers—something that is generally hard to achieve with in-house development.

For most organizations, software development is their core competency that must be continually improved. Their high-growth trajectory required the infusion of a large number of new engineers every year. Without tightly controlled processes, it would have been impossible to absorb new recruits into the development process quickly. Since the cost of manpower was not very high in India, it was possible for most companies to dedicate a team for its SPI effort. A survey of high-maturity organizations in India indicated that most companies had dedicated manpower for SPI equal to about 1 to 2 percent of their engineering manpower (Jalote 2001).

Most of the software companies in India are very young. Being followers in the software-development process, they could exploit the collective knowledge and experience of organizations the world over in implementing SPI. Most companies introduced quality systems very soon after they were formed. This ensured that the company had work standards to which each new entrant had to conform. After that, the company, people, and quality systems all matured together. As the people in the company have contributed actively to the SPI movements from the early days, it induced among the practitioners a sense of ownership for the quality system.

Software companies attract the best talent from engineering schools. Some of the CMM lead assessors have observed that the scores on the Myers-Briggs personality tests conducted as part of the capability appraisals often indicate that Indian engineers are different from their counterparts in the United States. Indian employees are ambitious and look for improvement in the way the organization works, which creates a need for process orientation. The average age of the Indian engineers is in the 20s and that of managers is late 20s to early 30s. Younger professionals are more receptive to change, as they have not invested in traditions and, indeed, want changes.

Indian culture is more family-oriented rather than individualistic. This prompts people to conform to established frameworks and systems. Professionals do not mind being measured. There are fewer privacy concerns, and in-house surveys have indicated that most engineers are more concerned about the nature of work and the overall work environment and not so much about being measured. The software background of top managers helps to secure backing from senior management for SPI initiatives.

Most of the facilitating factors are based in more general and societal contexts. Such factors are hard to emulate once the context changes. The government had little role to play in this movement. India does not have centers along the lines of the U.S. or European software engineering institutes. The Ministry of Information Technology in India did bring in the world's best Software Testing and Assessment of Software Maturity through licensing arrangements with the Software Engineering Institute at Carnegie Mellon University. Under this agreement, the Indian Standardization, Testing, and Quality Certification (STQC) Directorate of the Ministry of Information Technology undertook the job of certification, testing, and training of trainers and assessors in India.

The increasing importance of outsourced IT services from developed countries prompted many clients to voice concerns about data-protection practices of service providers. Issues of data confidentiality, integrity, and availability have come to the fore. The latest EU data-protection laws are designed to ensure that personal data of EU citizens are not sent to a country that has less stringent legal protection. Clients are also demanding adherence to security standards to ensure information security.

The government of India and NASSCOM are working closely to respond to these concerns. The government introduced clauses in its IT Act of 2000—covering privacy, digital signatures, and cyber crime—to meet EU requirements. More generally, the government strengthened software testing and assessment capabilities in India, in association with some of the leading organizations internationally. The Ministry of Information Technology set up the STQC directorate to train assessors and implement security standards. An Information Security Technology Development Council has also been set up to promote research in the area of information security.

Table 2.9 Profile R&D Effort in India's IT and Software
Service Sector

Year	Sample size	Firms reporting R&D (numbers)	R&D spending (Rs. millions)	R&D intensity[a]
1997/98	115	9	213.44	2.03
1998/99	155	14	244.28	2.82
1999/2000	238	16	703.29	6.82
2000/2001	217	17	763.19	4.91

Source: Parthasarathi and Joseph 2004: 97.
a. R&D expenditure as a proportion of sales.

Research and Development

R&D expenditure amounts have been and continue to be small
(table 2.9), with some increase in recent years. Low R&D expenditures
can partly be explained by the service (instead of product) focus,
which would require greater investment in R&D.

The bulk of R&D occurs in subsidiaries set up by multinationals.
As a cost-reduction strategy, a number of large and medium product
companies started captive development centers in India. Other companies have partnered with Indian firms to set up product-development centers, and still others are outsourcing to India functions such
as requirement specification, design, testing, and maintenance. The
availability of capable and low-cost Indian technical expertise, coupled with the deep financial resources of the multinationals, provide
for cost-effective R&D. However, as multinationals increasingly
distribute R&D or product development operations globally, they
become less likely to develop whole products in a single place such as
India.

Indian firms were reluctant to invest in product development,
because they lacked resources and expertise, and, more importantly,
because of the difficulties in designing products for distant and unfa-
miliar markets. Even when firms have the resources, they find it hard
to justify the high risks of product development. The risks are much
lower in providing services than in selling product, in part because of
the lower level of skill and financial risk.

In the software industry, product development is a small component of the overall costs of developing and promoting software products. Software firms may spend as much as 50 percent of revenues on
advertising and marketing and as little as 10 to 15 percent on product
development.

There are few examples of successful product development by Indian software companies. For the industry as a whole, only 1 to 5 percent of the software packages typically succeed in the market. Only recently have Indian companies reached a size and maturity to consider investing in R&D and marketing. One successful example to date are products developed by Indian companies for the banking sector.

Indian Software Companies As Learning Organizations

Tschang, Amsden, and Sadagopan (2001) examined the different ways in which upgrading takes place in the Indian software industry. They used the R&D classification of pure,[8] basic, and applied research[9] to differentiate different firms' technological abilities and functions. They found sufficient evidence of firms upgrading to the applied-research stage—itself an achievement, since it involves more conceptual work or longer-term efforts at research. The two highest levels of research—pure science and basic research—are almost nonexistent in India, especially in domestic firms. This shows that the nature of the industry is "applied," "service-oriented," or "incremental" in its innovation. The technologies developed are typically not breakthroughs, but are rather first implementations, involving "transforming, variating, and reapplying" known techniques to the software product under design.

The model of upgrading into products is perhaps the most difficult task, given the many reasons for failure. There is a paucity of success stories. The leaders of domestic firms who went into products, including those who left Wipro, Satyam, and other service companies to build their new enterprises, all noted that they had a different mentality and business objective in mind. Their goal was to build products or to create a fundamentally new service. Their plans would not have materialized if they remained in their former software service firms. Each new start-up may have also been trying to find a defensible or competitive niche within the Indian software industry.

Many firms are discouraged from trying the product market because of the distance from the final market and lack of sufficient resources or expertise at the outset. Service companies have resources, but find it hard to justify risk taking when they have such nice returns. Ultimately, even firms that break into the product market can stumble and fall. Ramco was an example. The shortage of this risk-taking attitude across the broader industry, coupled with resource and distance-from-market constraints, will make it difficult for more product firms to emerge. Unless all these factors change, it is unlikely that the Indian industry as a whole will change its complexion to one with more diverse models of upgrading.

The Indian service companies are clearly following a trajectory laid out by their original competencies, continuing to dominate the larger part of the services value chain all the way back to the requirements analysis and consulting stages.

The emergence of the specialized services model, such as Mindtree's contract R&D service, shows that the Indian industry does have its own style of promising entrepreneurial capabilities and the resourcefulness to develop a wider variety of areas.

Ultimately, the running of multinational subsidiaries on a cost-center basis (and the continued tradition of defining product requirements elsewhere) will constrain these subsidiaries from promoting new ideas or products locally. This pattern is different from domestic firms, which run as profit centers, giving them both heavier responsibility and greater scope for doing challenging work. In summary, both domestic firms and multinationals appear to be able to upgrade to applied research, but the business models themselves suggest that the domestic firms have broader scope to do products, if they so choose. Those local firms may engage in a broader range of R&D, though not necessarily the most advanced technologies.

The Indian product company Sassken built a large R&D arm to research technologies for making the "first implementation" of a communication standard (high-level and detailed design). That effort can be considered to involve both (i) "learning" in applied research and (ii) implementation in models of applied research.

The same kind of learning and concept modeling in applied research has been done at the CMC, one of India's earliest software companies set up by the national government. CMC had to design systems from scratch, many of which had social objectives, such as India's first railway reservations system, perhaps the most complicated systems endeavor ever undertaken in India.

Learning Through Alliances and Partnerships

Although a number of alliances have taken place between Indian firms and multinationals (table 2.10), India's firms have not benefited greatly in direct knowledge transfer from them. In a study by Gupta (2004), the perceived benefit of alliances in knowledge development and products skill is lower compared to the benefits of marketing know-how. Although IT firms have entered a large number of technology partnerships with foreign firms, they were primarily straightforward in nature (Gupta and Basant 2001).

Table 2.10 Examples of Alliances Formed by Indian IT Firms

Types of alliances	Examples
Services	
Staff augmentation	Aditi-Microsoft
Application development	GE-Satyam (JV)
Package implementation	TCS-SAP
Migrations	Compaq India—Persistant Systems
Remote maintenance	TIS-Silverline Technologies
ASP	Satyam-Computer Associates (JV)
IT enabled services	Wipro-Spectramind (Equity)
Nonservice industries	
Computer hardware	IBM-Wipro
Biotechnology	Satyam-CCMB
Verticals	
Engineering services	Van Dorn Demag—Infosys
Telecom and Internetworking	Nortel Networks—Infosys
Finance	Nordstorm—Infosys
Aviation	Swiss Air—TCS
Embedded systems and chip design	DCM Datasystems—Intel
Manufacturing	Oncourse-Geometric Software
Systems integration	Wipro-HP
CRM	Siebel-Infosys
Technology consulting	Answerthink-HCL
Alliance categories	
Marketing alliance, market access, new area	JASDIC-Infosys (Japan)
Marketing alliance, new domain	Wipro—Spectramind
Technology alliance implementation	SAP-Infosys
Technology alliance product development	Microsoft-Infosys (Hailstorm Technology Development
Technology alliance IP	Synopsis—HCL Technologies
Joint product-development alliance	Tata Infotech Ltd-WFS
Product marketing alliance	Vision Compass—Oasis
Product technology-compatibility alliance	Servion—Infosys
Standards	TCS (Internet Security Alliance)

Source: Basant 2003.

Basant and Chandra (2004) offer an interesting insight into different strategies used by alliance partners in India and China. They characterize an alliance in India with Nortel, a Canadian telecommunications equipment manufacturer, as having yielded limited benefits of knowledge transfer to Indian companies, which resulted in more

Indian contracts in the telecom sector. Nortel's alliance with five Indian subcontractors was focused on short-term export revenues by each company. On the other hand, Nortel's alliance in China was with a university for research that would benefit the Chinese domestic telecom market; its impact was more long-term.

D'Costa and Sridharan (2004) found what they term the creeping innovation capacity in a wide range of Indian firms of various sizes. Contributing to innovation are firm-level learning strategies that include systemization of knowledge, use of tools, and efforts at partnerships with research institutions. Many forms were incubated unwittingly in the import substitution regime, contributing to their strength in the post-liberalized era. Domain expertise, process standards, and short time to market have been pursued by all firms. Most large firms have begun small-scale partnerships for exploratory research with academia.

The Future

India, compared to its competitors, ranks high on several critical parameters, including level of government support, strong track record of quality and delivery, early-mover advantage of brand recognition, quality of labor pool, English-language skills, project-management skills, strong focus on processes, and a favorable time-zone difference with the United States that permits 24/7 internal operations. Some of the weaknesses that persist are slow growth in the domestic market and a lack of innovation and product orientation in the bulk of small and medium companies. The infrastructure needs improvement in many areas, such as roads, electricity, venture capital, and airports. Markets continue to be concentrated in North America and are therefore subject to nontariff barriers, such as visa denials. There has been some domestic political backlash against outsourcing in the United States and Europe. However, a comparison of India with competitors in software exports on strengths and weaknesses seems to suggest that India's current position is quite sustainable in the near future.

It is difficult to say whether India's success can be replicated in other countries. Any country hoping to emulate India's example would have to define a strategy that matches local capability to global opportunity and discover niches that can be exploited. The niche could very well be in terms of the market to be served on the basis of language competency. Late movers can take advantage of the demonstrated success of the offshore model and how it works.

Box 2.1 Key Factors That Explain the Success of India's
Software Industry

- A software industry can be built entirely on human capital. Requires
  limited infrastructure and up-front investment. Has good cash flows
  and is highly profitable.
- India had an early-mover advantage: repeated positive experience
  built trust in outsourcing and validated the Indian brand.
- Role of human capital, including software engineers, project managers, and corporate leaders.
- Early investments in engineering education and privatization of education created a large talent pool.
- Bodyshopping exposed a large population to new ways of working.
- Professionally trained entrepreneurs.
- Vigorous efforts at assimilating new technology and good management practices helped companies offer competitive costs for high
  quality and delivery performance.
- Selective support to industry in an otherwise constraining environment by a few enlightened bureaucrats and the role of NASSCOM in
  influencing policy.
- Lack of effective implementation of restrictive policies allowed
  market forces a significant play in the early phase. The economy was
  liberalized in later years.
- Highly entrepreneurial IT training and private education industry.
  Responded quickly to fill skill gaps and opportunities. Positive gov-
  ernment policies and lack of regulation meant few barriers.
- Large population created competition for engineering seats and jobs.
  Software industry faced no internal competition for technical talent.
  Competition from MNCs came when indigenous firms were prepared.

The availability of high-quality, trainable manpower and strong entrepreneurial and managerial talent is the only necessary condition for
the development of an IT industry (box 2.1). If countries cannot wait
for a high-quality technical-education system, it may still be possible
to mount focused training and certification programs in targeted niche
areas.[10] This would, of course, require the foundation of a good university education system that is producing easily trainable manpower.
Key infrastructure for offshore services, such as telecom, could be
created selectively through technology parks. Policy support and
incentives can also be provided selectively. Since trust is a key issue in
offshore work, the country's Indian diaspora and intermediaries can
play a critical role in the beginning.

Endnotes

1. Table on Indian software exports by country, NASSCOM 2004: 29. Also, according to NASSCOM, the number of $1 million plus customers in the IT exports segment increased from 331 in 2002/3 to 442 in 2003/4 (TNN, June 2004).
2. Infosys and Wipro fourth quarter results announced in April 2004. TCS crossed the $1 billion mark in 2003.
3. Information for the section above was obtained in a conversation with Dr. Nitin Patel, CEO Cytel Corporation, Boston. He was one of the first few employees hired by TCS.
4. Jain B. N., Indian Institute of Technology, Delhi. Presentation available online at www.iitd.ac.in/cgi-bin/nph-p/http/10.116.2.57/alumni/alumni.ppt.
5. "Sustaining Academic Excellence," Position Paper by IIMA Faculty, April 2, 2004. Presentation available online at http://www.iimahd.ernet.in/download/Presentation.ppt.
6. The prime minister's task force on IT, formed May 22, 1998, submitted 108 recommendations in an IT action plan aimed at increasing software exports to US$50 billion by 2008 and creating 1 million new jobs over five years. Recommendations included blanket approval for overseas acquisitions from export earnings; zero duty on all IT products by 2002 by advancing International Trade Administration schedules; broadening the definition of software to include the entire range of IT software as per WTO-ITA norms; and exemption for software developers and exporters from physical and customs bonding at software-technology parks, engineering development units, and export-processing zones ("IT Task Force Suggests Ways to Turn India into a IT Superpower," *Information Technology Review*, June 1998).
7. It should be noted that the SEI does not certify companies at maturity levels, nor does it confirm the accuracy of the maturity levels reported by the lead assessors or organizations. The list of Level 4 and 5 organizations is by no means exhaustive.
8. The category of pure science includes the development of mathematical algorithms, languages, or other computer-science theory. Typically, only the freshest start-ups in regions such as the United States or the main labs of the most established companies, such as Microsoft, will base their products on in-house basic research or university research into the pure sciences. Typical examples of the latter are the Web search engine companies, such as Lycos and Google.
9. For example, the implementation of a communications protocol stack (i.e., part of a computer-communications system) in a particular language or for a particular environment (with all its interface standards) requires

knowledge of the standards, how to create a logical model that meets the standard (applied research), and how to develop the model in a particular programming language in preliminary prototype form along the way to its final developed form.

10. For example Infosys Technologies is setting up a center in Mysore (a city in south India) to train 12,000 software engineers per year. ("Infy Plans Biggest Training Centre," *The Economic Times,* New Delhi, June 18, 2004. Available at: http://economictimes.indiatimes.com/articleshow/744132.cms.)

References

Arora, Ashish, Alfonso Gambardella, and Salvatore Torrisi. 2001. "In the Footsteps of the Silicon Valley? Indian and Irish Software in the International Division of Labour." Paper presented at the workshop on the Indian software industry in a global context, Indian Institute of Management, Ahmedabad.

Basant, Rakesh. 2003. "Economics Series." Working Paper No. 53. East–West Center, Honolulu, Hawaii.

Basant, Rakesh, and Pankaj Chandra. 2004. "Capability Building and Inter-Organization Linkages in the Indian IT Industry: The Role of Multinationals, Domestic Firms, and Academic Institutions." In Anthony P. D'Costa and E. Sridharan, eds., *India in the Global Software Industry: Innovation, Firm Strategies, and Development.* New York: Macmillan.

Bhatnagar, Subhas C., and Shirin Madon. 1997. "The Indian IT Industry: Moving Toward Maturity." In *Journal of Information Technology.* New York: Macmillan.

Bhatnagar, Deepti, and Mukund Dixit. 2004. "Stages in Multiple Innovations in Software Firms: A Model Derived from Infosys and NIIT Case Studies." In Anthony P. D'Costa and E. Sridharan, eds., *India in the Global Software Industry: Innovation, Firm Strategies, and Development.* New York: Macmillan.

Biao, Xiang. 2002. "Ethnic Translation of the Middle Class in Formation: A Case Study of Indian Information Technology." Paper presented during the International Seminar on ICTs and Indian Development: Processes, Prognoses, and Policies, December 9–11, Bangalore.

Dataquest. 2003. "Retention Rate." *Dataquest* 21 (16). Gurgaon: India, Cyber Media, Ltd.

D'Costa, Anthony P., and E. Sridharan, eds. 2004. *India in the Global Software Industry: Innovation Firm Strategies and Development.* New York: Palgrave Macmillan.

Gupta, Vivek. 2004. "Determinants of Incidence and Modes of Alliances: A Study of Indian Information Society." Fellow program in management thesis, Indian Institute of Management, Ahmedabad.

Gupta, Vivek, and Rakesh Basant. 2001. "Knowledge Acquisition by Indian Infotech Firms: Role of Global Internet Economy Alliances." Papers presented at the International Conference on Electronic Business, May 23–26, Beijing.

IIMA (Indian Institute of Management Ahmedabad) Faculty. 2004. "Sustaining Academic Excellence." Position paper presented April 2. [Retrieved from http://unpan1.un.org/intradoc/groups/public/documents/APCITY/UNPAN002944.pdf.]

Information Technology Review. 1998. [Retrieved from http://www.ipan.com/reviews/archives/0798it.htm]

Jalote, Pankaj. 2001. "The Success of the SPI Efforts in India." Working Paper 208016. Department of Computer Science and Engineering, Indian Institute of Technology, Kanpur. [Retrieved from http://www.cse.iitk.ac.in/users/jalote/papers/IndiaSPI.pdf]

Kumar, Nagesh. 2001. "Indian Software Industry Development: International and National Perspective." *Economic and Political Weekly.* Sameeksha Trust Publication, Mumbai.

NASSCOM (National Association of Service and Software Companies). 2001. *The IT Software and Services Industry in India.* New Delhi.

———. 2002. *IT Industry in India—Strategic Review.* New Delhi.

———. 2003. *IT Industry in India—Strategic Review.* New Delhi.

———. 2004. *IT Industry in India—Strategic Review.* New Delhi.

———. 2005. *IT Industry in India—Strategic Review.* New Delhi.

Parthasarathi, Ashok, and K. J. Joseph. 2004. "Innovation under Export Orientation." In Anthony P. D'Costa and E. Sridharan, eds., *India in the Global Software Industry: Innovation, Firm Strategies, and Development.* New York: Macmillan.

Srinavas, S., 1998. "The Information Technology Industry in Bangalore: A Case of Urban Competitiveness?" Working Paper No. 9. Development Planning Unit, London.

Tschang, Ted, Alice Amsden, and S. Sadagopan. 2001. "Measuring Technological Upgrading in the Indian Software Industry: A Framework of R&D Capabilities and Business Models." Working Paper (Final Draft). Asian Development Bank Institute.

van dijk, Meine-Pieter. 2002. "India-China: A Battle of Two New ICT Giants, Fought by Different Urban ICT Clusters." Paper presented during the International Seminar on ICTs and Indian Development: Processes, Prognoses, and Policies, December 9–11, Bangalore.

CHAPTER 3

Electronics in Taiwan—A Case of Technological Learning

John A. Mathews

THIS CHAPTER EXAMINES THE STEPS FOLLOWED by Taiwan, China, in building its electronics industry over the four decades from the 1970s to 2004. During the 1960s, Taiwan quickly diversified away from primary products to manufactured products, establishing a set of industries that absorbed the unemployed. By the 1970s, the island was ready to follow Japan's example and pursue knowledge-intensive sectors like electronics. But it did not follow Japan in every way. While building a distinct institutional system for diffusing technology, it also took extraordinary measures to build technological competence.

This study builds on literature that treats development as the enhancement of technological competence in tandem with macroeconomic stabilization.[1] Innovation is given a broad interpretation in the literature, covering not just the development of products and processes new to the world, but the adoption and adaptation of products and processes that are *new to a given country* under study. It is this aspect of innovation—of critical relevance to developing countries—that was used to great effect by Taiwan in the creation and building of its electronics industry. The institutional innovations made in that process all involved the timely capture of technologies; the building of capabilities in these technologies, such as in government-owned research and development (R&D) institutes; and the diffusion of those capabilities as rapidly as possible to the private sector, for example, through a sequence of targeted R&D consortia.

In view of the nature of innovation in developing countries, it may be more accurate to refer to the system of institutions and policies not as a "national system of innovation," but as a "national system of economic learning"—and the process involved not as one of innovation, but as one of managing technological diffusion.[2]

This chapter emphasizes the aspects of Taiwan's strategies that are most applicable to developing countries today. This is not to say that there is an easy path for development—there is not. But the strategies that worked for Taiwan, with appropriate institutional modifications and sufficient political will, could work for other countries. The lesson of Taiwan's successful building of an electronics industry is that, although each developing country faces a complex world of bewildering variety, the unprecedented flows of technology and capital available today can be harnessed to accelerate a country's development.[3] With help, firms can use temporary advantages, such as low costs, to insert themselves in global value chains and find strategic ways to complement the needs of established firms.

Of course, some of the actions taken by Taiwan, such as imposing outright temporary import bans on color television sets from Japan, are now outlawed under WTO rules and so are no longer available for developing countries that wish to abide by these rules. Complex rules of origin in the world trading system can also act as a constraint on individual country strategies. But the overall lesson of Taiwan's approach is that much can be accomplished with a few public institutions and clever strategies for concentrating and channeling knowledge. It is these lessons that are important for the developing world today.

Taiwan's Electronics Sector: Five Decades in Development

In its first decade of development, the 1950s, Taiwan followed a conventional import-substitution strategy, combined with major land-reform efforts to raise agricultural production. State-owned enterprises were established in key upstream sectors, such as oil, steel, electricity, and gas, to ensure a supply of raw materials and feedstocks for the small and medium-size firms that were allowed to flourish in the consumer- and intermediate-goods sectors.

In the 1960s, Taiwan's leaders continued to pursue import-substitution policies for the domestic economy while promoting a new export sector designed to earn foreign currency and supply external discipline to government-led programs for upgrading investment. Thus a dual structure was created, with different policies applying to the domestic and export-oriented sectors. The Statute for Encouragement of Investment, adopted in 1960, served as the umbrella for a variety of investment incentives intended to enhance export performance and upgrade technology and performance. Institutional innovations made at the time included export-processing zones, the manufacturing equivalent of nineteenth-century free ports.[4] Manufacturing GDP soared during the decade, from $370 million in 1960 to just over $1,400 million by 1969, a fourfold expansion that caused per capita national income to double from just under $200 in 1960 to just under $400 in 1969.

U.S., European, and Japanese firms competed fiercely in the 1960s to sell consumer items such as radios, calculators, TV sets, and electronic home devices, as well as communications products, computers, and components, particularly semiconductors. Seeking a competitive edge, U.S. electronics and semiconductor firms invested in low-cost manufacturing operations in East Asia, to which they first transferred production of components, then assembly of products (transistor radios, TV sets, and chips), and finally, test and assembly operations for both semiconductors and final electronic goods.

Taiwan's establishment of the world's first export-processing zone in 1965[5] opened the way for considerable investment, not only by U.S. firms, but by Japanese and European firms as well, in the production of electrical and electronic components and in the "back end" of the semiconductor cycle, which encompassed testing, packaging, and assembly. These activities tended to be isolated from the rest of the economy and had little potential in terms of industrial development, but they were valued for their export earnings.

In the production of TV sets, for example, U.S. multinationals led by Philco (1965), Admiral (1966), RCA (1967), and Motorola (1970) established export plants in bonded factories or in export-processing zones. Taiwan's government allowed such operations to be wholly owned—an unusual and distinctive approach in the 1960s. In the production of electronics parts, U.S. multinationals led by General Instrument (1964), then Philco and RCA, built plants to produce transistors, tuners, capacitors, and other parts. The strategy behind these U.S. investments was to take advantage of Section 807 of the U.S. Trade Act, which allowed duty-free reentry of U.S. components and parts embodied in final products that were assembled overseas (the first appearance of rules of origin, now so familiar in trade law and practice). Taiwan and Mexico rapidly became the major sites of such U.S. outsourcing, which was driven by global competition with Japan.[6]

The major policy instrument used by the Taiwanese government at this stage was a local-content obligation that forced foreign producers to upgrade the capabilities of local suppliers of parts and components. Local-content rules were imposed for black-and-white TV sets in 1965—at a level of 50 percent. This direct intervention by the government was designed to foster the development of a local parts and components industry. The local content level was raised to 60 percent in 1966 and to 70 percent in 1972—after the government helped create a local producer of cathode ray tubes (CRTs), Chung-Hua Picture Tube, a joint venture between Tatung and RCA. As locally manufactured CRTs became more widely available, the local-content rule was raised successively to 80 percent in 1973 and eventually to 90 percent in 1974. The rule for black-and-white TVs was phased out in 1983— by which time it had achieved its purpose. Similar rules were imposed for color TV sets, as described below.[7]

The 1960s also saw the emergence of the indigenous Taiwanese electronics industry. Initially, that industry served the domestic market, while foreign-owned firms served the export market. The dual-market structure was maintained through the 1960s. At the same time, small and medium Taiwanese firms organized themselves within the Taiwan Electrical Appliances Manufacturers Association, which represented the interests of the industry before government and ensured that multinationals did not monopolize the government's attention.

The Years of Consumer Electronics

In the 1970s, Taiwan's government encouraged the establishment of several new industries and opened the economy to foreign investment so as to acquire the technologies the island needed for further expansion of exports. In the first year of the decade, the government created a major institution, the Industrial Development Bureau within the Ministry of Economic Affairs, to coordinate investments. Under its guidance, major improvements were made in the machinery, textiles, petrochemicals, and motor industries. This decade also saw the government stimulate industrial development with major infrastructure projects (the Ten National Construction Projects), which also stimulated domestic demand for industrial output.

Electronics manufacturing picked up during the decade, mainly in the form of joint ventures with foreign partners (from Japan and the United States). A government requirement that foreign companies source components locally helped create a domestic parts and components industry. In addition to investment vehicles already in place, such as the National Chiaotung Bank, the government created new vehicles to diffuse technology, led by the Industrial Technology Research Institute (ITRI), formed in 1973 to provide public support for R&D.[8] ITRI's first laboratory, the Electronics Research Services Organization (ERSO), was dedicated to creating technological capabilities in the electronics sector.

To further promote the use of computers in industry, and thereby stimulate a national information technology (IT) industry, the government established the Institute for Information Industry in 1979. Manufacturing GDP expanded sixfold during the decade, from just under $2 billion in 1970 to $12 billion in 1979, leading to a fivefold expansion in per capita national income, from just under $400 in 1970 to $2,000 in 1979. The world was put on notice: a "miracle" was happening in Taiwan.

In the 1970s, some Taiwanese firms became more intensively involved in the production of goods such as transistor radios, black-and-white TV sets, and color TV sets. It was in color TVs that Taiwan's efforts to create an industry came to international attention—assisted by the arrival of domestic color TV broadcasting in Taiwan in 1969. Indigenous and joint-venture firms that had targeted only the domestic market in the 1960s began to venture into export markets, alongside the multinationals in the foreign-owned export sector. Companies such as Tatung took the lead. This company, which traced its origins back to the Tatung Iron Works of 1939, began producing electric fans in the 1950s, followed by refrigerators, air conditioners, and TV sets in the 1960s. By 1970, it had formed Chung-Hwa Picture Tube to build cathode ray tubes and TV sets. Today, Tatung is one of Taiwan's leading producers of flat-panel displays.

In the 1970s, the government allowed joint ventures to be formed, most with Japanese firms, as a means of bringing in technology from abroad, largely in the area of color TV production. The Taiwanese partners learned the technology of production quickly, and institutions such as the China Productivity Center were established to spread manufacturing knowledge even further.[9] In the next decade, Chung-Hwa Picture Tube and other companies translated their experience in producing color TVs to the production of new IT products, such as computers and monitors, without direct technology transfer from Japan. That accomplishment shows how far technological learning had progressed.

Taiwanese firms rapidly built up their exports of color TVs to the United States throughout the 1970s, to such an extent that imports had captured one-third of the U.S. market by 1976.[10] The "invasion" of the U.S. market by Taiwan, Japan, and other countries in the region sparked trade disputes that led to the imposition of "orderly marketing agreements." A three-year agreement with Japan, enforced in 1977, led to a surge in Taiwanese and Korean imports into the United States, with latecomer firms from these countries taking advantage of the limits on Japanese exports, at a time when incumbent firms in the United States were cutting back production. Taiwanese and Korean latecomers would put this early experience of the advantages of world-wide downturns in providing an opportunity for market entry to good use in later downturns, in electronics as well as in semiconductors.

The United States imposed orderly marketing agreements on Korea and Taiwan during 1979–82. The drastic drop in U.S. imports of color TVs from East Asia toward the end of the 1970s was a direct result of these trade embargoes and not a result of some loss in competitiveness.

At the same time, Taiwan was taking steps to protect its nascent color TV industry. Until 1984, Taiwan imposed high tariffs on imported color TVs, *and banned imports from Japan.* The government's pointed intervention allowed Taiwanese firms to benefit from domestic sales while building up their export performance.

Orderly marketing agreements in consumer electronics probably had little effect on the competitiveness of U.S. manufacturers of consumer electronics. By the mid-1970s the only U.S. producers left in the market were Zenith and RCA, both of which had extensive investments in production facilities in Taiwan. But the agreements probably did accelerate the trend (already under way) toward the disintegration of the value chain and promote the growth of new industrial sectors specializing in electronics parts and components (box 3.1). Parts and components, which could be imported into the United States without restriction, were brought in first by Japanese firms with production facilities in the United States, and later by Taiwanese firms. Both Tatung and Sampo had established production bases in the United States by the end of the 1970s—initiatives directly traceable to the imposition of the orderly marketing agreements of 1979.

By the late 1970s, Taiwanese firms had become exporters of consumer electronics and communications equipment, as the export statistics cited above reveal. Although firm-level data are hard to come by, it is probable that the contributions of large firms, such as Tatung, were important as catalysts, but that the majority of production-for-export work was accomplished by small and medium-size firms. This feature of the Taiwanese industrial structure meshed well with the growing disintegration of the value chain in electronics production worldwide.

By 1979, Taiwan exported 80 percent of its electronics production—a figure unsurpassed by other countries. (In the same year, Korea exported 70 percent of its electronics output.) The bulk of the exports went to the U.S. market. By 1984, Taiwan's government deemed the color TV industry ready for international competition and removed tariffs.

Box 3.1 Profiting from the Disintegration of the Value Chain

"The division of labor is limited by the extent of the market." Adam Smith's famous dictum explains why an expanding global market for electronics products led inexorably to the creation of specialized markets for parts and components—markets that would not have been able to appear with a smaller overall market for the final products.

Taiwan's electronics industry can be linked to the growth of the global market for electronics goods. Its firms took advantage of the trends toward vertical disintegration by winning contracts to manufacture components for large, established players, such as RCA, which put their own brand on the products. By contrast, Japan nurtured the development of large, wholly integrated electronics firms, typified by Matsushita and Sony, which also bought in a large proportion of their components from other Japanese firms.

The process known as "disintegration of the value chain" was accelerated in the 1980s with IBM's decision (in 1982) to build its personal computer on the basis of open standards, leading to the phenomenon of "IBM-compatible" PCs.

By upgrading their activities within global value chains, firms can gradually gain entry to a global industry. In the case of the electronics industry, the path followed various steps from original equipment manufacturing (OEM) to customers' specifications, to own design and manufacturing (ODM) to customers' outline specifications, and, finally, to manufacturing and selling under one's own brand. The OEM pattern originated in the clothing industry, but it was rapidly taken up in the electronics industry, particularly in Taiwan, with very considerable effects.

The case of Proview illustrates the possibilities of OEM in the electronics field. Established in 1989, Proview produced computer monitors on an OEM basis for Xerox, Gateway, Samsung, and Philips. It built its first plant in Taiwan but quickly moved to China—setting a trend that many other Taiwanese firms have followed. In the early 2000s, Proview began phasing out CRT monitors in favor of flat-panel displays.

Proview distinguishes itself by offering an extraordinary customer warranty: full replacement for any machine found to be faulty. The company found that it was cheaper to offer replacements than to employ technicians for after-sales work in the United States and Europe. Proview's experience is an example of how a latecomer firm can turn its latecomer status from a liability into an advantage by building technological capabilities that enable the firm to keep up with technological trends created elsewhere.

As a 1983 report from the U.S. Office of Technology Assessment on the global electronics industry put it: "With an economy that has been growing at an annual rate of about 8 percent, unemployment at less than 2 percent, and a persistent trade surplus with the United States, Taiwan's electronics industry is well positioned for further expansion" (OTA 1983: 386). This was in fact a prescient conclusion: the glory days of Taiwan's electronics industry still lay in the future.

The IT Stage

The 1980s saw a further intensification of the guided national industrialization strategy, this time with a focus on electronics, IT, and other high-technology industries. Capital-intensive industries were established, and the government introduced major infrastructure innovations to accommodate them, including the new Hsinchu Science-based Industry Park (HSIP), established in 1980. The first semiconductor venture was spun off from ITRI/ERSO in 1980, as United Microelectronics Corporation (UMC). Many more followed—notably TSMC, created in 1986 as a joint venture with Philips. During this decade, manufacturing GDP tripled from $15 billion in 1980 to just over $50 billion in 1989, bringing per capita national income to $8,000 at the end of the decade—a fourfold expansion. Taiwan was on the brink of becoming a "developed" nation.

In 1981, electronics still accounted for only 6.8 percent of Taiwan's manufacturing production—although, as noted, a large part of that was already dedicated to exports. Taiwan's National Economic Development Plan for the years 1980–89, issued by the Council for Economic Planning and Development (CEPD) in 1980, named electronics as the highest priority for development.[11] The government's plan to double output in electronics was a conscious effort to emulate successful Japanese strategies and to compete with perceived Korean efforts. The plan, driven by a new focus on IT products, covered all sectors of the electronics industry, called for building a semiconductor industry in Taiwan, and urged imitation of new electronics developments introduced by Japan, such as video cassette recorders (VCRs), which even U.S. producers were wary of doing.

Locally owned Taiwanese firms began to produce integrated circuits (ICs), or chips, in 1982, with small-scale activities undertaken by UMC, the ITRI spin-off. The first ICs were simple ones intended for use in watches, calculators, and other consumer products such as greetings cards. But they were a foot in the door. Meanwhile, ITRI was building its own semiconductor capabilities in its laboratory devoted to large-scale integration, which soon evolved to the next technological level: very large-scale integration. These laboratories were built with substantial government funding in the mid-1980s.

High tariffs were never imposed in the case of ICs. The government expected Taiwanese firms to reach international levels of productivity and quality rapidly. Policy pronouncements stressed automated manufacturing as a means of raising productivity. Displaced workers were expected to undergo rigorous retraining to be able to use the new computer-based and automated production systems.

The government sought to spark further private investment in semiconductors by creating TSMC, another ITRI spin-off, as a joint venture with Philips, which already had extensive production facilities in Taiwan. TSMC, launched in 1986 during a downturn in the world semiconductor industry, caught incumbent firms off guard. It was dedicated from the outset to the novel strategy of providing third-party IC fabrication services in very large scale integration (VLSI) facilities—at first to other Taiwanese firms for firms that could not afford to build their own chips (thus sparking an IC design industry) and then to overseas firms.

Thus Taiwan pursued a familiar pattern through its gradual entry into the world semiconductor industry, conquering each step in the value chain. It started with back-end processing (testing and assembly) in the 1960s and 1970s, then moved into various phases of fabrication and eventually arrived at IC design (figures 3.1 and 3.2).

Flowering

The 1990s saw the flowering of high-technology industries, as Taiwan became the "Silicon Valley of the East" (Mathews 1997).

Figure 3.1 Value Chain in the Semiconductor Industry

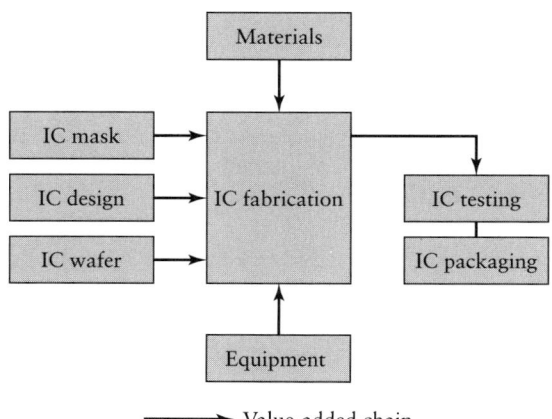

Source: Mathews and Cho 2000.
Note: IC = integrated circuit.

Figure 3.2 Revenues from the Manufacturing of Integrated
Circuits in Taiwan, 1989–98

US$ billion

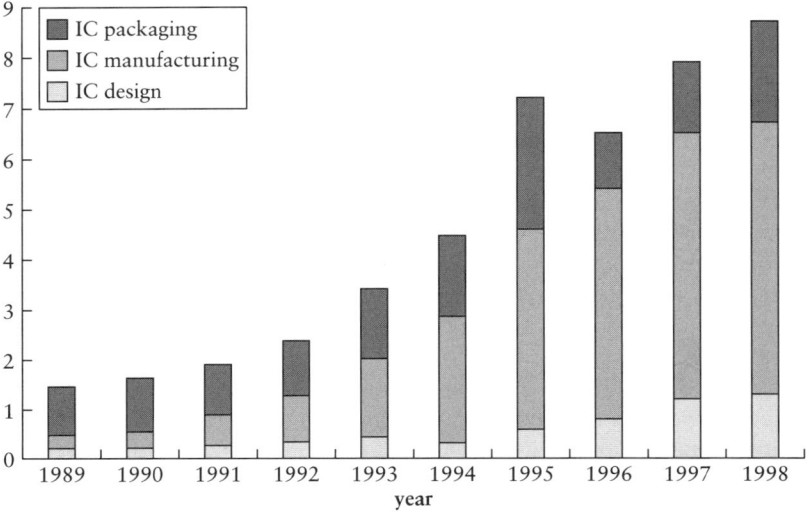

Source: Electronics Research Services Organizations/Industrial Technology Research Institute.

Continuing its guidance of the economy in 1991, the government
enacted the Statute for Upgrading Industries, which offered more
incentives to firms for enhancing exports, upgrading technology, participating in ITRI-led R&D consortia through which technology
could be diffused to the private sector, and relocating to new industry
parks.[12] The Industrial Development Bureau sustained the policy of
targeted development by issuing, in 1993, "Measures for Boosting
Economic Development" and "Strategies and Measures for Development of Ten Emerging Industries," with included industries being IT,
communications, flat-panel displays, and semiconductors. Manufacturing GDP grew from just under $50 billion in 1990 to $80 billion
in 1997—and then recovered quickly after the 1997/98 Asian financial crisis to reach just under $80 billion again by 1999. Per capita
national income rose above $12,000 by 1994; it recovered to that
level again by 1999.

By 1990, the electronics industry in Taiwan accounted for no less
than 12.2 percent of the country's manufacturing output and for a
substantial proportion of the country's exports. With further efforts

targeted toward IT products, components, semiconductors, and industrial electronics, the sector continued to expand, contributing 25 percent of Taiwan's manufacturing value added by the end of the 1990s and giving Taiwan a large share of the world market for various IT and electronic products. At this point, consumer electronics had declined in favor of components and semiconductors—consistent with a worldwide trend toward semiconductors accounting for ever consistently increasing value of electronic systems.

But Taiwan still trailed behind Japan, the world leader in electronics, in several sectors of the industry, particularly in equipment for electronics manufacturing, where Japan commands formidable leadership. For example, Japanese firms are world leaders in surface-mount technology that is used in the production of printed circuit boards that are at the heart of all electronic products. Taiwanese firms (and even U.S. firms) show no signs of penetrating this leadership cluster anytime soon.[13] The persistence of the gap between leader and fast follower remains as a reminder that the task of catch-up is never easy.

By the end of the century, Taiwan had recorded the astonishing feat of raising national income nearly a hundredfold, from about $100 in 1951 to more than $10,000 in 2000. Manufacturing GDP was the primary engine of growth, expanding by 400 times from less than $200 million in 1951 to just under $80 billion in 2000. In the process, Taiwan became the world's sixteenth-largest economy, with worldwide trade totaling $288 billion in 2000. It has become a major IT manufacturer, producing 5.5 percent of the world's semiconductors and most of the world's computer terminals and peripherals.

There are lessons here for all developing economies. The principal one is that to emphasize manufacturing as the engine of industrial development, a country may have to flout the principle of comparative advantage. Formulated in the early nineteenth century by David Ricardo, the notion was wrong at the time and is even more so today—from the perspective of a developing country. Taiwan inherited no comparative advantage in manufacturing, but it created competitive advantages and used them to drive the development process. Early in the process, Taiwan exploited low labor costs—a temporary competitive advantage. But its leaders knew they could not rely on such advantages forever, and so created the infrastructure and institutions to translate a temporary competitive advantage based on low costs into a more durable one based on manufacturing strength. The electronics industry, broadly defined, played a major role in this process in Taiwan.

The Electronics Sector As an Engine of Growth

Starting in the 1970s, the electronics sector became the engine of growth in Taiwanese manufacturing and exports. Electronics' share of manufacturing value added grew throughout the 1980s, overtaking metal machinery, chemicals and food, and textiles and related products in the mid-1990s. By 2000, electronics was the island's leading sector, accounting for 35 percent of manufacturing value added (figure 3.3).

As it grew, the electronics sector passed through several important phases. By 1986, when electronics accounted for more than 10 percent of value added in manufacturing, it had already surpassed textiles in importance. Total production passed the $25 billion mark in 1992. In that year, consumer electronics accounted for 13.1 percent of output; IT products for 36.2 percent; communication electronics for 6.9 percent; and electronic parts and components (including semiconductors) for 43.8 percent. The early predominance of consumer electronics gave way in the later 1980s and 1990s to IT, semiconductors, and

Figure 3.3 Electronics: The Engine of Manufacturing in Taiwan, 1981–2002

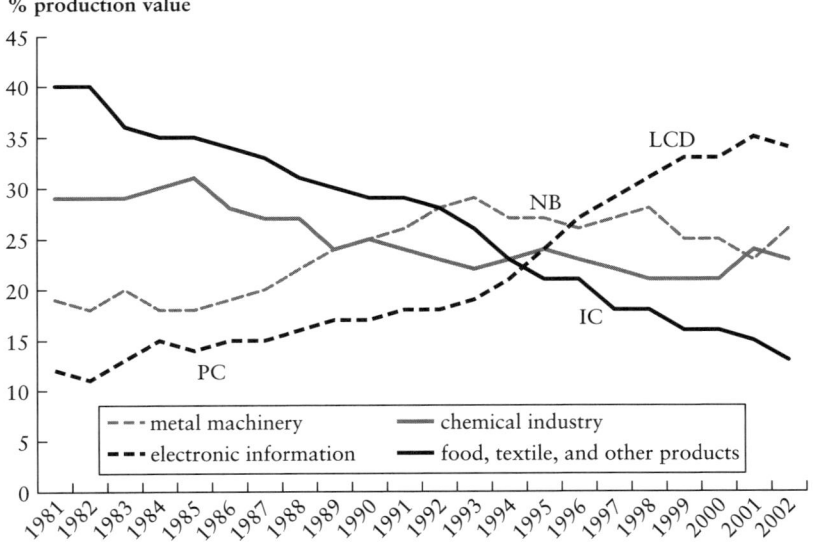

Source: Industrial Development Bureau, Ministry of Economic Affairs.
Note: PC = personal computer, IC = integrated circuit, NB = notebook, LCD = liquid crystal display.

Table 3.1 Taiwan's Export Performance in Electronics, 1976–2001
(Share of export)

	1976–88		1989–2001	
	Taiwan (China)	*World*	*Taiwan (China)*	*World*
Consumer electronics	2.5	1.2	1.8	1.1
IT products	2.9	1.4	8.5	2.8
Parts and components	2.1	1.9	9.6	3.4
Telecommunications	6.2	1.5	3.7	2.6
Semiconductors	2.4	1.3	8.6	3.4
Total electronics	16.1	7.3	32.2	13.3

Source: World Bank.

components. By 1998, electronics accounted for more than 25 percent of manufacturing value added in Taiwan and had come to dominate the island's manufacturing.

But the real strength of the electronics sector in Taiwan is revealed in export figures. Taiwan's electronics exports made up an average of 16.1 percent of total exports in the period from 1976 to 1988, but rose to 32.2 percent of total exports in the period from 1989 to 2001 (table 3.1). In the world as a whole, electronics exports accounted for 7.3 percent of exports in the first period and 13.3 percent in the second period. Thus Taiwan's performance in both periods was more than twice that of the rest of the world. Taiwan was also a major *importer* of electronics goods, as inputs into its IT products and consumer electronics. In the first period, imports of electronics accounted for 10.9 percent of total imports in Taiwan, and in the second period for 21.7 percent of total imports—again exceeding world figures by a strong margin.

Taiwan's performance was strong in all sectors of electronics. Exports of consumer electronics, IT products, telecommunications, and semiconductors were all at least twice the world averages in the first period, from 1976 to 1988; only electronics parts and components, where Japanese firms have been and remain dominant, did not reach that height. In the second period, Taiwan's export share in IT products grew to 8.5 percent, as against the world average of 2.8 percent—making Taiwan three times more export-competitive in these products. Its share of parts and components also rose dramatically during the second period to account for 9.6 percent of total exports, as against the world average of 3.4 percent. Its share of semiconductors in the second period rose to 8.6 percent of total exports, as compared with the world average of 3.4 percent.

Key Factors in the Development of the Electronics Sector in Taiwan

The creation of an electronics industry in Taiwan was a joint effort between government agencies set up for the purpose of steering the island's development and a private sector that was itself largely created to meet specific economic goals. The Taiwanese government's Sixth Four-Year Plan (1973–76) first identified electronics as a sector to be developed as part of a second round of import substitution in which key technological imports from Japan (chiefly parts and components) were targeted for replacement by domestic production. The emphasis on electronics has been repeated in every subsequent development plan, making electronics industries eligible for incentives for adopting and upgrading technology.

In this section, we consider the systems and processes that Taiwan's government instituted to encourage the island's firms to keep up with competitors and eventually draw abreast of world leaders in electronics.[14]

An Institutionalized Technology Strategy

Several institutions established as parts of Taiwan's overall technology strategy are particularly relevant to the electronics industry.

As previously noted, the Industrial Development Bureau was constituted in 1970 as part of the Ministry of Economic Affairs to formulate policies, strategies, and measures for industrial development; develop and manage industrial parks; and devise financial and tax measures for industrial development. It continues to oversee Taiwan's industrial development strategy, selecting industries and technologies to be promoted, based on their expected contribution to the island's economy.

This case of "picking winners" needs to be clearly distinguished from similar efforts in more advanced countries, however. For "fast followers," such as Taiwan, the basic technologies used in targeted sectors have already been developed and are in commercial production. The country need not concern itself with developing risky new technologies, but rather with securing access to the technologies it needs and choosing appropriate policies to seed industries in the home market.

Taiwan's leaders were adept at using expert committees to channel government funds to targeted industries. They were also adept at ensuring that companies could not become "rent seekers" that targeted government assistance as a business strategy. All incentives offered were performance-based, so that if, for example, a company was receiving a tax credit for its export performance, then its tax credit would be discontinued if it failed to achieve its export targets. Above all, there were to be no bail-outs for companies that failed; if, with all the incentives available, a company still threatened to go under, then the Taiwan authorities allowed this to happen—as in the case of several failures in the semiconductor industry in the mid-1980s.

All-important political support for the project to lift Taiwan's technological competence, and in particular to continue with the relatively expensive promotion of an electronics and IC industry, was generated by a science and technology congress convened by the minister for industry, Dr. K. T. Li, in 1978. Taiwan's cabinet considered and accepted the recommendations set forth in the congress report, including the formation of a science and technology advisory group convened by Dr K. T. Li and reporting directly to the premier. This group, which consisted largely of foreign experts who would meet for several days at a time to debate potential technology options and strategies, was influential in the development of Taiwan's high-technology industries generally, and its electronics and semiconductor industries in particular. The group was divided into two technical review boards: one for the semiconductor industry and the other for the electronics industry at large.[15]

By far the most significant of the recommendations that emerged from the 1978 congress was the creation of high-technology infrastructure within a so-called science-based industry park. The task of creating such a park was entrusted to the National Science Council, which after much searching (in crowded Taiwan), eventually appropriated land near Hsinchu from the Taiwanese military. Taiwan has had many industry parks, both before and since, but the Hsinchu variant was designed to create an "industrial ecology" in which high-technology industries could flourish. The Hsinchu Science-based Industry Park (HSIP) is described at greater length later in the chapter.

These institutions were complemented by public and public–private financial institutions that played the role of development banks. The National Chiaotung Bank, the China Development Corporation (CDC), and the Bank of Communications are national institutions created to mobilize the financial capital needed to build new industries. The theory behind such "development banks" dates back to the nineteenth-century German political economist Friedrich List. It was taken up and analyzed by the economic historian of "late" development Alexander Gerschenkron.[16] It can be argued that all successful cases of industrial development in the twentieth century have been associated with the mobilization of savings and finance by such development banks. Conversely, wherever such banks have been wanting (as in Hong Kong), technological development has stalled at an early stage.

Technology Development in the Electronics Sector

A comprehensive account of the development of the entire electronics industry in Taiwan is beyond the scope of this chapter. Four cases will give the background needed for the remainder of the discussion.

Color TV Sets   Taiwan's success with color TV sets in the 1970s derived from its earlier experience with black-and-white sets and transistor radios in the 1960s. Considered extremely complex products in the 1970s, color TVs were the subject of fierce international competition between U.S. and Japanese multinationals, as well as European firms, such as Philips and Grundig. Seeking a competitive edge over their Japanese and European rivals, U.S. firms began outsourcing much of the value chain to lower-wage countries, such as Taiwan, where they could obtain good conditions—notably protected production in export-processing zones, high-skilled labor at low cost, and full ownership of their plants and operations. Their major obligation to the Taiwanese government was local-content regulations, imposed initially in 1970 at a level of 40 percent, then raised successively to 50 percent and 60 percent as local industrial capacity for producing key components, particularly CRTs, improved. Taiwan's government considered the rules necessary as a means of forcing Japanese suppliers of parts and components to transfer technologies to their Taiwanese partners and local parts makers. In some cases, the government took steps to ensure that a local parts maker was present to permit foreign firms to meet their local-content obligations. This was the case with Chung-Hua Picture Tube, formed as a joint venture between Tatung and RCA in 1971 to produce CRTs. It took Chung-Hwa Picture Tube 10 years to become profitable, but, by 1980, it had become an internationally successful and competitive producer. It has since become one of Taiwan's leading producers of flat-panel displays.

Semiconductors   Taiwan first acquired semiconductor capabilities in 1973, through ITRI. In 1976, an ITRI laboratory entered into a technology-transfer agreement with RCA, thereby acquiring initial capabilities in semiconductor fabrication and design. The technology transferred was considered obsolete by RCA, but it served as training material for ITRI, which then diffused the skills to the private sector by spinning off UMC in 1980. UMC has since become a world-class firm, in part through alliances with advanced firms. In 1986, ITRI entered into a technology-transfer agreement with Philips to form TSMC, giving Philips new fabrication capacity and privileged access to the Taiwan market. To avoid competing directly with Philips, TSMC elected to produce chips only for third parties, inventing in the process the notion of the "silicon foundry." This has proven to be remarkably successful, and TSMC has continuously enlarged and deepened its technological capacities by absorbing the technological specifications of its customer firms as it takes orders to produce their chips.

The strategy of the latecomer is to close the gap between in-house technological capabilities and the world frontier. That strategy has dominated Taiwanese thinking about semiconductors, where the state of technological sophistication can be captured in terms of the line-widths used in etching circuits onto the silicon substrate. Taiwan's initial technology transfer from RCA was seven microns, which had been reduced to two microns by 1985, when the world frontier was at just over one micron (figure 3.4). By 1995, the Taiwanese firms had

Figure 3.4 Taiwan and the Technology Gap in Semiconductors, 1975–95

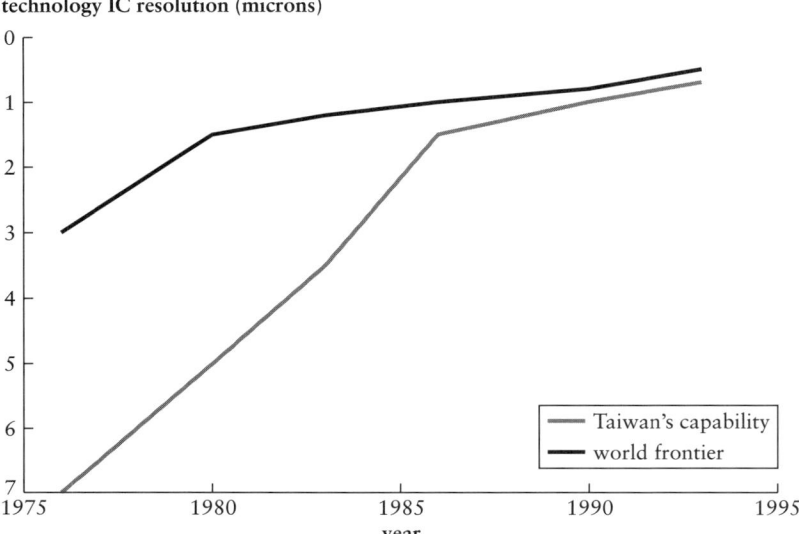

Source: ERSO.
Note: IC = integrated circuit.

just about caught up, with submicron technology comparable to that employed in the world's leading firms. Such "technology gaps" need to be tracked obsessively by latecomers engaged in catch-up—as they were by Taiwan in its process of catching up in electronics.

Dynamic Random Access Memory (DRAM)   Taiwanese firms mastered DRAM technology in the 1990s. Earlier efforts to enter the industry by firms such as Quasel, Mosel, and Vitelic, all failed. Quasel disappeared as a company; Mosel and Vitelic merged in the 1990s to become a strong DRAM producer. Before the 1990s, the supporting infrastructure and skills—in a word, the absorptive capacity of the Taiwan semiconductor industry—were not able to support the demands of advanced DRAM fabrication.

Absorptive capacity was enhanced by the activities of semiconductor firms in various non-memory devices, which rapidly deepened their experience in the 1980s. At the same time, the Industrial Development Bureau and ITRI ensured that as many steps as possible in the semiconductor value chain were being included (such as IC design, mask production, and supply of specialist materials and equipment). But the decisive contribution to the industry's absorptive capacity was made by ERSO, with its Submicron Project. The ITRI affiliate built a major pilot fabrication plant, the seed of the Vanguard International Semiconductor Corporation.

In the 1990s, one Taiwanese firm after another announced that it would produce DRAMs.[17] Nan Ya Technology, Umax, Winbond, and Mosel-Vitelic entered DRAM production through technology-transfer agreements, most with Japanese firms seeking reliable "second sources" of supply for their products. By the late 1990s, Taiwanese firms were established as serious and highly competitive DRAM producers, putting great pressure on their Korean, Japanese, and U.S. counterparts.

Taiwan's capacity to produce DRAMs was developed by the public sector. Over time, the initiative for leveraging technology and knowledge moved from the public sector to the private sector, as firms gained in sophistication and the industry matured.

Flat-Panel Displays   Taiwan grew in the second half of the 1990s to become the world's second-largest supplier of notebook computers, based partly on innovative developmental consortia in the 1980s. But it was unable to penetrate into the key component segment of flat-panel displays, whose manufacture was dominated by Sharp, Fujitsu, Toshiba, and Matsushita. This meant that much of the value added in notebook computers was lost to Japan, a drain on Taiwan's balance of trade. Efforts by ITRI/ERSO to launch a liquid crystal display (LCD) industry in Taiwan in the early 1990s were frustrated by resistance from semiconductor incumbents and insufficient absorptive capacity. But in the second half of the 1990s, several Taiwanese firms committed themselves to LCD production, as it became clear that thin-film transistor LCDs were becoming the technological standard for notebooks and as Japanese firms sought to outsource and second-source much of their LCD work to Taiwanese firms. By the late 1990s, Taiwan was moving to establish itself as the world's third-largest supplier of thin-film transistor LCDs, after Japan and Korea. All three were considerably ahead of U.S. and European firms in terms of mass production.[18]

Technology Diffusion in the Electronics Sector

ITRI spread knowledge to the private sector through two principal channels: spin-off enterprises and R&D alliances. Two of the spin-offs have already been described. Here we focus on the R&D alliances, which since the early 1980s have been instrumental in keeping Taiwanese firms abreast of new technologies in electronics, semiconductors, and IT.

A series of collaborative R&D ventures emerged in Taiwan in the 1980s and 1990s, bringing together firms, public sector research institutes, and trade associations, with catalytic financial assistance from the government. Unlike many collaborative arrangements between established firms in the United States, Europe, and Japan, where mutual risk reduction is frequently the driving force, those forged in Taiwan had as their chief goal technological learning and industry creation. Formed hesitantly in the 1980s, Taiwan's R&D alliances flourished in the 1990s as institutional structures got established that encouraged cooperation among firms to raise their technological capacity. Most of the alliances have been in the broad IT sector, covering personal computers, work stations, servers, and multimedia, as well as a range of consumer products, telecommunications, and data-switching systems and products. Alliances have also emerged in other sectors, including automotive engines, motorcycles, electric vehicles, and in services and finance.

A good example of the early R&D consortia was the one formed to produce the PowerPC. In June 1995, when IBM introduced a new personal computer based on its PowerPC microprocessor, Taiwanese firms exhibited a range of products based on the same processor just one day later. That achievement rested on a carefully nurtured R&D consortium involving IBM and Motorola, joint developers of the PowerPC, as external parties (Mathews and Poon 1995; Mathews 2002a). The incentive for IBM and Motorola to be involved was to promote the development of a market for PowerPC products in Taiwan—such as laptop PCs and (it was envisaged) other products such as set-top boxes. The incentive for Taiwanese firms was to have a technological capability in a product that could well have become a real alternative to Intel and Microsoft architectures. The PowerPC microprocessor did not take off and the "Wintel" standard remained in place.

Behind many of these successes lie remarkable institutional structures that favor collaborative product development, Taiwan's version of the R&D alliance. Taiwan's current dominance of portable PCs such as laptops and notebooks, for example, rests at least in part on a public–private sector consortium that rushed a product to world markets in 1991. Taiwan's strong performance in data switches and other products now dominant in computer networks also rests on a consortium that worked with ITRI to produce a switch to meet the ethernet standard, in 1992–93. By doing so, Taiwan firms could market their communications products in all markets that followed the Ethernet standards.

These successes were followed by many more such alliances in digital communications and multimedia (table 3.2). Some consortia have been more successful than others—but all seem to have learned organizational lessons from early cases where the government contributed all the funds, and where research tasks were formulated in generic and overly ambitious terms so that companies could take advantage of them. The more recent alliances have been more focused, more tightly organized and managed, and more likely to involve participant firms directly in co-developing a core technology or new technological standard that can be incorporated by companies, through adoption and adaptation, in their own products.

The relatively small budgets of these consortia are remarkable. In all, the consortia have required no more than NT$4 billion over 15 years, with government input of no more than NT$2.3 billion (around $100 million), equivalent to just one year's government subsidy of the 10-year Sematech program in the United States.[19] The figures reveal just what a "David and Goliath" struggle it has been for Taiwan to take on firms in high-technology industries. They also underscore the significance of Taiwan's achievements, which owe as much to organizational finesse and learning as to subsidies.

Industrial Cluster Formation

Taiwan's goal was not to build state-owned firms, as in Brazil or India, or conglomerate firms, as in Japan and Korea, but rather to create industrial clusters—before the term became fashionable.[20] Thus for each

Table 3.2 Representative R&D Alliances in Electronics in
Taiwan (China), 1983–97

Alliance	Years	Number of companies (second stage)	Total budget (millions of NT$)
Electronics and information technology			
PC 100 (IBM PC XT-compatible)	1983–84	5(9)	40
PC 400 (IBM PC AT-compatible)	1984–85	3	24
Workstation (SunSPARC-compatible)	1989–91	2(3)	150
Notebook PC	1990–91	46	100
Graphics terminal	1991–93	34(9)	25
Palm PC	1991–92	16	50
Pentium server	1991–93	2	50
Taiwan NewPC (PowerPC)	1993–97	40	250
Consumer electronics and communications			
Ethernet switch	1993–96	5(8)	75
Digital loop carrier	1992–94	3(4)	60
Liquid crystal display	1995–97	4	230
High-definition TV	1994–96	11	250
Interactive TV	1995–97	21	200
V5 network access standard	1996–	12	150
High-speed loop access system	1996–	14	120
Software/services			
Java-based Internet products	1996–	24	250
Electronic commerce	1996–	61	300

Sources: ITRI; industry interviews.
Note: NT$1 = US$0.031; US$1 = NT$32.

new wave of technology acquisition and diffusion, the state in Taiwan
sought to develop new players or to induce existing firms to invest in
new enterprises. This distinct pattern of new firm formation, pioneered and perfected in Taiwan, differs remarkably from the process
followed in Japan and Korea, where existing conglomerates were encouraged by state incentives to enter each successive phase of a new
industry, usually as a group. Taiwan's process, by contrast, created
new industries through new venture formation and invited the entry of
new firms. Entirely a Taiwanese innovation, it has served the country
very well indeed.

The beginning of the semiconductor industry provides a good example of Taiwan's process. Acting in the spirit of a Gerschenkronian "collective entrepreneur," the government sought to catalyze, or spark, the development of a private sector in 1980 but could not find Taiwanese companies willing to make the necessary commitments. Instead of inviting a group of established players, such as Tatung and Sampo, to cooperate in building semiconductor facilities, the government instead invited firms to invest in a new venture, to which resources (technology, skilled staff, initial capital, initial plant and equipment) were transferred from ERSO. Thus UMC was launched. Allowed to compete internationally on its own, the company built its own product portfolio. TSMC was similarly launched a few years later as a private venture, with initial capital provided largely from public investment funds and Philips.[21] Later spin-offs include the Taiwan Mask Company and Vanguard, which received ERSO's VLSI fabrication technology. These spin-offs had the desired effect of stimulating investments by established firms in a new industry and of generating further spin-offs.

If firms are to come together and capture the benefits of clustering, then land and facilities need to be provided. This is precisely the role played by the Hsinchu Science-based Industry Park, established in 1980 as a specialized home for the electronics, IT, and semiconductor industries. Hsinchu has been a spectacular success, providing land and facilities for an enormous number of firms in electronics, IT, and semiconductors. The cluster benefits include shared facilities, shared access to skilled staff trained in nearby universities (Tsing-Hua University and National Chiaotung University), and shared access to the technology services of ITRI, whose main campus is next door to Hsinchu. By 1995, the 180 firms located in the park employed 42,257 persons. Of these, more than 60 percent held a first postsecondary qualification, while more than 30 percent held university degrees.

Hsinchu offers firms an attractive working environment and living conditions (far superior to the crowded conditions of cities like Taipei and Kaohsiung) as well as proximity to technical expertise. It continues to provide high-quality employment for a growing number of well-educated staff. Because it is government-owned, it also offers firms that settle there a range of special benefits—among them low-interest government loans, R&D matching funds, tax benefits (including special exemptions from tariffs, commodity, and business taxes), government purchases of technology abroad for transfer to participating companies, government equity investment of up to 49 percent of enterprise capitalization, and access to government laboratories and test facilities located in the park.

Human Resource Development

A constant concern in Taiwan's campaign to acquire technological competence has been the building of an adequate skills base. Investments in education were high through the 1950s and 1960s, with an emphasis on training technical cadres and sending students abroad for postgraduate study. The Science and Technology Congress convened in 1978 called for a major focus on skills development. In 1983, the Commission on Research, Development and Evaluation, formed by the cabinet as a result of the forum, drafted a "Scheme to Strengthen the Cultivation and Recruitment of High Technology Experts" that was later approved by the cabinet. This policy document laid out the nation's need for high-technology talent, compared that need with the current sources of supply, and discussed options for bridging the gap. The study estimated the cost of the program and identified sources of finance.

The task of building up a large intellectual base in a very short time was achieved in two phases. The first focused on bringing back to Taiwan those who had studied abroad and were pursuing successful technical careers, particularly in the United States. Many of these people, having studied at the best American universities, had taken up positions of responsibility in management and engineering in U.S. corporations. Measures were taken to attract them back to Taiwan. New institutes were established in which they could take leading positions, for example within the framework of ITRI. Facilities were created in the Hsinchu Science-based Industry Park to help them form companies of their own. In selected areas of high technology, start-up capital was provided by the government. In total, 19,000 scientists and engineers returned from abroad between 1950 and 1988, with their families being accommodated in superior housing and their children going to bilingual schools.

This short-term measure was backed by the longer-term strategy of rapid expansion of Taiwan's system of technical education. In 1952, the island had four universities and four junior colleges with a total enrollment of 10,037 students; of these, 2,590 studied engineering. By 1989, Taiwan had 42 universities and 75 polytechnics or colleges—a massive expansion in a very short period of time. The programs begun in 1978 have been remarkably successful—turning Taiwan into an "Intelligent Island."

The role of human capital in the Taiwanese electronics case is extremely important, for at least two reasons. First, as early as the 1950s, and in contrast to the situation in most developing countries, Taiwan already had a highly skilled workforce. This was not the outcome of government policies (which came much later). Emigrants who left China for Taiwan in 1949 brought with them high technical and entrepreneurial skills. These emigrants provided the initial skills base for Taiwan, enabling the construct of increasingly sophisticated industries. Second, the availability of a skilled workforce, delivered by university-level engineering institutions under relentless pressure from the government, reinforced the competence of the Taiwanese government, helping it to conceive and implement strategies for technology development. These are issues that go to the core of any country's industrial-development strategy today.

Patenting

The passage from imitation to innovation is measured in R&D expenditures and by the rate of patenting in the electronics industry. While R&D expenditures are well covered in the existing literature, the recent patent performance of Taiwanese firms and institutions is less well known. One of the clearest indications of innovation is the rate of use (or take-up) of patents issued by the U.S. Patent and Trademark Office (USPTO).[22] Recent studies of "national innovative capacity" have linked the rate of patenting with economic variables, such as R&D expenditures and the proportion of scientists and technologists employed in a sector or country. These studies have shown that East Asian firms and institutions, particularly those in Taiwan, have made astonishing strides in recent years.[23] Taiwan ranked third in the world in per capita take-up of U.S. patents in 1997–2001 (table 3.3).

The experience of East Asia in patenting with the USPTO is shown in table 3.3, as compared with G7 countries and a reference group of comparable countries, including Finland, Israel, and Ireland. The table reveals just how rapid has been the rise of East Asia as an innovative force, shifting from imitation to innovation. In terms of utility patents taken out in the United States over the past five years, per capita, Taiwan ranks third in the world behind the United States and Japan, as can be seen. Korea ranks as number 8, with 6.6 patents per capita per year, averaged over the past five years, while Singapore is rising fast, to reach 11th on a per capita basis. China has yet to make an impression.[24]

Table 3.4 shows the number of U.S. patents taken out in each of the five years from 1997 to 2001 by East Asian organizations (firms and institutions), almost all of which operate in the electronics, IT, or communications sectors, with a particular concentration in semiconductors. The pattern of concentrating on certain sectors is consistent with the latecomer catch-up strategy.

Table 3.3 Per Capita Take-Up of U.S. Utility Patents in Selected Countries
(Averages for 5- and 30-year periods)

Country	Patents per year			Patents per capita			Annual growth rate (percent)		
	1968–97	1992–97	1997–2001	1968–97	1992–97	1997–2001	1968–97	1992–97	1997–2001
United States	44,850	56,683	79,717	15.5	21.5	28.6	4.9	7.9	9.7
Japan	11,216	22,433	29,949	10.3	17.9	23.7	8.6	6.5	10.2
Taiwan (China)	437	1,535	3,778	2.3	7.3	17.2	26.2	21.4	27.8
Israel	183	400	757	4.2	7.2	12.4	12.5	15.4	17.3
Finland	181	370	609	4.2	7.2	11.8	11.4	10.0	13.6
Germany	5,806	6,895	9,387	9.2	8.4	11.4	2.7	4.3	13.0
Canada	1,380	2,119	3,121	4.9	7.2	10.2	6.5	7.5	11.2
Korea, Rep. of	267	1,134	3,113	0.7	2.5	6.6	39.1	36.1	20.4
France	2,432	2,881	3,662	4.3	5.0	6.2	16.4	3.7	8.5
United Kingdom	2,492	2,427	3,469	4	4.2	5.9	2.7	6.7	10.8
Singapore	16	59	174	0.6	1.7	4.4	44.9	26.2	33.7
Italy	855	1,215	1,548	1.7	2.1	2.7	4.4	4.5	9.2
Hong Kong (China)	31	72	162	0.6	1.2	2.3	14.2	23.0	35.6

Source: USPTO; World Development Indicators 2003.
Note: The data for Germany include West Germany before 1990.

107

Table 3.4 U.S. Patents Obtained by Firms in Taiwan (China), Korea, Singapore, and China, 1997–2001

Firm	1997	1998	1999	2000	2001	1997–2001
Taiwan (China)						
UMC	149	174	266	430	584	1,603
TSMC	130	218	290	385	529	1,552
ITRI	153	218	208	198	221	998
VISC	53	120	112	131	112	528
Winbond	24	59	115	115	126	439
Mosel-Vitelic	15	32	38	66	68	219
Korea						
Samsung Electronics	584	1,305	1,545	1,441	1,450	6,325
Hyundai Electronics	154	212	242	294	533	1,435
LG Electronics	113	215	229	220	248	1,025
Daewoo Electronics	215	319	273	120	54	981
LG Semiconductor	119	235	311	255	42	962
ETRI	58	120	130	124	72	504
KIST	29	44	41	35	35	184
Singapore						
Chartered	30	39	44	79	135	327
China						
WSMC	0	0	6	61	37	104

Source: USPTO.
Note: Only companies that obtained at least 100 U.S. patents during the period are included in the table.

Lessons: Managed Diffusion and Economic Learning

What are the lessons of Taiwan's success in building an electronics industry? Two stand out: (i) the diffusion of innovation throughout the economy *as a managed process,* and (ii) the creation of a national system of economic learning.

Diffusion of Innovation in a Developmental Context

Innovations can benefit economies only as they spread, or diffuse, to a large number of firms.[25] This was certainly the case with the buildup of technological prowess in electronics in Taiwan. The processes of diffusion are generally held to follow two major pathways, namely market-induced imitation and organizationally induced technology transfer (usually within the context of the product cycle).[26]

But the customary framework fails to account for the integration of Taiwan's latecomer firms into the high-technology segments of the electronics sector. In fact, most of Taiwan's successful firms have not been innovators in the usual sense of the word, nor have they been "recipients" of diffusion or technology transfer. Instead, they have been the instigators of the process of diffusion. Successful latecomer firms in Taiwan, such as UMC, TSMC, and Acer, devised sophisticated strategies to acquire and internalize technology. Those strategies in themselves soon became a source of competitive advantage. Meanwhile, institutional structures created in Taiwan accelerated the process of diffusion, assuming many of the functions of the market.

Diffusion As a Process   Technological diffusion was actively managed in Taiwan. From this perspective, diffusion is not driven solely by strategic decisions and calculations taken by the originators of something new. Rather, it is a complex process in which technological leverage and strategic management play critical roles.[27] In this sense, diffusion is triggered as much by decisions of the adopters (who assimilate, accommodate, adapt, and improve) as by the original innovators. "Diffusion," with its connotations of passive transfer, is thus a misnomer; Taiwan's experience shows diffusion to be a multipolar process of active dissemination and leveraging of resources in which decisions to adopt and adapt are primary and account for the extent to which diffusion actually occurs (Mathews 2001).

Diffusion Management As a Source of Competitive Advantage   The strategic calculations of the latecomer firm engaging in leveraging practices are quite different from those normally depicted in discussions of strategy and the enhancement of "sustainable competitive advantages" by firms.[28] Whereas the conventional discussion is couched in terms of the firm's identification of sources of competitive advantage and then framing strategy to enhance and defend those advantages, this makes no sense from the perspective of the latecomer firm. For the latecomer lacking resources and advantages other than temporary cost advantages, the approach to strategy is first to identify the resources that are most available and most susceptible to leverage and then to tap those resources and improve on them. This is resource leverage in a developmental context.[29]

Institutions for the Management of Diffusion   The Taiwanese latecomers found that the management of diffusion called for institutions quite different from those developed in the advanced countries to support R&D-led innovation. The institutions of diffusion management are concerned with accelerating the uptake of technologies by firms, spreading new techniques, and hastening the enhancement of organizational capabilities (organizational learning) through devices such as engineering research associations (Japan) and developmental consortia. The creation of such an institutional framework means that firms do not have to leverage and learn on their own. The results of earlier experiences with collaborative dissemination can be used to improve the outcomes—in a process that can be described as "economic learning." Such a process calls for a national system of economic learning, in contrast to the more conventional national system of innovation.[30]

Whatever labels we wish to attach to them, Taiwan's three organizational innovations—diffusion as a process, diffusion management as a source of competitive advantage, and institutions for the management of diffusion—represent a major departure in the understanding of innovation and its propagation, away from knowledge generation in individual firms and toward the management of technological diffusion as a strategic process of economic upgrading. The shift in theoretical perspective involved here is profound, for several reasons. First, it implies that the industries being created cannot be conceived simply as firms, as in much industrial economics analysis, but must be seen as clusters of firms together with their institutional supports. Even where this is given due weight, a conventional view of industry evolution is concerned with patterns of innovation and their supporting institutional structures. By contrast, the approach pioneered in Taiwan is to see the process in terms of patterns and dynamics of diffusion and its management; the emphasis is on how innovations can be leveraged and turned into technological capabilities and competitive products as rapidly as possible.[31] Three perspectives on the propagation of innovation—that derived from the U.S. perspective (emphasizing product innovations), that derived from the Japanese perspective (emphasizing process innovations), and a novel perspective from East Asia, emphasizing the management of technology diffusion—are summarized in table 3.5.

Institutional Elements of a National System of Economic Learning

The building of the electronics industry in Taiwan may be summarized under several headings, which may also serve the purpose of generalizing the discussion to make it applicable to other sectors in other developing countries.

Industry Targeting The first step in a latecomer's strategy of technological catch-up is to select industries and technologies that can feasibly be developed. This means creating institutions that can (1) marshal available technical advice and (2) assess industries in terms

Table 3.5 Propagation of Innovations: Competitive Postures

(Basis of competitive) posture	Product innovation	Process innovation	Diffusion management
Competitive focus	Product	Process	Access to technologies; diffusion
Competitive tools	Intellectual property rights; first mover advantages	Costs; quality	Resource leverage
Competitive vehicles	Firms	Firms	Firms; government research institutes; consortia
Dynamic capabilities	Product enhancement: R&D	Process enhancement: quality/time improvement	Combinative capabilities; learning
Strategic goals	Sustainable competitive advantage	Temporary competitive advantage	Transient competitive parity
Sources of competitive advantage	First mover advantages	Quality/time enhancement	Fast followership
Institutional framework	Atomistic competition	Limited competition	Accelerated diffusion within consortia; MNCs
Lead countries	United States	Japan	Korea, Taiwan (China), Singapore

Source: Author.

of technological goals to be achieved, financial hurdles to be overcome, market access, and the capacity of local industries to provide components and parts. Taiwan's Industrial Development Bureau is an exemplary institution from this perspective, as is Singapore's Economic Development Board.

Technology Leverage In Taiwan, Korea, and Singapore, pivotal to the successful development of the semiconductor industry was the presence of a public sector R&D institute that scanned global technological developments, quickly mastered new techniques, and diffused them to the private sector. In Taiwan the lead agency for electronics was ITRI/ERSO. Comparable institutions are the Korea Institute for Industrial Economics and Trade and Singapore's Institute of Microelectronics.

Financial Leverage Also present in Taiwan, Korea, and Singapore was a development bank or investment vehicle that identified and financed investments to enhance the catch-up effort—in Taiwan, the China Development Corporation; in Korea, the state-owned or regulated development banks; in Singapore, the Development Bank of Singapore. These institutions mobilize domestic savings or form international bank syndicates that issue depositary receipts and other debt instruments.

Provision of Land and Infrastructure for Knowledge-Intensive Firms All the cases of electronics and semiconductor industry formation in East Asia have involved provision of suitable infrastructure. Taiwan has the Hsinchu Science-based Industry Park (1980) and its successor, the Tainan Park (late 1990s). Singapore has a science park in Jurong and a new semiconductor fabrication park in Woodlands. Institutions have been designed to allocate and administer the infrastructure— such as the Hsinchu Science-based Industry Park Administration in Taiwan and the Jurong Town Council in Singapore. In the 1990s, Taiwan extended the park concept to the creation of a multistory incubator building on the ITRI campus designed to house the operations of new technology-intensive firms, most of which were formed by ITRI personnel.

Vehicles to Attract Investment Economic-development agencies, such as Singapore's Economic Development Board and Malaysia's Penang Development Corporation, have been central to the process of attracting and monitoring investment in high-technology activities. The East Asian agencies have been so successful in attracting investment by multinational companies that they have been emulated in Scotland, Wales, and Ireland, where development agencies lately have played major roles in attracting foreign investment to areas formerly stricken with industrial decline. There is no Taiwanese equivalent, as Taiwan did not place such emphasis on attracting foreign capital. But most countries today clearly have to do so.

Incentives and Discipline for Industrial Upgrading   Behind every case of successful industrial upgrading in East Asia lies an institution, such as the Industrial Development Bureau in Taiwan that combines incentives to create new firms or technologies with processes for continual technological upgrading and enhancement. Performance monitoring is an essential aspect of such programs.

Skills Upgrading and Technical Training   Technological capabilities rest on an infrastructure of specialist industrial training that ensures that firms will have access to skilled staff and engineers as they need them. Markets alone will not provide the necessary infrastructure. Singapore created specialist training centers in alliance with selected multinational corporations. In the 1990s, it has underpinned its expansion of wafer fabrication activities with a specialist technical training program funded through the Economic Development Board.

Market Shaping and Creation   In all the cases of semiconductor industry formation in East Asia, new markets had to be created to complement those in advanced countries, which acted as the export drivers for the nascent industries. Institutions like the Institute for Information Industry in Taiwan helped to create a public and private sector market for IT products by driving the associated standards for IT use in government agencies. In Singapore, a similar role was played by the National Computer Board.

Export Promotion   The collective enhancement of export performance through institutions, such as the China External Trade Association in Taiwan, the Korea Trade Investment Promotion Agency, and the Trade Development Board in Singapore, has also been a significant source of institutional support for firms in East Asian countries seeking to enter new markets.

Public R&D Coordination   The role of public institutions in promoting targeted R&D in advance of the private sector, and eventually as supplementing private R&D, is essential to the technological development process. It is coordinated by institutions such as Taiwan's National Science Council.

Industry Self-Organization   Bodies such as the Taiwan Electrical and Electronics Manufacturers Association, the Taiwan Semiconductor Industry Association, and the Korea Semiconductor Industry Association provide a means for communication between government and firms. They also help produce consensus over new directions for the industry. The capacity of an industry to organize for more than merely defensive ends helps it upgrade and sustain its competitiveness.

Patents and Protection of Intellectual Property Rights As a developing country moves from imitation to innovation, the importance of intellectual property rights increases. Growing firms must become sophisticated in their treatment of patents and copyright issues. The Taiwan Intellectual Property Organization and other institutions offer coordination and protection of intellectual property rights at the national level.

The number of such institutions can clearly multiply, posing a danger of "institutional inflation" unless governments prune institutions as they become less relevant. While institutions concerned with technology capture are very important early on, those concerned with intellectual property rights become more important later. But the most important feature in this list is the institution devoted to coordinating, or "steering," the efforts of all the others. This is the lead agency.

Lead Agency Because industry adaptation and adjustment depend on coordination, a lead agency is needed to set strategic direction and coordinate the activities of other agencies and associations. Prominent examples of the lead agency are Japan's MITI, Taiwan's Council for Economic Planning and Development, Singapore's Economic Development Bureau, and the former Economic Planning Board in Korea.

Each on its own, the foregoing agencies are unremarkable and easily replicated. What is remarkable in East Asia is the total system of economic learning formed by their interaction and mutual support (figure 3.5). It is the adaptability of the total system that counts, and its ability to improve adaptation over time as experience is gained and stored in appropriate institutional form, generating what might be called institutional capacity.

Generalizability of the Taiwanese Model

To what extent can Taiwan's approach to technology-diffusion management be generalized? The key lesson of Taiwan's success in electronics lies in the way Taiwanese firms managed to insert themselves into existing value chains—whether as OEM contractors in the early consumer electronics sector, or again as OEM contractors in the IT sector, or as assembly and test houses for IC products, and then as wafer foundries offering contract fabrication services to established IC firms. For every step in the industry-building process, Taiwan's firms developed strategies to complement the strategies being pursued by established firms.

Figure 3.5 National System of Economic Learning in East Asia

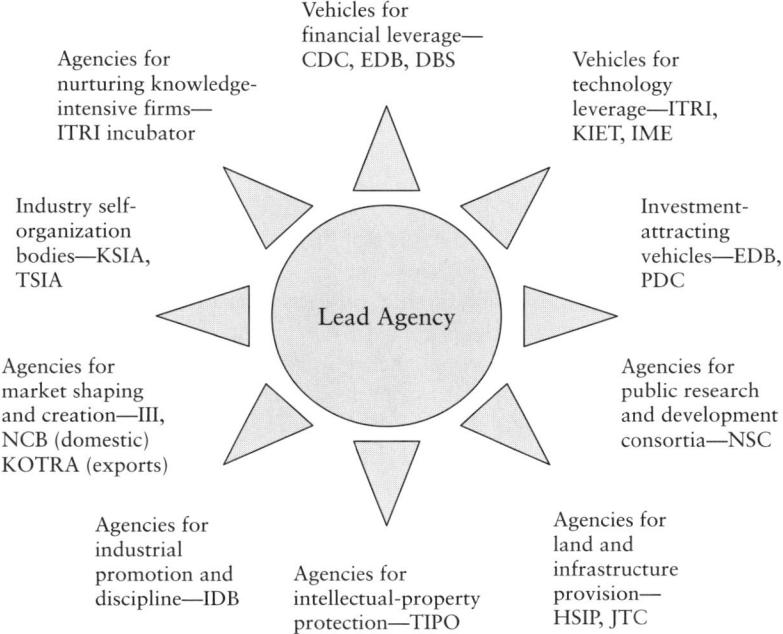

Source: Author.
Note: CDC = China Development Corporation (Taiwan), DBS = Development Bank of Singapore, EDB = Economic Development Board (Singapore), HSIP = Hsinchu Science-based Industry Park (Taiwan), IDB = Industrial Development Bureau (Taiwan), III = Institute for Information Industry (Taiwan), IME = Institute for Micro-Electronics (Singapore), ITRI = Industrial Technology Research Institute (Taiwan), JTC = Jurong Town Corporation (Singapore), KIET = Korea Institute for Electronics and Telecommunications, KOTRA = Korea Overseas Trading Association, KSIA = Korea Semiconductor Industry Association, NCB = National Computer Board (Singapore), NSC = National Science Council (Taiwan), PDC = Penang Development Corporation (Malaysia), TIPO = Taiwan Intellectual Property Office, TSIA = Taiwan Semiconductor Industry Association.

Taiwan has been less successful in some sectors of electronics, such as hard-disk drives, and indeed in other industrial sectors. In the auto industry, for example, the Taiwanese established several motor vehicle companies, linking them through technology leverage to Japanese automotive producers. But those enterprises have remained tied to the Japanese firms, producing under the Japanese brand for two decades, with little prospect of breaking free to design and produce cars of their own, as would be the expected outcome of the leverage strategy. With certain qualifications, one might say that the leverage strategy has not served the Taiwanese particularly well in automotive products.[32]

Likewise in aerospace, the attempts of the Taiwanese to launch aircraft projects through technology transfer from European and U.S. producers have yet to meet with success. The reasons for these shortcomings are many. While undoubtedly there were failings on the Taiwanese side, in terms of technological or marketing deficiencies, there were also powerful factors rooted in the very nature of the industry that worked against their success. Whereas the semiconductor industry, electronics, and IT are generally characterized by many competitors as industries with fast product life cycles and companies with rapid turnover, none of these conditions applies in the automotive or aerospace sectors. These latter sectors, on the contrary, are characterized as industries with few players that require huge investments in long product-development cycles and very low company turnover—all of which make it difficult to secure technology on terms other than those dictated by the incumbents. In other countries of Southeast Asia, such as Indonesia, attempts to enter high-technology sectors, such as aerospace, without the accumulated technological capabilities and absorptive capacity that Taiwan can demonstrate, have been even more controversial and problematic. The Indonesian case demonstrates quite vividly the limits and obstacles to catch-up by latecomers in technologically complex fields, such as aircraft manufacture.[33]

Of the countries in the developing world today, China and, to some extent, India appear to be the most successful at applying the lessons of technology leverage. They are drawing on the stock of knowledge of the advanced world and applying it in accelerated fashion to their own development.[34] China, in particular, seems to have studied the Taiwan model very closely and is applying it very successfully in sector after sector. Meanwhile, the East Asian models need to be updated—as discussed at some length in recent World Bank studies.[35]

The point made in these and other studies is that the success of a strategy of technology leverage really depends on the cultivation of high technical and managerial skills both in the private sector and in government; organizations like ITRI cannot be managed efficiently without such skills. China and India have succeeded precisely because they have the capabilities to manage smart institutions and a technically skilled workforce. Another way of saying this is to insist that technology-leverage strategies depend for their success on the prior absorptive capacity of the country. That capacity needs to be raised through relentless pressure on the social institutions (schools, institutes, universities) that develop human resources. Likewise, firms must be induced to upgrade their industrial and technical capacities.

The principal difference between the world Taiwan faced in the 1960s and what Latin American and Central Asian countries face today is a world system that is tightly regulated by the World Trade Organization

(WTO). Tightly worded agreements associated with and forming part of the WTO system extend to trade-related intellectual property rights (TRIPS) and investment-related policies, such as local-content regulations (TRIMS: Trade Related Investment Measures). Such WTO instruments limit the policy options open to developing countries.[36]

Final Remarks

Earlier theories of industrial development, such as the product life-cycle theory and early versions of global commodity chains, stressed the point that firms in latecomer countries were caught up in decisions taken elsewhere, by firms in the advanced countries, and frequently trapped in positions from which they could not extricate themselves. But the evidence makes clear that latecomer firms need not remain passive in the face of global developments. They are free to make strategic choices, but their choices are made within an institutional setting. Some institutions encourage firms to take the risks of innovating their technological capabilities. Others induce them to take the easy road of lowest-cost production. Singapore's Economic Development Board has become famous for its unrelenting pressure on multinational corporations to continually upgrade the technological level of their operations and to expand the scope of their functions to the benefit of domestic firms. Complementary institutions provide incentives to local firms to become involved with multinationals as local partners or suppliers. In short, the institutional framework biases the choices that firms make. In Taiwan, cooperative R&D consortia between competitor firms succeed because they are structured to enable participating firms to isolate their competitive efforts from their collaborative efforts. This is an outcome of institutional design.

Countries on the development ladder must devote as much care to the growth and nurturing of these institutions as they do to the firms operating within the context that they create. The point has been made that Taiwan fashioned a system that offered incentives to firms that succeeded in raising their productivity or their export levels—both clear and objective benchmarks—and punished firms that did not achieve such goals by withdrawing the tax credits and other financial incentives. In this way, Taiwan's leadership maintained a tight focus on policy goals and on performance toward those goals rather than allowing incentive programs to become mere rent-seeking systems. Institutions were fashioned to encourage performance-oriented behavior. Institutional learning can take place, from country to country and within countries, as the institutions created adapt to new circumstances and acquire experience. Ultimately, technology leverage will succeed only with adequate preparation and investment in absorptive capacity by institutions designed for the purpose. That is what characterizes the Taiwan experience and makes it so eminently generalizable in the twenty-first century.

Endnotes

1. For a recent summary of this literature, see Lall and Urata (2002). Classic contributions include Dahlman (1994), Kim and Dahlman (1992), Lall (1990, 1992, 1997, 2000); Enos (1989, 1992); Lin (1994); Enos, Lall, and Yun (1997); and Hobday (1995). A first draft of the present study appeared as Mathews (2004), and a revised and abridged version as Mathews (2006b). This chapter has been refereed and further revised.
2. On national systems of innovation, see Nelson (1993) or Lundvall and others (2002). On the notion of a national system of economic learning, see Mathews (2001, 2002b, 2003).
3. See UNIDO (2002) and UNCTAD (2003) for ample documentation of these flows of capital and technology.
4. The first export processing zone, opened in 1966 under the guidance of industry minister K.T. Lee, was at Kaohsiung. In such zones, companies could import components and export finished products free of duties.
5. The zone was managed by the Industrial Development Bureau. On its origins, by the man who invented the concept, see Li (1988).
6. See Levy and Kuo (1991) and Chen and others (1998) for discussions of this U.S. outsourcing strategy, and Sato (1997) for a discussion of the dual-market approach in Taiwan in the 1960s.
7. See Chen and others (1998) for a discussion of the local-content rules in the electronics sector.
8. The moving spirit behind the formation of ITRI was Minister of Economic Affairs Yun-Tsuen Sun. Sun had to fight a long bureaucratic battle to ensure that ITRI would be established as an independent entity. It was founded initially as a merger between three existing national R&D institutions. In 1974, a fourth laboratory was added for electronics, the precursor of ERSO. For further details on the origins of ITRI, see Xue, Hsu, and Perkins (2000) and Mathews and Cho (2000).
9. The China Productivity Center was established in the mid-1950s as an independent extension service for the manufacturing industry (on the model of agricultural extension services) to provide Taiwanese industry with productivity-improving advice in automation, quality control, and other subjects.
10. Color TV sets were the first major electronics exports from Taiwan to the United States, following the example set by Japan; indeed, Taiwan was the second East Asian country to pursue an export-led strategy in electronics, in advance of the Republic of Korea.

11. The Council for Economic Planning and Development was established as a cabinet-level economic planning institution in 1977, taking over from the Economic Planning Council and earlier bodies. Its 11 members include a minister without portfolio designated by the premier, the governor of the Central Bank of China, the minister of finance, the minister of economic affairs, the minister of transportation and communications, the chairman of the Council of Agriculture, the secretary-general of the Executive Yuan, and the director-general of budget, accounting, and statistics.

12. For example, under the Statute for Upgrading Industries in 1991, as amended in 1995, firms investing in productivity-enhancing automation would be able to offset the investment against tax liabilities; likewise R&D expenditure would qualify for a tax credit of between 5 and 20 percent against the tax payable in a given year. Low-interest loans would be made available in certain key strategic industries, including electronics.

13. On Japanese leadership in electronic manufacturing, and U.S. concerns over the matter, see the report by the Japanese Technology Evaluation Center (JTEC) of Loyola University in 1995.

14. There is, by now, a substantial literature devoted to Taiwan's experience in the electronics sector. This paper draws on Chang, Shih, and Hsu (1994); Chen (1992); Chen and others (1998); Hobday (1995); Levy and Kuo (1991); Lin (1994); Mody (1990); San and Kuo (1993); Sato (1997); Schive and Yeh (1980); Tung (2001); Wu and Tsen (1997); and Tsai and Wang (2005).

15. The idea for forming the science and technology advisory group is attributed to Pat Haggerty, former CEO of Texas Instruments, who died before the group was formed. The group was chaired by Fred Seitz, later president of Rockefeller University. It also included Bob Evans of IBM, who in 1995 became president of the new Vanguard International Semiconductor Corporation in Taiwan. Dr K. T. Li convened the group and remained an honorary adviser through the mid-1990s.

16. See Gerschenkron (1962) for the classic exposition of the latecomer strategy in economic development, and Hobday (1995) and Mathews (2002b) for its application to the latecomer firm. Mathews (2005b) delves into the intellectual origins of the latecomer concept, which sets the context for Taiwan's strategy in creating an electronics industry.

17. The process started with the TI-Acer joint venture of 1991 (taken over by Acer in 1998 as Acer Semiconductor Manufacturing Inc.), and proceeded through several tie-ups with Japanese firms, for example Nan Ya Technology with Oki; Umax with Mitsubishi (to form Powerchip Semiconductor); Mosel-Vitelic with Oki; and Winbond with Toshiba.

18. See Linden, Hart, Lenway, and Murtha (1998) for a description of the Taiwanese experience, and Wong and Mathews (1998) in the same special issue of *Industry and Innovation* for an introduction that sets the scene for the analysis of this important global industry. Mathews (2005a)

provides an updated discussion in the context of the industry's cyclical dynamics.

19. On Sematech's budget and its appropriation in the mid- to late-1990s, see Sematech (1997).

20. See Porter (1998) on clusters. Porter has never used Taiwanese cases in his writings to date, despite their relevance.

21. The China Development Corporation took 48 percent of the equity, while Philips took 28 percent, with other private investors taking up the balance of 24 percent.

22. The USPTO is itself a product of American catch-up efforts; it was the first government agency established by the federal government in the 18th century, and its charter is embedded in the U.S. Constitution.

23. See Hu and Jaffe (2001), Jung and Imm (2002), and Hu and Mathews (2005) for recent examples.

24. See Hu and Mathews (2005) for an analysis of the patenting performance of five East Asian countries in terms of their uptake of patents from the USPTO. This paper opens the way to an anticipated series of studies in which latecomer countries will be able to measure their transition from imitation to innovation, utilizing the patenting experience within the USPTO as a benchmark. It is anticipated that this framework will be applied to more and more developing countries, starting with China and India, but also encompassing middle-ranking but highly innovative non-OECD countries such as Finland, Israel, and Ireland, as well as countries in central and eastern Europe, Central and South America, Australasia, and, eventually, Africa.

25. See Rogers (1995) and Westney (1988) for pioneering studies of the processes of diffusion and Schnaars (1994) for an analysis of how firms develop competitive advantages based on imitation rather than innovation. Kim and Dahlman (1992) provide an important application to the industrialization process, using Korea as their example.

26. The argument is that technologies are imitated in competitive fashion by firms, or that technologies are transferred by incumbent firms as they mature; see Vernon (1966) for an early exploration of such product-cycle considerations.

27. Technology leverage is here used in the sense of a process that delivers more than is put into it. In the strategy literature, resource leverage refers to a firm's capacity to gain resources over and above those provided by the company itself. The U.S. Internet router company, Cisco, for example, is well known as a company that leverages other firms' R&D by acquiring the firms and their technology. In the same way, developing-country firms can leverage resources from incumbents by entering into contractual relations with such firms. For example, they can secure OEM contracts and, from them, leverage knowledge about quality control and design, over and

above the revenues earned. These extra knowledge resources can then be used to develop an in-house design capacity that can lead from original equipment manufacturing under contract to so-called own-design-and-manufacture contracting.

28. Porter (1990) takes it as axiomatic that competitive advantages are built on a foundation of innovation.

29. See Mathews and Cho (2000) for an elaboration of the concept of developmental resource leveraging.

30. See Mathews (2003) for an elaboration of the notion of national system of economic learning. San (1993) provides a comprehensive description of the institutions concerned, under the more conventional label of "national innovation system."

31. Recent treatments of technological catch-up from this perspective include Lall (2000). For a closer focus on the firm level, see Teece (2000) and Mathews (2002b).

32. There are important qualifications. In the mid-1990s, for example, the Taiwanese launched a collaborative R&D exercise to jointly develop a four-stroke automotive engine (after an earlier attempt in the 1980s to develop a two-stroke engine ended in failure). Technology in this instance was leveraged from the U.K. firm Lotus under the auspices of ITRI's Mechanical Engineering Laboratory. This has resulted in several participant Taiwanese automotive firms becoming engine producers, through a consortium, the Taiwan Engine Company, now a successful exporter of engines to China.

33. See McKendrick (1992) for an assessment of Indonesia's catch-up efforts in the aircraft industry, where the accumulation of impressive technological capabilities was not matched by managerial competencies.

34. See Hu (2002) and Zhao (1995) for arguments along these lines in relation to China.

35. Yusuf (2003) is concerned in particular with the capacity of East Asia's economies to move toward a more open system of innovation, in transition from the imitation practices that have worked so well in the past.

36. On the situation facing Latin American and Caribbean countries, as seen by a group of World Bank economists, see de Ferranti and others (2002). On TRIPS and the development issue generally, see Reichman (2000).

References

Chang, Pao-Long, Chintay Shih, Chiung-Wen Hsu. 1994. "The Formation Process of Taiwan's IC industry—Method of Technology Transfer." *Technovation* 14 (3): 161–171.

Chen, Tain-Jy. 1992. "Technical Change and Technical Adaptation of Multinational Firms: The Case of Taiwan's Electronics Industry." *Economic Development and Cultural Change* 40: 867–881.

Chen, Tain-Jy, Been-Lon Chen, and Yun-Peng Chu. 1998. "The Development of Taiwan's Electronics Industry." Paper presented at the conference on Rethinking the East Asian Development Paradigm, sponsored by the Sasakawa Peace Foundation Project, Tokyo.

Dahlman, Carl J. 1994. "Technology Strategy in East Asian Developing Economies." *Journal of Asian Economics* 5 (Winter): 541–572.

de Ferranti, David, G. E. Perry, I. Gill, J. L. Guasch, and N. Schady, with W. F. Maloney and C. Sanchez Paramo. 2002. *Closing the Gap in Education and Technology*. World Bank Latin American and Caribbean Studies Series. Washington, D.C.

Enos, John. 1989. "Transfer of Technology." *Asian-Pacific Economic Literature* 3 (1): 3–37.

———. 1992. *The Creation of Technological Capacity in Developing Countries*. London: Pinter.

Enos, John, Sanjaya Lall, and Mikyung Yun. 1997. "Transfer of Technology: An Update." *Asian-Pacific Economic Literature* 11 (1): 56–66.

Gerschenkron, Alexander. 1962. *Economic Backwardness in Historical Perspective*. Cambridge, Mass.: The Belknap Press of Harvard University Press.

Hobday, Mike. 1995. "East Asian Latecomer Firms: Learning the Technology of Electronics." *World Development* 23 (7): 1171–1193.

Hu, Albert G. Z., and Adam B. Jaffe. 2001. "Patent Citations and International Knowledge Flow: The Cases of Korea and Taiwan." Working Paper 8528. National Bureau of Economic Research, Cambridge, Mass.

Hu, Angang. 2002. "Knowledge and Development: The New Catch-Up Strategy." In Bhajan Grewal, Lan Xue, Peter Sheehan, and Fiona Sun, eds., *China's Future in the Knowledge Economy: Engaging the New World*. Beijing: Tsinghua University Press, and Melbourne: Victoria University, Centre for Strategic Economic Studies.

Hu, Mei-Chih, and John A. Mathews. 2005. "National Innovative Capacity in East Asia." *Research Policy* 34: 1322–1349.

JTEC (Japanese Technology Evaluation Center). 1995. *Electronic Manufacturing and Packaging in Japan*. Chicago: Loyola University. [Retrieved from http://www.wtec.org/loyola/ep/toc.htm]

Jung, Sungchang, and Keun-Young Imm. 2002. "The Patent Activities of Korea and Taiwan: A Comparative Case Study of Patent Statistics." *World Patent Information* 24 (4): 303–311.

Kim, Linsu S., and Carl J. Dahlman. 1992. "Technology Policy for Industrialization: An Integrative Framework and Korea's Experience." *Research Policy* 21 (October): 437–452.

Lall, Sanjaya. 1990. *Building Industrial Competitiveness in Developing Countries*. Paris: Development Centre of the Organisation for Economic Co-operation and Development.

———. 1992. "Technological Capabilities and Industrialization." *World Development* 20 (2): 165–186.

———. 1997. *Learning to Industrialize: The Acquisition of Technological Capability by India*. London: Macmillan.

———. 2000. "Technological Change and Industrialization in the Asian Newly Industrializing Economies: Achievements and Challenges." In Linsu S. Kim and Richard R. Nelson, eds., *Technology, Learning, and Innovation: Experiences of Newly Industrializing Economies*. Cambridge, U.K.: Cambridge University Press.

Lall, Sanjaya, and Shujiro Urata, eds. 2002. *Competitiveness, FDI, and Technological Activity in East Asia*. Cheltenham, U.K., and Cambridge, MA: Edward Elgar.

Levy, Brian, and Wen-Jeng Kuo. 1991. "The Strategic Orientations of Firms and the Performance of Korea and Taiwan in Frontier Industries: Lessons from Comparative Case Studies of Keyboard and Personal Computer Assembly." *World Development* 19 (4): 363–374.

Li, Kuo-ting. 1988. *The Evolution of Policy behind Taiwan's Development Success*. New Haven: Yale University Press.

Lin, O. 1994. "Development and Transfer of Industrial Technology in Taiwan, ROC." In O. C. C. Lin, C. T. Shih, and J. C. Yang, eds., *Development and Transfer of Industrial Technology*. Amsterdam and New York: Elsevier.

Linden, Greg, Jeffrey Hart, Stephanie Ann Lenway, and Thomas P. Murtha. 1998. "Flying Geese as Moving Targets: Are Korea and Taiwan Catching Up with Japan in Advanced Displays?" *Industry and Innovation* 5 (1): 11–34.

Lundvall, Bengt-Ake, Björn Johnson, Esben Sloth Andersen, and Bent Dalum. 2002. "National Systems of Production, Innovation, and Competence Building." *Research Policy* 31 (2): 213–231.

Mathews, John A. 1997. "Silicon Valley of the East: Creating a Semiconductor Industry in Taiwan." *California Management Review* 39 (4): 1–29.

———. 2001. "National Systems of Economic Learning: The Case of Technology Diffusion Management in East Asia." *International Journal of Technology Management* 22 (5/6): 455–479.

———. 2002a. "The Origins and Dynamics of Taiwan's R&D Consortia." *Research Policy* 31 (4): 633–651.

———. 2002b. "Competitive Advantages of the Latecomer Firm: A Resource-Based Account of Industrial Catch-Up Strategies." *Asia Pacific Journal of Management* 19 (4): 467–488.

————. 2003. "Competitive Dynamics and Economic Learning: An Extended Resource-Based View." *Industrial and Corporate Change* 12 (1): 115–145.

————. 2004. "Understanding the 'How To' of Technological Change: The Case of Electronics in Taiwan." Working Paper 2004-21. Macquarie Graduate School of Management, Sydney, Australia.

————. 2005a. "Strategy and the Crystal Cycle." *California Management Review* 47 (2): 6–32.

————. 2005b. "The Intellectual Origins of Latecomer Industrial Development." *International Journal of Technology and Globalization* 1 (3/4): 433–450.

————. 2006a. *Strategizing, Disequilibrium and Profit.* Stanford, CA: Stanford University Press. Forthcoming.

————. 2006b. "How Taiwan Created an Electronics Industry." In Y. C. Yeung, ed., *Handbook of Research on Asian Business.* Cheltenham, U.K., and Cambridge, MA: Edward Elgar. Forthcoming.

Mathews, John A., and Dong-Song Cho. 2000. *Tiger Technology: The Creation of a Semiconductor Industry in East Asia.* Cambridge, U.K.: Cambridge University Press.

Mathews, John A., and T. S. Poon. 1995. "Innovation Alliances in Taiwan: The Case of the New PC Consortium." *Industry in Free China* 84 (6): 43–58.

McKendrick, David. 1992. "Obstacles to 'Catch-Up': The Case of the Indonesian Aircraft Industry." *Bulletin of Indonesian Economic Studies* 28 (1): 39–66.

Ministry of Economic Affairs, Industrial Development Bureau. 1997. "The Development and Upgrading of Manufacturing Industries in Taiwan." *Industry and Innovation* 4 (2): 277–302.

Mody, Ashoka. 1990. "Institutions and Dynamic Comparative Advantage: The Electronics Industry in South Korea and Taiwan." *Cambridge Journal of Economics* 14 (3): 291–314.

Nelson, Richard R., ed. 1993. *National Innovation Systems: A Comparative Analysis.* New York: Oxford University Press.

OTA (Office of Technology Assessment). 1983. *International Competitiveness in Electronics.* Washington, D.C.: Congress of the United States.

Porter, Michael E. 1990. *The Competitive Advantage of Nations.* New York: Free Press.

————. 1998. "Clusters and the New Economics of Competition." *Harvard Business Review* (Nov.–Dec.): 77–90.

Reichman, J. H. 2000. "The TRIPS Agreement Comes of Age: Conflict or Cooperation with the Developing Countries?" *Case Western Reserve Journal of International Law* 32 (3): 441–470.

Rogers, Everett. 1995. *Diffusion of Innovations, Fourth Edition.* New York: The Free Press.

San, Gee. 1993. "National Systems Supporting Technical Advance in Industry: The Case of Taiwan." In Richard R. Nelson, ed., *National Innovation Systems: A Comparative Analysis.* New York: Oxford University Press.

San, Gee, and Wen-Jeng Kuo. 1993. "Technological Dynamism behind Taiwan's Successful Export Performance: An Examination of the Electronics and Textiles Industries." A Report to the United Nations Conference on Trade and Development, October, Chung-Hua Institution for Economic Research, Taipei.

Sato, Yukihito. 1997. "Diverging Development Paths of the Electronics Industry in Korea and Taiwan." *The Developing Economies* 35 (4): 401–421.

Schive, Chi, and R. S. Yeh. 1980. "Direct Foreign Investment and Taiwan's TV Industry." *Economic Essays* 9 (2): 261–291.

Schnaars, Steven P. 1994. *Managing Imitation Strategies: How Later Entrants Seize Markets from Pioneers.* New York: The Free Press.

Sematech. 1997. *A World of Opportunity: 1996 Annual Report.* Austin, Texas.

Teece, David J. 2000. "Firm Capabilities and Economic Development: Implications for the Newly Industrializing Economies." In Linsu S. Kim, and Richard R. Nelson, eds., *Technology, Learning, and Innovation: Experiences of Newly Industrializing Economies.* Cambridge, U.K.: Cambridge University Press.

Tsai, Kuen-Hung, and Jiann-Chyuan Wang. 2005. "An Examination of Taiwan's Innovation Policy Measures and Their Effects." *International Journal of Technology and Globalisation* 1 (2): 239–257.

Tung, An-Chi. 2001. "Taiwan's Semiconductor Industry: What the State Did and Did Not Do." *Review of Development Economics* 5 (2): 266–288.

UNCTAD (United Nations Conference on Trade and Development). 2003. *World Investment Report 2003: FDI Policies for Development—National and International Perspectives.* Paper presented at the United Nations Conference on Trade and Development, September 4, Geneva.

UNIDO (United Nations Industrial Development Organization). 2002. *Industrial Development Report 2002–2003: Competing through Innovation and Learning.* Vienna.

Vernon, Ray S. 1966. "International Investment and International Trade in the Product Cycle." *Quarterly Journal of Economics* 80 (2) May 1966: 190–207.

Westney, D. Eleanor. 1988. *Imitation and Innovation: The Transfer of Western Organizational Patterns to Meiji, Japan.* Cambridge, MA: Harvard University Press.

Wong, Poh-Kam, and John A. Mathews. 1998. "Competing in the Global Flat Panel Display Industry" (Introduction to special issue). *Industry and Innovation* 5 (1): 1–10.

Wu, Rong I., and Ming-Sheng Tsen. 1997. "The Development of Information Industry in Taiwan." Paper presented at the conference Foundation for Advanced Studies in International Development, Tokyo.

Xue, Li-Min, Chen-Kuo Hsu, and Dwight H. Perkins. 2000. *Industrialization and the State: The Changing Role of Government in Taiwan's Economy, 1945–1998.* Cambridge, MA: Harvard Institute for International Development.

Yusuf, Shahid. 2003. *Innovative East Asia: The Future of Growth.* Washington, D.C.: World Bank.

Zhao, Hongxin. 1995. "Technological Imports and Their Impacts on the Enhancement of China's Indigenous Technological Capability." *Journal of Development Studies* 31 (4): 585–602.

CHAPTER 4

Electronics in Malaysia: Export Expansion but Slow Technical Change

Rajah Rasiah

SINCE 1987 THE ELECTRONICS INDUSTRY HAS been the leading export earner in Malaysia. In 2000 it accounted for a quarter of employment, a third of fixed assets, and a third of value added in manufacturing (Malaysia 2003). From barely any activity in 1968 (Rasiah 1995), electronics became the cornerstone of manufacturing growth, accounting for more than 80 percent of manufactured exports in 2000 (Malaysia 2001). However, the industry is still dominated by low-value-added activities and low R&D intensities.

This chapter will explain the policies that stimulated Malaysia's rapid expansion in electronics and that kept the industry from upgrading to higher value added over 30 years of rapid growth. The failure to upgrade is a point of serious concern in view of rising domestic costs and the emergence of low-cost producers in China, Indonesia, the Philippines, and Thailand.

The connections between Malaysia's still-emerging electronics cluster and global markets and value chains shape the dynamics of production technology in the firms making up the cluster. Access to global markets through multinational corporations (MNCs) has facilitated rapid growth in the industry, but the lack of a system to support upgrading has restricted operations to low-value-added assembly and test operations.

Explaining technical change and its effect on upgrading in firms requires an audit of the institutions that support learning and innovation. Links between firms and those institutions drive the processes of upgrading, differentiation, and division of labor in dynamic clusters such as Malaysia's electronics sector. Relevant high-tech institutions supply human capital, create the policy framework needed to stimulate firms' participation in risky and uncertain activities (such as

R&D), provide financial incentives, promote adherence to international standards, regulate intellectual property rights, and disseminate research findings.

The framework adopted here examines the role of policy instruments and how those instruments, and the institutions they create, affect firms. We focus on government-led strategies to engender growth and structural change, how government policy can create or transform structural conditions to induce desired conduct in firms (Best 2001; Porter 1990), and how high-tech institutions help stimulate innovative activities in economies on a national scale (Freeman 1989; Nelson 1993).

Following a brief historical review, the chapter discusses the policy environment facing the electronics industry. We then examine the critical drivers of export expansion in the electronics industry and the industry's successful role in stimulating investment, employment, and production growth in Malaysia.

Origin and Early Development

Import-substitution industrialization in Malaysia began officially with the Pioneer Industry Ordinance of 1958. However, it was not until the late 1960s that the instruments foreseen in the ordinance were implemented. The Malaysian Industrial Development Authority (MIDA) opened in 1964 (as the Federal Industrial Development Authority) to spearhead industrial promotion. Japanese-owned Matsushita Electric, which began assembling household consumer electronics in Malaysia in 1965 to supply the domestic market, was Malaysia's first electronics firm (Rasiah 1993). However, major growth began in 1972 with the opening of free trade zones (FTZs). Clarion and National Semiconductor were the first export-oriented electronics firms to build factories in the country—in 1971 in Penang.

First Wave

Several MNCs opened electronics facilities in Penang and the Klang Valley in 1972–75, which helped boost employment and value added in the industry. Electronics employment and value added grew on average by 44.7 percent and 46.4 percent per year in 1971–79. American and Japanese components firms, especially semiconductors, led this wave of investment—National Semiconductor, Intel, Motorola, Hitachi, and Advanced Micro Devices were among the firms that relocated in this period. European investment (Philips and Siemens) came later.

This wave was also characterized by the assembly of highly labor-intensive components. Although the Industrial Relations Act of 1967 did not prohibit the formation of trade unions, the government allowed "pioneering" firms to keep unions out.[1] Interviews suggest that electronics firms are generally reluctant to allow trade unions owing to volatile fluctuations in demand.

The first phase of Malaysia's electronics industry included almost no local firms—except for a few small ones such as Penang Electronics, established in 1970. Foreign direct investment (FDI) dominated the small manufacturing sector, but FDI levels declined from 1975 until the 1980s, because of the OPEC-induced oil crises. The decline in FDI was exacerbated by the enactment of the Industrial Coordination Act, which implemented the ethnic restructuring provisions of the National Economic Policy of 1971 (NEP; Malaysia 1976). Nevertheless, the NEP did not fetter FDI inflows because export-oriented firms located in export-processing zones were not required to introduce ethnicity-based equity conditions.

Large reserves of potentially trainable labor with at least secondary education were important in encouraging FDI (Lim 1978; Rasiah 1993). But Malaysia was just emerging from a bloody ethnic crisis in 1969 and the infrastructure outside the Klang Valley was inadequate to facilitate air freight shipments of inputs and output. To compensate, the government improved the airport facilities and created FTZs and licensed manufacturing warehouses (LMWs) to provide the requisite infrastructure and political security to enable labor-intensive activities. The incentives were instrumental in offsetting the risks associated with relocating to a new and potentially dangerous site. Personal visits from state leaders helped raise the attractiveness of the sites.

Operations in Penang were moribund in the 1970s. To ensure repairs of machinery and other support functions foreign firms either built in-house facilities (Intel, AMD, and Hewlett-Packard) or opened complementary subsidiaries (National Semiconductor started Dynacraft and Micro Machining). Most inputs—including machinery parts—were imported (Rasiah 1993).

Second Wave

In the first half of the 1980s, growth was slow because of the global slowdown and inward-oriented government. As a result, it was only after 1986 that the second wave of electronics firms relocated operations to Malaysia. Annual employment and value added in the industry grew by 2.0 percent and 11.4 percent on average in 1980–85. The second wave of firms boosted growth in employment

and value added to annual rates of 21.6 percent and 21.1 percent in
1985–90. Firms responded to the reintroduction of generous tax incentives, to the maturation of Malaysia's experienced labor force, and
to rising costs in Japan, Korea, Taiwan, and Singapore caused by the
appreciation of their currencies following the Plaza Accord and with-
drawal of privileges under the Generalized System of Preferences in
February 1988.

Some firms had started operations in Melaka, Negeri, Sembilan,
and Joho in the early 1980s, but the second wave helped expand electronics assembly operations in these states, and in Kedah and Perak.

The second wave of new investment included the relocation of giant
consumer electronics firms—among them Hitachi, Sony, Toshiba, and
Samsung. The synergies created by these firms, as well as the high costs
of operations in Singapore, attracted disk drive and computer companies in the early 1990s. Most of the disk drive companies shifted out
of Malaysia by the late 1990s in response to the tightening labor market. China and Thailand eventually became their main target, though
Seagate retained a plant in Penang. Financial incentives offered to all
firms in 1998 on the basis of investment dissuaded several MNCs from
leaving Malaysia.

Despite rising production costs and the threat of China by the end
of the 1990s, the electronics industry remained the main contributor
to employment and value added in Malaysia. Value added grew on average by 21.3 percent per year in 1990–97, and electronics made up
80.1 percent of manufactured exports in 2000 (Malaysia 2003).

It can be argued that the industry remains viable. Employment rose
strongly until 1997, after which it began to decline (figure 4.1). Nevertheless, exports and imports have continued to expand.

In response to the economic crisis of 1985–86 (real GDP shrank by
1.1 percent in 1985 and grew by only 1.2 percent in 1986[2]) the government formulated a very generous Industrial Master Plan (IMP) to
retain foreign firms. Incentives were renewed even for labor-intensive
firms whose pioneer status and investment tax credits had expired.
The IMP also included complementary industries upon the recom-
mendation of the expert team from the United Nations Industrial
Development Organization. Initiatives were launched to increase linkages and technological deepening—among them allowance of double
deductions of approved expenses to promote human resource development and investment in R&D activities. A subcontract exchange
program and the vendor development program were introduced to encourage large firms to support local suppliers (Malaysia 1988; Rasiah
1995). Largely because of network cohesion in areas outside Penang
and weak institutional support for upgrading, these programs had

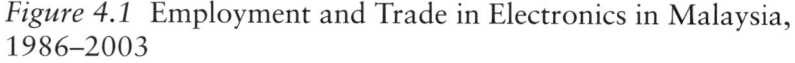

Figure 4.1 Employment and Trade in Electronics in Malaysia,
1986–2003

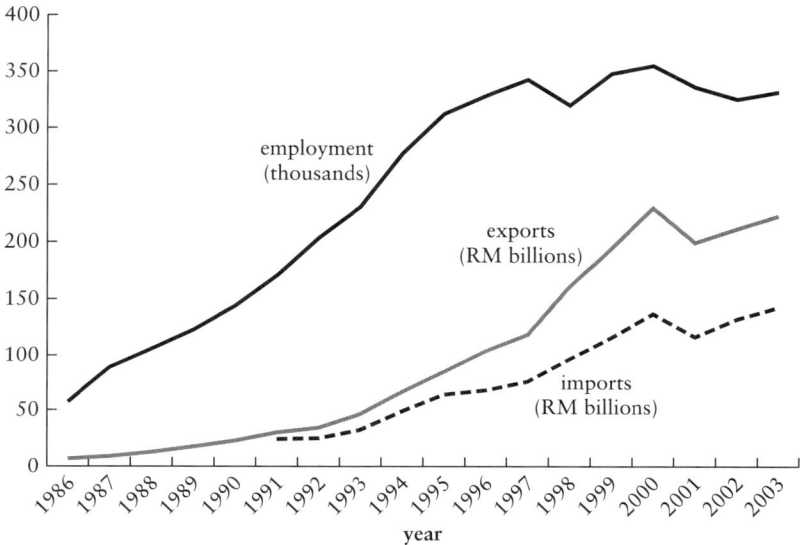

Source: Compiled from unpublished data supplied by the Malaysian Statistics Department (2004).

little impact on local firms' participation in the supply chain of electronics MNCs (Rasiah 1996b, 1999).

Policy Environment

Government policy on industrial promotion in Malaysia can be grouped into four phases: a first round of import substitution, a first round of export orientation, a second round of import substitution, and a second round of export orientation (table 4.1). This section examines the government's role in promoting the growth of manufacturing in general and of the electronics industry in particular.

Import Substitution I

Promotion of industry in Malaysia began with the Pioneer Industry Ordinance of 1958, which was designed to encourage import-substituting industrialization (Hofmann and Tan 1980).[3] Matsushita Electric relocated some operations to Malaysia, but the small domestic market discouraged others. The handful of electric and electronics

Table 4.1 Industrial Strategies and Trade Orientation in
Malaysia, 1958–2000

Phase	Trade orientation	Period of dominance	Policy instruments
1	Import substitution	1958–1972	Pioneer Industries Ordinance, 1958
2	Export orientation	1972–1980	Investment Incentives Act, 1968 Free Trade Zone Act 1971
3	Import substitution	1981–1985	Heavy Industries Corporation of Malaysia (HICOM) 1980
4	Export orientation	1986–2005	Industrial Master Plan 1986 Promotion of Investment Act, 1986 Action Plan for Industrial Technology Development (APITD) 1990 Industrial Master Plan 2, 1996

Source: Compiled by author.

firms that opened plants simply carried out minor assembly work because import controls were imposed only on final consumer goods and
because no policy was in place to stimulate upgrading and domestic
capacity building (Edwards 1975; Rasiah 1995: chapters 4 and 5).

By 1965, rising unemployment and industrial stagnation in the face
of worsening economic inequality led the government to expand its industrial policy to include export promotion (Jomo 1990b). However,
in contrast to Singapore—which shifted completely to export orientation following its secession from Malaysia in 1965—Malaysia's export
orientation coexisted with import substitution. Although only fiscal
instruments were used to control trade, import substitution was per-
petuated except in the tariff-free export-processing zones that were
effectively cut off from the principal customs area (Rasiah 1996a).

The key phases in Malaysia's industrial policy are presented in
table 4.1.

Export Orientation I

Incentives for export-oriented firms in Malaysia were introduced in the
Investment Incentives Act of 1968. However, the ethnic bloodshed of
1969 and Malaysia's unproven capacity for large-scale export-oriented
activities raised uncertainty. The first major wave of export-oriented
firms relocated to Malaysia only after the creation of FTZs in 1972,

following enactment of the Free-Trade Zone Act the previous year. The government launched LMWs to offer firms the same incentives in sites where FTZs were unfeasible or undesirable. Incentives and personal visits by government leaders helped raise the confidence of the top management of the MNCs and induce export-oriented electronics firms to begin assembly and test operations in Malaysia. FDI has dominated ownership in Malaysia's electronics industry since 1968, in contrast to Korea and Taiwan. Indeed, FDI levels in the ownership of fixed assets in the electronics industry have always exceeded those in the manufacturing sector.

The Industrial Coordination Act (ICA) was passed in 1975 to coordinate the application of the NEP, which was designed to alleviate poverty and inequality while eliminating the identification of economic function by ethnicity. Through the NEP, the government set out to raise corporate equity ownership by the indigenous Bumiputeras from 2.5 percent in 1971 to 30 percent in 1990. While the politicians have exulted in its success in raising Bumiputera participation in the economy, critics have rightfully argued that it has created ethnic polarization in government-controlled sectors such as the civil service, where more than 90 percent of the staff in 2000 were Bumiputeras (Jomo 1986).[4] In addition, although the 30 percent corporate equity target was not met in 1990, inequality also rose sharply in the Bumiputera community over the NEP period (Ishak 1995). The ICA was administered so as not to discourage FDI. Exporting firms were not required to have local equity. Although some firms took on some local equity, most export-oriented multinationals did not. Foreign capital continued to dominate the ownership of fixed assets in the electronics industry even after 1975.

Import Substitution II

A global slowdown in semiconductor sales and government efforts to shift incentives to domestic heavy industries caused new investment in electronics to contract in 1980–85, which in turn contributed to the economic crash of 1985. High commodity prices after 1979 encouraged the government to invest extensively in heavy industries, as well as in the construction of the North–South highway and Penang Bridge. But prices for Malaysia's key export commodities—rubber and palm oil—fell in the first half of the 1980s. Tin exports collapsed after 1980, and oil prices fell sharply in 1985. At the same time, the appreciating Japanese yen raised the cost of the debts the government had contracted to finance infrastructure development.

Although macroeconomic accounts focus on heavy investment in domestic heavy industries, falling primary commodity prices, and the appreciation of the currency in 1981–85 as the major causes, the electronics industry faced a severe downturn following a massive glut and collapse in prices (Rasiah 1988). Several large multinationals had already threatened to leave Malaysia if expiring financial incentives were not renewed. For example, in 1985, Intel announced plans to revive assembly in Manila, while AMD began construction on a plant in Thailand (Rasiah 1993).[5] These developments inspired the generous incentives contained in the IMP of 1986. Incentives for exporting firms, including those in electronics, were reintroduced, and older firms enjoyed a new round of tax incentives.

Export Orientation II

Burgeoning debt-service problems, a slowdown in GDP (the country grew by only 0.2 percent in 1985), and rising unemployment (from 6.0 percent in 1980 to 8.6 percent in 1985) forced the government to reverse its approach to FDI. In addition to devaluing the ringgit in 1986, the government renewed financial incentives to export-oriented firms with the IMP launched the same year. Coupled with the exodus of firms from Japan, Korea, and Taiwan following the Plaza Accord of 1985 and the exclusion of Asia's newly industrializing economies from the Generalized System of Preferences in February 1988, the IMP (1986–96) brought Malaysia a major resurgence of FDI. The plan renewed pioneer status, the investment tax allowance (ITA), and certain allowances for stimulating training and R&D.[6]

High-tech activities enjoyed incentives when R&D expenditures by Malaysian operations were at least 1 percent of gross sales for three years or when science and technical graduates made up at least 7 percent of the workforce (Rasiah 1995: appendix 4.2).

However, incentives for R&D did not have many takers because the allowances focused on product and tangible process technologies at a time when firms were concentrating their R&D on process engineering. Also, pioneer status and ITA made it unnecessary for most firms to seek the R&D incentives, especially since investment classified as "strategic" qualified for total tax exemption under the pioneer and ITA provisions. Strategic investments were defined as those that exceeded RM 100 million, involved integrated and high-tech manufacturing activities, stimulated backward and forward linkages, or occurred in approved R&D facilities.

In the government's attempt to stimulate R&D, several other incentives were applied. These included:

- ITA of 50 percent on qualifying capital expenditure incurred within 10 years. Under the ITA, firms could seek tax exemptions equivalent to 50 percent of all approved capital expenditures over the 10-year period.

- Double deduction for R&D expenditures—including spending on services from approved R&D institutes and direct cash contributions to approved R&D institutions.
- Deduction for spending on buildings used for R&D activities and other R&D-related capital expenditures, such as machinery and equipment. Firms' R&D expenditure on buildings was deductible from taxable income.

The government also launched several funds to stimulate participation by Malaysian firms in R&D activities. The Technology Acquisition Fund was designed to help Malaysian firms seek strategic technology from foreign sources. The Industry R&D Grant Scheme and the Commercialization of R&D Fund were supposed to encourage firms to adapt, create, and commercialize new technologies.

Over the Sixth (1991–95), Seventh (1996–2000) and Eighth Malaysia Plans (2001–5), the government allocated RM 1 billion in each of the periods to support basic research in institutions of higher education. Since then, the Intensification of Research in Priority Areas has become a major source of R&D funding in Malaysian institutions and R&D organizations.

These generous incentives stimulated considerable interest among universities and R&D organizations to undertake research. However, implementation problems restricted their effectiveness.[7] First, the mechanisms used to evaluate the usefulness of projects involved little participation of relevant representatives from the private sector. Second, academic and private researchers have argued that the criteria for evaluation did not reach acceptable standards owing to the lack of a mechanism to appraise evaluations. Third, the confidentiality of assessments was often compromised, so that individual agents hardly sought excellence as the basis for approving grants. Because applicants often knew the assessors, there was a tendency toward collusion. Fourth, MNCs were not brought into the activities—thereby excluding a critical source of knowledge. (Singapore and Ireland, by contrast, benefited from participation by multinationals in schemes to promote R&D.) Fifth, the failure to appraise completed projects kept the mechanism from improving. Sixth, grants to stimulate R&D were offered only to successful local firms already operating in the country, even though they had been intended to include start-ups and individual innovators who operated outside the domain of firms. (In Ireland and Singapore, foreign multinationals and individual innovators enjoyed access to such grants.)

During the IMP period, the government also tried to stimulate upgrades in process technology and human resources. Capital spending on process technology related to tangible changes in machinery and equipment qualified for a reinvestment allowance (in the form of a tax exemption of 60 percent or more), while the double-deduction tax incentive was introduced to stimulate training. That instrument, designed to stimulate training in smaller firms, was considered largely redundant by larger firms that enjoyed access to the more lucrative pioneer and ITA allowances. However, the paucity of training institutions in areas where firms with fewer than 100 employees were concentrated and the difficulty that small firms had in releasing employees for training meant that few small firms made use of the incentive (Rasiah and Osman 1998). Instead, multinationals were the main beneficiaries.

The Malaysian Institute of Microelectronic Systems (MIMOS) was established in 1985 in the office of the prime minister for the strategic purpose of stimulating microelectronic R&D activities in local firms. However, rules governing public servants and organizations constrained its activities. Not only did MIMOS's governance structures prevent it from acting quickly on policy issues, but also the hiring of high-ranking civil servants was delayed by government procedures. The salary structure of public employees also discouraged attempts by MIMOS to attract knowledgeable employees from private firms. For example, a former MIMOS employee reported that the institute was unable to entice a senior engineer from Advanced Micro Devices because it could not offer sufficient remuneration (Rasiah 1995: chapter 6).

The government privatized MIMOS in the 1990s in a partially successful attempt to resolve the salary problem. The institute was a key member of a consortium that attracted a key expert from Silicon Valley and used him to incubate wafer fabrication in Malaysia. The expert later became vice president at First Silicon, a contract wafer-fabrication plant established in Kuching in 2000. However, MIMOS still has not achieved the success of the Industrial Technical Research Institute (ITRI) in Taiwan. Through its incubator strategy, ITRI has successfully spun off several firms, including TSMC in 1986 (see the chapter by Mathews in this volume). In contrast to ITRI, which offered shares and other benefits to offset the disadvantages of leaving the United States, MIMOS's inability to promote loyalty, trust, and opportunities for self-actualization appear to have restricted the number of experienced Malaysians returning home. In fact, the Malaysian government managed to draw home only 231 experienced and Malaysian professionals between 2001 and 2003.[8]

Following the Action Plan for Industrial Technology Development of 1990, the government created several intermediary organizations to encourage innovative activities with higher value added. The action plan included efforts to diversify production, strengthen training and R&D, and penetrate markets in higher-value-added activities. The execution of the blueprints led to the creation of the Human Resource Development Fund (HRDF) in 1992, the formation of the Malaysian Technology Development Corporation (MTDC) and the Malaysia Industry Government High Technology (MIGHT) body in 1993, the introduction of the Second Industrial Master Plan (IMP 2) in 1996, and the opening of the Multimedia Super Corridor (MSC) and the Multimedia Development Corporation (MDC) in 1997.

Under the HRDF legislation, manufacturing firms with 50 or more employees were required to pay 1 percent of their payroll to the Human Resource Development Council, which could be reclaimed upon presentation of evidence of approved training expenses. As noted earlier, MNCs in the electronics industry were seen as the prime beneficiaries of the double tax deduction given for training expenses in 1988–92, and interviews showed that their training strategies were not influenced by the double-deduction incentive. The high-tech foreign multinationals in the electronics industry invested in training to meet rising skills requirements arising from global competition (Rasiah and Osman 1998). Because firms that engaged in low-value-added activities were investing little in human resource development, the HRDF provided the "shaking" needed to get them to train. Manufacturing firms with fewer than 50 employees still enjoyed the double deduction under the act that created the HRDF. The following additional incentives were introduced to encourage training:

- Deduction of cash contributions to nonprofit technical or vocational training institutions
- Exemption from import duties, sales tax, and excise duties for imported machinery, equipment, and materials used for training purposes
- ITA for new investment in training equipment or training capacities

Reviews of the HRDF by electronics firms showed that it forced smaller firms and firms in inward-oriented industries to train. However, its implementation faced two major problems (Rasiah 1996b, 1999). First, training institutions, especially outside the western industrial corridor, were ill-equipped to handle state-of-the-art training, where they existed at all. Second, interfirm networks outside Penang were not sufficiently cohesive to coordinate demand and supply effectively. Hence, while foreign electronics firms enjoyed cutting-edge training practices even before the introduction of the HRDF—Intel and Motorola had their own training centers—local firms did not get similar training support. Penang was the exception. The Penang Skills Development Center (PSDC) was formed in 1989, when the state government offered a huge building to support industry-oriented training. The cluster in Penang attracted strong participation by both foreign and local firms. In 1998, PSDC had 81 members who employed about 75,000 workers (Best and Rasiah 2003: 50). The Selangor Human Resource Development Center and the Johor Skills Development Center failed to produce similar results. These institutions are fairly new, and they lack the interfirm and firm–institution relationships, economic and social, needed to attract strong participation.

The government launched four different programs to develop linkages between multinational firms and small and medium-size enterprises (SMEs). The first program adopted the umbrella concept. Beginning in 1984, local SMEs in food processing were identified and given marketing and financial support. The government helped promote the sales of these firms. Second, the subcontractor exchange program (SEP) was launched in 1988 to match local firms with MNCs. Third, a vendor development program (VDP) was introduced the same year to encourage MNCs to foster local firms. Under this program a number of local firms were identified for support from MNCs. And fourth, the industrial technical assistance fund (ITAF) provided incentives for firms whose products contained 50 percent domestic content (a maximum of 30 percent value added). In 1995, the government consolidated all public institutions representing SMEs into the Small and Medium Industry Development Corporation (SMIDEC), which fell under the authority of the Ministry of International Trade and Industry (MITI). Some firms benefited from these programs. But given that the successful suppliers of electronics firms emerged in Penang largely without direct government support, it can be said that they had but a small impact on the development of suppliers in Malaysia.

The government also took steps to establish Malaysia as a regional distribution center like Singapore and Hong Kong. Firms incorporated in Malaysia with capital of RM 0.5 million and annual sales of at least RM 100 million were eligible to apply for incentives under the International Procurement Center when they handled goods and services through Malaysian airports and seaports. The incentives allowed qualifying firms to:

- Hire expatriates
- Open foreign-currency accounts with licensed commercial banks to keep export revenue in foreign currencies

- Enter forward contracts on exports using foreign exchange
- Enjoy exemptions from ICA equity-ownership conditions on wholesale and retail trade in Malaysia
- Import and export goods and services duty free

Tax exemptions based on export volume and subsidized credit for export refinancing had been available since the launch of the IMP in 1986.

The government launched the Kulim and Bukit Jalil high-tech parks in the 1990s. To attract strategic high-tech firms engaged in R&D activities to the parks, it offered pioneer-status tax incentives. Electronics firms became the main beneficiaries of this initiative, although the rate of take-up was relatively low compared to that of the FTZs and LMWs.

MTDC was intended to attract venture capitalists to support innovative firms. Although it acquired equity in UNICO and Globetronics in 2000 to stimulate expansion of those firms, MTDC failed to play the bigger role expected of venture capitalists, namely to finance potentially innovative entrepreneurs who lack the capital to start firms. Hence, it has not attracted scarce human capital with the potential to produce profitable products. The dynamic frameworks used to support the new high-tech entrepreneurs of Silicon Valley, Route 128, Taiwan, and Israel should be adapted to the Malaysian environment if venture capital is to play its role in supporting the creation of new firms in the high-tech electronics industry.

MIGHT has played an important role in coordinating consultation among CEOs, academics, and government officials in high-tech industries to help encourage technological deepening and competitiveness. However, it does not focus sufficiently on improving networks to drive differentiation and division of labor in the country or on promoting firms' participation in R&D activities. Taiwan's ITRI, by contrast, managed to attract considerable participation by firms (after initial gaps) by engaging firms directly in their planning. MIGHT needs to engage the real entrepreneurs on the ground rather than by placing in firms retired officials with civil service or university backgrounds.

IMP 2 was launched in 1996 to drive industrialization toward higher-value-added activities (Malaysia 1996). The "manufacturing ++" strategy of the plan sought to stimulate value added both horizontally (expanding operations beyond assembly to include designing and research upstream, and marketing and sales downstream) and vertically (raising the value added of assembly activities). Since 1986 the government had tried to attract consumer and industrial electronics firms to diversify the industry, but their numbers remained much smaller than components electronics. More aggressive efforts have been made since 1996 to add missing segments to industrial clusters. Among the missing segments promoted through incentives offered in IMP 2 were R&D and marketing to enable firms to become integrated manufacturing complexes, and electronics wafer fabrication to extend the value chain upstream. In addition, the regulatory framework governing the more lucrative pioneer status and ITA was changed in 1996 to limit incentives to high-tech activities such as training and R&D.

Despite these changes, two major constraints have blocked the emergence of cluster synergies. The first is the lack of human capital to drive technological deepening. The second is the lack of sufficient coordination between government and firms to support industrial upgrading outside Penang.

IMP 2 contained measures to induce Malaysian-owned firms to participate in cutting-edge activities, including original equipment, own design, and own brand manufacturing, but the lack of a stick to instill discipline has limited their impact (Rasiah 2002a; Best and Rasiah 2003). On the positive side, Sapura, OYL Electronics, and Malaysian Electronics Company (MEC) managed to obtain special incentives to sell in domestic markets. However, Sapura shifted its focus to auto parts once the special contract it enjoyed with the Malaysian government to sell telephone sets and other telecommunications services ended (Alavi 2000), and MEC closed down its consumer electronics assembly operations following competition from imports.

IMP 2 had very little impact on the development of the supplier firms in the electronics industry. Firms that grew from demand generated by MNCs, especially in Penang, benefited from better state–industry coordination. Local firms in Penang hired former employees of MNCs to use their knowledge and connections with these companies to expand operations. Some of these firms had already begun operations before the special incentives were started, so even if the incentives facilitated expansion, they did not initiate start-ups in Malaysia. UNICO, Globetronics, Shinca, Trans Capital, Prodelcon, Metfab, and BCM—all located in the dynamic Penang cluster—are some of the companies whose top management had worked in MNCs and that have expanded strongly. The Penang cluster has also produced several successful ancillary firms that supply metal engineering, machinery, plastic engineering, and components to electronics firms, while also exporting on their own (Rasiah 1994; Narayanan and Wah 2000). Eng Technology, LKT Engineering, SEM Engineering, Polytool, Atlan, and Wong Engineering are examples. LKT Engineering has production operations in Costa Rica to supply Intel operations there. Eng

Technology has subsidiaries in China, Thailand, and the Philippines. Interviews by the author showed that only 10 percent of the firm's value added in 2001 was recorded in Malaysia. Atlan had subsidiaries in Indonesia. The microchip firms of Carsem and Unisem located in Perak have also strongly expanded contract assembly. Although these firms did not enjoy the same cluster synergy benefits of Penang firms, they benefited from hiring former employees of electronics MNCs in Malaysia.

The MSC and MDC were launched with sophisticated infrastructure created between Sepang and Kuala Lumpur, including the two modern, "wired" cities of Putra Jaya (Malaysia's administrative capital since 1998) and Cyber Jaya. However, as with the IMP 2 efforts to deepen high-tech activities, the MSC has suffered from serious human resource and network constraints. The network problems included poor connectivity and coordination between firms and institutions. In addition, providing incentives only to IT firms violated the concept of clustering (in which differentiation and division of labor require extension into complementary industries).[9]

The environment for stimulating R&D activities was also poorly developed. Incentives were generous but more suitable for mature innovators than for start-ups and MNCs seeking to offset the risks of relocating developmental R&D activities. The implementation of R&D grants also bypassed start-ups and potential individual innovators. In contrast to Singapore, Ireland, and Israel, where grants and labs helped stimulate relocation of MNCs' activities, Malaysian incentives and grants tended to reward innovators after the fact. The fact that tax holidays were available based on different criteria—such as level of investment and strategic classification—made the R&D incentives redundant for many firms.

Although national policy limited states' developmental activities, some states used their relative autonomy to play a stronger role than others. Penang state—through its development corporation—played a direct role in wooing foreign firms to relocate and participate in local supplier and training activities (Rasiah 1996b, 2002b; Lim 1998).

Overall, Malaysia managed to attract export-oriented electronics firms in assembly and test activities. Its success in this regard is comparable with the world's best examples. Despite failing to stimulate successful start-ups, government incentives managed to spur further expansion in some firms. However, in contrast to Singapore and Ireland, where similar firms made the transition to higher-value-added activities, Malaysia has yet to achieve similar results. For one thing, the policy framework has meshed poorly with potential entrepreneurs. The mechanisms for identifying potential entrepreneurs and innovators, drawing them into the policy process, and establishing a legal framework to connect venture capitalists with innovators have been poorly developed, with the result that entrepreneurial energies have not yielded significant new firms.

Critical Drivers of Growth

Malaysia's unemployment rate stood at 8.1 percent in 1971 when the government took serious steps to attract export-oriented MNCs to Malaysia. Although English was widely spoken and literacy high, substantial government efforts were required to prime the flow of investment and ensure that the flow continued. Rising unemployment and inequality in the 1960s—which caused the defeat of several ruling-party candidates in the 1969 national elections and led to ethnic bloodshed—pushed the government to promote labor-intensive, export-oriented manufacturing. In addition to restoring peace, the government had to improve infrastructure (notably air transport) before MNCs would consider relocating to the country.

A combination of policy instruments, intermediary organizations, and external forces shaped Malaysia's electronics industry as it grew from the 1970s. Unlike the experience of Korea and Taiwan, where local firms quickly gained global dominance as original equipment, own design, and own brand manufacturers, Malaysia's electronics manufacturers remain foreign affiliates that rely extensively on the technology of their parent companies. Domestic efforts have yet to induce foreign or local firms to invest substantially in R&D.

Malaysia has been successful in attracting strong foreign investment, creating jobs, engendering diversification, promoting exports, and forging significant production-related linkages in the state of Penang. But on the whole, production has been limited to low-value-added activities. Efforts to stimulate the transition to higher-value-added activities have not borne fruit.

Basic Infrastructure, Political Stability, and Investment Coordination

Malaysia is a small, open, middle-income country that was able to create the conditions to attract FDI in electronics manufacturing. Although the country has been politically stable for many years and by the 1990s boasted basic infrastructure comparable to that of developed economies, neither was true when the first wave of electronics firms opened operations in the country in 1972–74.

Financial incentives helped offset the risks associated with foreign sites. But they were not enough. The Investment Incentives Act of 1968 introduced tax exemptions for export-oriented firms, but it took five years more for a critical mass of firms to relocate to Malaysia. The Malaysian government opened FTZs in 1972, following legislation the previous year that helped provide necessary basic infrastructure such as power, water, security (from theft), and trouble-free access to airports. The FTZs and LMWs made possible predictable duty-free operations, while the Malaysian Industrial Development Authority (MIDA, formed in 1964 as the Federal Industrial Development Authority) helped investors resolve basic infrastructure problems.

Confidence among investors that Malaysia's leaders would deliver on their promises was as important as any initiative. Officials from the Ministry of Industry (including the minister) and the head of the Penang state government knocked on the doors of flagship firms such as National Semiconductor, Intel, AMD, and Hitachi to persuade them to relocate operations. Rasiah (1988) documented evidence of the dynamic role of Lim Chong Eu, Penang's chief minister from 1969 to 1990, in visiting firms and hosting visits to Penang by their chief executive officers. Intel, AMD, Hewlett-Packard, and others reported that the dynamism showed by Penang's leadership was a major factor in their decision.

The government took special measures to ensure that its ethnic restructuring policy did not collide with its export-oriented FDI strategy. The Industrial Coordination Act (ICA) was enacted in 1975 to implement the principles of the New Economic Policy (NEP) of 1971, notably to eliminate ethnic economic segregation and increase the participation of indigenous Malays in the economy. Foreign firms that exported most of their output were exempted from ICA. Although the exemption averted a potentially harmful exodus of firms from Malaysia, it also deepened the dualism of the manufacturing sector (Edwards and Jomo 1993; Rasiah 1995).

ICA also discouraged multinationals located in FTZs and LMWs from selling in the domestic market, although the domestic market was too small to affect most of the firms. The law was designed to promote ethnically based ownership conditions among firms selling in the domestic market, but its effect was to discourage the development of domestic suppliers. The limits on domestic sales were realized after 1989 when the state government of Penang made serious efforts to persuade disk drive and computer assembly firms to relocate alongside manufacturers of integrated circuits, capacitors, resistors, disk-drive components, and opto-electronic components. The same problems were evident when the government began encouraging domestication by tying incentives to local sourcing after 1991.

In the 1980s, the government averted a potential exodus of MNCs by renewing financial incentives as recommended in the IMP of 1986. Ironically, the government played a role in encouraging the exodus when it attempted in the early 1980s to shift the country's industrial focus away from exports led by foreign MNCs to domestic heavy industries (see table 4.1). The import-substitution years from 1981 to 1985 saw a decline in investment in the electronics industry. Malaysia's treasury had already been hit by falling export prices of rubber, palm oil, and petroleum; the crash of tin (Jomo 1990a); and a huge surge in the value of the yen that increased the cost of loans used to finance infrastructure construction and support heavy industries. The economic slowdown and nonrenewal of expiring financial incentives for foreign firms soon began to discourage reinvestment in the electronics industry, already hit by a global slowdown. AMD was among the firms that threatened to relocate (to Bangkok) if their expiring tax incentives were not renewed (Rasiah 1988).

MITI's quick processing of applications and the open door policy practiced by MIDA and the state economic development councils encouraged firms to resume investments in assembly and test operations in Malaysia. The focus on attracting FDI was so intense that Penang, Kuala Lumpur, and Selangor had to take steps after 1993 to persuade firms to consider locating in other regions in Malaysia. Investors in labor-intensive industries were encouraged by the state development corporations in these states to relocate in the other parts of Malaysia. The federal government reintroduced regional incentives for labor-intensive firms to shift operations to East Malaysia and the states of Sarawak and Sabah in 1995. However, the initiative lost steam when the financial crisis and the emergence of China threatened to attract foreign MNCs away from Malaysia. Ericsson, Omega, and Intersil were among the firms that wholly or partially relocated from Malaysia to China during this period. To stem the outflow, the government in 1998 restored financial incentives for all export-oriented firms regardless of their chosen location in the country.

Malaysia's successful strategy to attract export-oriented assembly and test operations despite rising production costs meant that the share of electronics in manufactured exports continued to rise—from 0.7 percent in 1968 to 71 percent in 1997 (table 4.2). The trade balance improved as well, from −88 percent in 1968 to a positive balance of 10 percent in 1990 before falling to −2.6 percent in 1997. Exports as a share of production fell slightly from 16 percent in 1968 to 11 percent

Table 4.2 Output and Trade Ratios for Electrical Machinery
in Malaysia, 1968–97
(Percent)

	1968	1973	1979	1985	1990	1997
Share in manufactured exports [$xi/\Sigma xi$]	0.7	2.1	28.4	44.6	53.4	71.0
Trade balance [$(xi - mi)/(xi + mi)$]	−87.7	−72.3	−3.1	−3.7	10.0	−2.6
Share in manufactured output [xi/yi]	16.0	11.0	72.8	99.0	94.0	81.9

Sources: Computed from Malaysia, *External Trade Statistics,* 1969, 1974, 1980, 1986, 1991, 1998.

Note: x, m, and y refer to exports, imports of industry *i* respectively

in the first oil crisis year of 1973 before rising sharply to 99 percent in 1985. Quite clearly, expansion in the industry was driven by global markets. The export-intensity of output fell gradually to 82 percent in 1997. Sustained expansion in the industry ensured growth in employment. The focus on labor-intensive, low-wage operations also caused wages to decline by 2.6 percent during 1971–79 and 3.5 percent during 1985–90.

Network Cohesion

Dynamic clustering, which is defined by strong connectivity and coordination among institutions and firms (Rasiah and Lin 2005), has anchored interactive learning, innovation, and performance of participating firms in the leading industrial districts of the world. Cohesively integrated clusters enjoy continuous upgrading, innovation, and differentiation and division of labor. Network cohesion[10] in industrial clusters like Silicon Valley and Route 128 in the United States (or Emilia Romagna and Marche in Italy) supports new firm creation and new growth opportunities. The participation of MNCs in clusters has helped stimulate differentiation and new firm creation in Israel, Ireland, and Singapore. MNCs have also been instrumental in stimulating the growth of local firms in Taiwan, Republic of Korea, and, to a lesser extent, the Malaysian state of Penang. The growth of local suppliers of precision metal tools and engineered plastics helped MNCs upgrade and source locally in Penang. MNC operations also helped encourage the development of contract manufacturing firms in electronics.

The strength of the systems in regional clusters differs significantly, even within a single country. The Penang cluster has enjoyed far stronger network cohesion than have other regions in Malaysia. Dynamic intermediary organizations such as the Penang Industrial Coordination Councils (PICC) brought together critical stakeholders to initiate and drive interactive and interdependent coordination among firms, to create production links, and to start the Penang Skills Development Center (PSDC) in 1989. Although the PICC was not formally established until 1990, its evolution as a network from the 1970s helps explain why Penang shows a stronger differentiation and division of labor than do other locations in Malaysia (Rasiah 2002b).

Strong relationships between the Penang Development Corporation (and the state government) and firms (MNCs as well as local firms) helped forge strong systemic coordination in the Penang cluster. Outside Penang, by contrast, one saw truncated operations with few links beyond the allocation of land and coordination of trade. The Klang Valley, where key ministries and MIDA are located, suffers from a lack of network cohesion, while the state economic development corporations of Selangor, Negri Sembilan, Melaka, Johor, Kedah, Perak, and Sarawak have limited their support activities to the provision of land. Firms rely on the police and other organizations for security and basic infrastructure (Rasiah 2002a, 2002b; Best and Rasiah 2003). The hands-off approach to coordination adopted by state development corporations outside Penang has reduced their capacity to strengthen firms' relations with institutions.

These contrasting roles—the proactive but intermediary role of the Penang Development Corporation in Penang and the hands-off role of other state development corporations after firms obtained their operating licenses—have produced other contrasts. Despite efforts of state governments to copy Penang since the late 1990s, the failure of economic agents to interact has limited the growth of system synergies.

Network cohesion in Penang has generated strong interactive relationships between state institutions and firms. The PSDC—considered by firms to offer world-class training in generic skills—was started as a direct consequence of relationships that arose from initiatives to resolve collective-action problems. Enhanced information flow among firms and institutions and support from the state government—which built trust and helped solve collective action problems—also stimulated new firm creation, which in turn strengthened intra- and interindustry linkages in Penang. Penang has successfully spawned suppliers in precision machine tools and plastic injection molding. LKT Engineering, Eng Technology, Unico, and BCM are some of the sophisticated firms that started directly from the cluster synergies in Penang. LKT

Table 4.3 Aspects of Institutional and Systemic Coordination, Penang and Klang Valley, 2005

Systemic features	Penang	Klang Valley
Chambers of commerce	Strong	Weak
Number of MNCs	High	High
Network cohesion	Strong	Weak
Skills development and training	Strong	Weak
Matching of supplier firms with MNCs	Strong	Weak
Basic infrastructure support	Strong	Strong
Security	Strong	Strong
Meetings between state, MNCs, and supplier firms	Active	Passive
Production of high-tech human capital	Low	Low
Links with standards organizations	High	High
Industry–public R&D ties (e.g., MIGHT)	Weak	Weak
Industry–university ties	Weak	Weak
R&D labs (e.g., with MIMOS)	Weak	Weak
Publication of documents on product and process technology of suppliers	Strong	None
Access to foreign high-tech human capital	Restricted[a]	Restricted[a]

Source: Adapted from Rasiah (2002b).

a. IT firms enjoying Multimedia Super Corridor (MSC) status can import foreign engineers and scientists, but this opportunity is still underused. A strip of land between the KLIA airport and Kuala Lumpur received MSC status in 1997. Penang was offered MSC status in 2005.

Engineering and Eng Technology have already become global service providers. The lack of such systemic coordination inhibited the development of similar linkages in other regions of Malaysia. Compared with the Klang Valley—which had electronics employment comparable to that of Penang in 1972–2002—stronger systemic and institutional coordination have stimulated greater differentiation in technologies, firms, and industries, as well as greater division of labor (table 4.3).

Whereas Penang spawned many local firms, most of them suppliers to MNCs, the remaining regions in Malaysia generated only a handful. State-created rents gave rise to telecommunication component and product assembly at Sapura in Kuala Lumpur, and the acquisition of licenses allowed OYL Electronics to manufacture consumer electronics appliances. Malaysian Electronics Company (MEC) began assembling appliances such as fans in the 1990s before folding operations in 2001. Chip assembly also emerged in the state of Perak to meet low-end assembly and test demand. Carsem and Unisem carry out these activities in the town of Ipoh in Perak.

Local firms in Penang moved into three distinct activities in (or linked to) the electronics industry. They entered electronics manufacturing from the 1980s, initially as subcontractors that assembled printed circuit boards and conducted burn-in tests. Cluster-differentiating synergies then gave rise to the development, in Penang, of local firms in component electronics, interindustry supplies in metal engineering and robotics, plastic molding, and packaging.

The first two of these activities are related. The first involved intraindustry electronics activities—e.g., subcontract chip assembly, surface mount technology, monitor and keyboard assembly, burn-in and test services, and low-end IC assemblies. The second was related to interindustry activities—e.g., plastic injection molding, precision metal tooling, and related parts for a wide range of electronics firms. Suppliers grew thanks to demand from foreign MNCs. The expansion of local firms in these first two activities stemmed directly from industrial-district synergies, as effective network cohesion (connection and coordination among the different economic agents) and strong intermediary links between the state government and industry associations, particularly Penang, stimulated differentiation and division of labor.

Interindustry linkages emerged because MNCs sought quick, flexible, accessible, and cheaper alternatives to imports from abroad, and the state government strongly supported them with matching and promotional initiatives, and by resolving certain collective action problems—such as the need for training. Most of the local entrepreneurs who started precision metal and plastic engineering firms gained their experience in MNCs (Rasiah 1994, 1995, 1996b). Local firms in metal precision engineering include Semiconductor Equipment Manufacturer (SEM), Polytool, Eng Hardware, and Wong Engineering. Local firms in plastic injection molding include Sanda Plastics, Precico, and LKT. Local firms in electronics components and subassemblies include Globetronics (integrated circuits), Shinca (subassemblies), and Unico (subassemblies and components).

The third activity arose from subcontract chip manufacturers that started firms to carry out assembly and test operations. Globetronics, started by former employees of Intel, has become an important subcontract chip assembler in Penang. Although the synergy for all three activities arose from the activities of MNCs seeking government incentives to relocate in Malaysia, direct government support through incentives and grants only came later.

Institutional Support for High-Tech Synergies

The Malaysian government created excellent conditions to support assembly and test activities in electronics manufacturing, but as the labor market tightened the government failed to construct a potent institutional and systemic framework to stimulate a transition to innovative activities by firms. Herein lie the problems that have stalled further expansion even in Penang; problems that threaten to temper Malaysia's position in global electronics and illustrate the dangers of taking a passive role in the promotion of R&D activities.

The government's productive role in coordinating technology transfer—from *ex ante* approval (vetting), monitoring, and *ex post* appraisal using experts from the industry—was vital in Japan, Korea, and Taiwan (Johnson 1982; Amsden 1989; Mathews and Cho 2000). But the Technology Transfer Unit (TTU) set up in 1975 under Malaysia's Ministry of Trade and Industry merely records technology transfer agreements between firms. Technology-transfer agreements in the country's electronics industry are dominated by agreements between Malaysian subsidiaries and the firms' affiliates abroad. In contrast, the government in Japan, Korea, and Taiwan helped local firms reduce their royalty payments under technology-licensing agreements and promoted the diffusion of technology.

Initiatives to set up institutions and to create links between institutions and firms began in the late 1980s with the Industrial Master Plan of 1986 but quickened with the launching of the Action Plan for Industrial Technology Development in 1990. Institutions and instruments were created or strengthened in the 1990s to support upgrading to higher-value-added activities in firms and participation in innovative activities (as discussed earlier). Unfortunately, the institutions were insufficiently connected and coordinated with firms to encourage firms' movement up the value-added chain. Despite frequent meetings involving MIMOS, MTDC, MIGHT, MDC, high-tech park officials, and firms, few activities reflect the interests of the latter. In Taiwan, industry participation and close coordination among stakeholders in the allocation of government grants, incentives, and other promotional instruments helped reduce failures (Rasiah and Lin 2005), but in Malaysia links between firms and public bodies were lackluster.

MIMOS is Malaysia's counterpart to ITRI in Taiwan (Mathews and Cho 2000; Lin 2003). But whereas ITRI has spun off numerous high-tech firms around key product technologies developed through incubation, MIMOS has managed to assist just a few firms, none of which has yet achieved success in export markets. Among its latest spin-offs, First Silicon is a joint-venture wafer fabrication plant that began producing 0.25 and 0.5 micron wafers in Kuching in 2000. It has lost money since its opening.

To stimulate R&D, the government began allowing double deduction from taxes of approved R&D expenses in 1988. Since 1990 it has classified certain industries as high-tech and strategic, thereby qualifying them for "pioneer status" incentives. Most firms interviewed in 1995, however, considered the incentives as an insufficient inducement for undertaking R&D activities. The electronics industry reported high R&D spending in 1995 (Rasiah 1996b), but studies show that the expenditures were directed primarily at adapting process technologies (Rasiah 1996b, 1999; Ariffin and Bell 1999; Ariffin, Bell, and Figueiredo 2003). The exceptions, such as Motorola and cordless phones (Rasiah 1996b; Hobday 1996) and Matsushita split-level air conditioners, are extensions of existing products.

Linkages between universities and electronics firms also have been weak. Despite the Intensification of Research in Priority Areas (IRPA) initiative in the Sixth Malaysia Plan (1990–1995), which set aside RM 1 billion to support R&D in postsecondary institutions, local universities' participation in industrial R&D has been limited to peripheral activities (Best and Rasiah 2003).

Electronics firms in Taiwan and Korea enjoyed higher R&D intensities in 2000 than firms in Malaysia (table 4.4), where foreign firms enjoyed marginally higher intensities than local firms (table 4.5), largely due to access to superior R&D support from their parent companies. Clearly government policy in Malaysia has failed to induce firms to engage in R&D activities as actively as firms in Taiwan and Korea. Whereas the government provided grants to motivate firms to begin R&D activities in Korea and Taiwan, Malaysia's government did no more than offer double-deduction incentives. Three reasons explain why: (i) the incentive mechanisms used have been redundant for many MNCs, as most already enjoyed favorable tax treatment on other criteria, (ii) the incentives for innovation offered in Malaysia are typically useful only for firms already engaged in R&D activities (and not for stimulating the initiation of such activities), and (iii) government initiatives have not been matched with sufficient levels of human capital.

To offset shortcomings in skill intensities, firms in Malaysia invested more in training than did firms in Korea and Taiwan. In Penang foreign MNCs and local firms established the Penang Skills Development Center (PSDC). In 1992 the government introduced its Human Resource Development Fund along with related legislation to pressure manufacturing firms with 50 or more employees to increase training. In 2000, the average intensity of human resource development efforts in foreign and local firms in Malaysia exceeded that of firms in Korea and Taiwan (tables 4.4 and 4.5).

Weak institutional support has weakened the participation of Malaysia's foreign and local electronics firms in R&D activities, compared with electronics firms in Korea and Taiwan (tables 4.4 and 4.5).

Table 4.4 Technological Intensities of Electronics Firms in Malaysia, Taiwan (China), and Korea, 2000

		Malaysia	Taiwan (China)	t	Malaysia	Korea	t
SI	All	0.310	0.687	−8.668*	0.310	0.635	−6.361*
	Foreign	0.309	0.629	−5.444*	0.309	0.665	−5.302*
	Local	0.317	0.718	−4.856*	0.317	0.619	−3.008*
HR	All	0.470	0.386	2.016**	0.470	0.314	4.016*
	Foreign	0.475	0.319	2.829*	0.475	0.361	2.072**
	Local	0.454	0.421	0.408	0.453	0.288	2.343**
PT	All	0.487	0.447	1.037	0.447	0.358	3.083*
	Foreign	0.479	0.452	0.507	0.479	0.358	2.507**
	Local	0.333	0.505	−2.543*	0.333	0.357	−0.747
R&D	All	0.088	0.546	−10.737*	0.088	0.212	−4.265*
	Foreign	0.103	0.423	−6.928*	0.103	0.225	−2.807*
	Local	0.033	0.610	−6.584*	0.033	0.205	−3.583*

Source: Rasiah 2004b.

Note: Samples include 46 (36 foreign and 10 local) firms from Malaysia, 38 (13 foreign and 25 local) firms in Taiwan, and 43 (15 foreign and 28 local) firms in Korea. See Rasiah (2004b) for the formulas used to compute the SI, HR, PT, and R&D variables. SI = skill intensity; HR = human resources; PT = process technology.

* *t*-statistics significant at 1 percent; ** *t*-statistics significant at 5 percent

Table 4.5 Two-Tailed *T*-Tests of Technology Intensity by Country and Ownership, Electronics Firms, 2000

	Foreign	Local	t		Foreign	Local	t
SI				*PT*			
Korea	0.665	0619	0.487	Korea	0.358	0.357	0.08
Taiwan (China)	0.629	0.718	−1.189	Taiwan (China)	0.452	0.505	−0.96
Malaysia	0.317	0.309	0.135	Malaysia	0.479	0.333	2.29**
HR				*RD*			
Korea	0.361	0.288	1.21	Korea	0.225	0.205	0.48
Taiwan (China)	0.319	●0.421	−1.48	Taiwan (China)	0.423	0.610	−2.38**
Malaysia	0.475	0.454	0.33	Malaysia	0.103	0.033	1.41

Source: Rasiah 2004b.

Note: Samples include 46 (36 foreign and 10 local) firms from Malaysia, 38 (13 foreign and 25 local) firms in Taiwan, and 43 (15 foreign and 28 local) firms in Korea. See Rasiah (2004a) for the computation of these variables. SI = skill intensity; HR = human resources; PT = process technology.

* *t*-statistics significant at 1 percent; ** *t*-statistics significant at 5 percent

While very successful in offering cutting-edge manufacturing skills, the government–firm coordination framework anchored by the PSDC has not expanded to stimulate R&D activities, owing to federal policy constraints. Low levels of skills (including engineering skills)—a consequence of weaknesses in overall human capital development policies and of restrictions against using imports of experts from abroad to fill gaps, as is done in the United States, Ireland, Singapore, and Israel—have restricted upgrading in the industry. A policy preference for importing unskilled labor from Indonesia and Bangladesh has only reduced labor-market pressure on firms to upgrade. Special exemptions are provided in the MSC, allowing approved firms to import foreign experts. Unfortunately, not only is this policy confined to a narrow corridor, but also it is badly implemented.

Electronics firms in Malaysia showed higher process-technology intensities than did Korean firms in 2000, but this is largely a consequence of skills-displacing automation to compensate for shortages in skilled labor. Foreign firms scored better on process technology than local firms in Malaysia in 2000 (table 4.4). While the mean for local firms in Malaysia was lower than that of local firms in Korea and Taiwan, the opposite was obtained for foreign firms. Foreign multinationals in Malaysia are obviously using cutting-edge machinery, process, and quality control techniques in assembly and test operations.

Unlike in most other developing economies, several of the institutions needed to spur value-added upgrading have already been created in Malaysia. Paralleling the Science and Technology Projects (STP) launched within Taiwan's ITRI (Rasiah and Lin 2005), the Malaysian government introduced the IRPA program, endowing it with RM 1 billion in funding for the two five-year periods between 1990 and 2000 (Rasiah 1999). The government also launched the Malaysian Institute of Microelectronics Systems in 1985. However, the lack of an interactive and interdependent relationship between firms and these new institutions has limited learning and innovation. The necessary relationship can be spawned by requiring that grants to the institutions be jointly determined and implemented with private firms. Given that grants rather than financial incentives are the primary stimulant for infant firms to participate in R&D, funding for these institutions should be made conditional on take-up by firms.

Human Capital

We discuss human capital separately to explain Malaysia's inability to make the transition to higher-value-added activities in the electronics industry. Two major sources of human capital have been critical in

Table 4.6 Educational Enrollment Index in Selected Countries, 1970–95

	1970	*1975*	*1980*	*1985*	*1990*	*1995*	*2000*
Net primary	$n = 60$	$n = 57$	$n = 78$	$n = 78$	$n = 72$	$n = 89$	$n = 107$
China	n.a.	n.a.	n.a.	n.a.	1.2	1.2	1.2
Hong Kong (China)	1.2	1.1	1.2	n.a.	n.a.	1.1	1.1
Indonesia	n.a.	0.9	1.1	1.2	1.2	1.1	1.3
Ireland	1.3	1.1	1.1	1.1	1.1	1.1	1.1
Israel	n.a.	n.a.	n.a.	n.a.	n.a.	n.a.	n.a.
Japan	1.3	1.2	1.3	1.3	1.2	1.2	1.2
Korea, Rep. of	1.3	1.2	1.3	1.2	1.3	1.1	1.2
Malaysia	1.2	n.a.	1.2	n.a.	1.2	1.1	1.2
Philippines	n.a.	1.2	1.2	1.2	n.a.	1.2	1.2
Singapore	1.3	1.2	1.3	n.a.	n.a.	1.1	1.1
South Africa	0.9	n.a.	n.a.	n.a.	n.a.	n.a.	0.8
Thailand	n.a.	n.a.	n.a.	n.a.	n.a.	n.a.	1.2
United States	n.a.	n.a.	n.a.	1.2	1.2	1.1	1.1
Gross secondary	$n = 132$	$n = 137$	$n = 153$	$n = 156$	$n = 152$	$n = 145$	$n = 149$
China	0.8	1.2	0.9	0.7	0.8	1.0	1.2
Hong Kong (China)	1.1	1.3	1.2	1.3	1.4	1.1	1.2
Indonesia	0.5	0.5	0.6	0.8	0.8	0.8	0.6
Ireland	2.3	2.3	1.7	1.8	1.7	1.8	1.8
Israel	1.8	1.7	1.4	1.5	1.5	1.4	1.7
Japan	2.7	2.4	1.8	1.7	1.7	1.6	1.6
Korea, Rep. of	1.3	1.5	1.5	1.7	1.6	1.6	1.7
Malaysia	1.1	1.2	0.9	1.0	1.0	0.9	1.0
Philippines	1.4	1.4	1.2	1.2	1.3	1.2	1.2
Singapore	1.4	1.3	1.2	1.1	1.2	1.2	1.3
South Africa	n.a.	n.a.	n.a.	n.a.	1.3	1.5	1.3
Thailand	0.5	0.7	0.6	0.6	0.5	0.9	0.9
United States	2.6	2.2	1.8	1.8	1.6	1.5	1.4
Gross tertiary	$n = 118$	$n = 120$	$n = 144$	$n = 141$	$n = 129$	$n = 110$	$n = 121$
China	n.a.	0.1	0.1	0.2	0.2	0.2	0.4
Hong Kong	1.1	1.1	0.8	n.a.	n.a.	n.a.	n.a.
Indonesia	0.4	0.3	0.3	n.a.	0.5	0.5	0.4
Ireland	1.8	1.8	1.4	1.6	1.6	1.8	1.9
Israel	2.8	2.6	2.2	2.3	1.9	1.8	1.9
Japan	2.7	2.9	2.3	2.0	1.7	n.a.	1.7
Korea, Rep. of	1.1	1.0	1.1	2.4	2.2	2.3	2.2
Malaysia	n.a.	n.a.	0.3	0.4	0.4	0.5	0.8
Philippines	2.6	1.8	1.8	1.8	1.6	1.3	1.4
Singapore	0.9	0.9	0.6	1.0	1.0	1.5	1.7
South Africa	0.6	n.a.	n.a.	n.a.	0.7	0.8	0.9
Thailand	0.5	0.4	1.1	1.3	0.0	0.9	0.8
United States	7.2	6.0	4.2	4.3	4.2	3.6	3.7

Sources: Computed from World Bank (2004), national databases of Indonesia, Malaysia, South Africa, and Thailand.

Note: Figures calculated using the formula $x_i[\Sigma(x_1 \ldots x_n)]^{-1}n$, where x_i is the enrollment percentage of country i, and n is the number of countries reporting data. n.a. = data not available.

stimulating upgrading through the value chain and the growth of firms in dynamic clusters. The first is formal education in schools, training institutes, and universities. The second involves the tacit knowledge required to run firms. The second type of human capital comprises technical as well as entrepreneurial knowledge.

Education and Training Institutions Malaysia offered a large reservoir of literate and trainable labor. The prohibition of unions until 1989, when in-house unions were allowed, and subsequent repression to keep them passive has attracted firms to expand labor-intensive activities in both the low-wage (printed-circuit-board assembly) and the moderate-wage (microprocessor assembly) segments. The quality of Malaysia's primary and secondary education systems explains why labor-intensive, export-oriented assembly operations expanded strongly after the early 1970s (table 4.6). However, the tightening labor market, especially since 1993, has discouraged new investment in this area.

Although the renewal of incentives and permission to recruit foreign human capital (extended to IT firms in 2005) has persuaded most firms to remain, China's expansion and attractive low wages have discouraged new MNCs from relocating to Malaysia. Although tertiary enrollments in Malaysia have risen compared with most countries where data are available, they are still short of the critical mass available in competing economies occupying higher segments of the electronics value chain (United States, Japan, Ireland, Singapore, Taiwan, Korea). Even China enjoyed a higher index of tertiary enrollment than Malaysia in 2000.

Despite rising wages the country has retained a significant portion of labor-intensive activities through automation. However, Malaysia's progress in the development trajectory has raised wages so much that low-cost labor is no longer a major attraction for new firms. Average annual wages in the electronics industry in Malaysia rose by 8.4 percent in 1990–97, despite unskilled labor imports from Indonesia and Bangladesh. At the same time, electronics assembly has become increasingly automated in response to precision requirements driven by miniaturization and the need for a clean atmosphere for assembly operations.

Penang state managed to engender the conditions for institutional coordination to stimulate off-firm training of generic precision skills, which formed the basis for the PSDC in 1989. This institution has received laudatory reviews for offering cutting-edge training to firms' employees, as well as individuals seeking training in the state (Best and Rasiah 2003). However, efforts to replicate it in other parts of

Malaysia have yet to materialize, although the Johore State Development Corporation and the Selangor Human Resource Development Corporation have attempted to create similar training synergies.

The federal government has taken pains to expand the supply of science and technical graduates in the country. The Private University Bill of 1995 authorized private universities in Malaysia; enrollments in science courses rose in the country's universities; additional scholarships were introduced to support bright Malaysians seeking education abroad in promoted disciplines; and financial incentives were extended to training institutions. These instruments helped raise the supply of scientists and engineers in the population from 85 per million in 1992 to 154 in 1998 (World Bank 2002). However, Malaysia's figure in 1998 compared poorly against Korea's of 1993. As shown in table 4.5, all firms in Malaysia, foreign and local, had significantly lower skill intensities than firms in Taiwan and Korea. The differences are statistically highly significant. The much higher skill intensities in Korea and Taiwan reflect the strong education and human resource development policies pursued in those countries.

Immigration regulations have also stifled short-term solutions to human capital deficits. In contrast to the United States, Singapore, Israel, and Ireland, which recruit foreign human capital to substitute for domestic deficits, Malaysia has freely allowed foreign experts only in the IT sector and only when they relocate in the MSC. Although Penang is among the few regions in the country to qualify for MSC status in 2005, six firms reported that it might be too late to prevent high-tech firms from relocating to China.

The lack of human capital has prevented value-added upgrading of electronics firms in Malaysia. At the same time, federal policy instruments have not offered the push necessary for firms to participate in R&D activities (Rasiah 1999, 2002b; Best and Rasiah 2003). The combination of severe deficits in human capital and ineffective support instruments has prevented firms from investing in new product and process development. The shortcomings of Malaysia's institutional support for R&D activities in firms can be seen in the share of R&D scientists and engineers in the population (table 4.7). Malaysia not only lagged far behind Taiwan and Korea, but it also fell short of China in 1997–2000. While R&D grants and labs have been critical catalysts for firms in most emerging regions (e.g., Ireland, Israel, and Singapore), the Malaysian government continues to offer only tax exemptions for approved R&D expenditures. Hence, the Penang cluster has not achieved R&D intensities comparable to the United States, Japan, Korea, Taiwan, Singapore, Ireland, and Israel.

Table 4.7 R&D Scientists and Engineers per Million People, 1981–2000

	1981	1985	1990	1997–2000
	$n = 31$	$n = 36$	$n = 28$	$n = 65$
China	n.a.	n.a.	n.a.	0.2
Hong Kong (China)	n.a.	n.a.	n.a.	n.a.
Indonesia	n.a.	0.1	n.a.	0.0
Ireland	0.6	0.8	0.7	1.6
Israel	n.a.	n.a.	n.a.	2.6
Japan	2.8	3.5	2.8	3.6
Korea, Rep. of	0.4	0.8	0.8	1.5
Malaysia	n.a.	n.a.	0.0	0.1
Philippines	0.1	n.a.	n.a.	n.a.
Singapore	0.3	n.a.	0.7	1.6
South Africa	n.a.	0.3	n.a.	n.a.
Thailand	n.a.	n.a.	n.a.	0.1
United States	2.1	2.7	n.a.	2.5

Source: Computed from World Bank 2004.

Note: Figures calculated using the formula $x_i[\Sigma(x_1 \ldots x_n)]^{-1}n$ where x_i refers to R&D scientists and engineers per million people in country i, and n the number of countries reporting data. n.a. = data not available.

Tacit Knowledge Malaysian workers who gained experience working in MNCs built up Malaysia's stock of human capital with "tacit knowledge," providing a pool that might support new firms. While intense training and learning by doing took place in all electronics firms, it was highest in chip assembly and test firms, where workers had to have knowledge of statistical process control and often underwent recertification examinations. Some entrepreneurial firms such as Intel and Motorola kept a balance of new and old personnel by encouraging employees to seek jobs in complementary supplier firms. In this way, some mid-level managers relocated to other firms. But the prime source of new managers for MNCs has been other MNCs, which recruit individuals seeking higher salaries or those laid off during downturns. Local start-ups gained tacit knowledge and experience from such corporate reshufflings.

The role of MNCs as invisible colleges that impart tacit knowledge was encouraged by industry–government coordination councils and by close rapport between the state government and firms promoted by the Penang Development Corporation (PDC). The PDC occasionally has matched firms with the MNCs, and has even induced some early entrepreneurs to leave MNCs to start their own firms. However, its role was never sufficiently institutionalized.

Conclusions

Overall Malaysia fared extremely well in attracting foreign multinationals in the electronics industry. Except in 1981–85, government policy targeted foreign multinationals by promising security, improvements to basic infrastructure, and financial incentives. That policy helped to retain existing firms in the late 1990s despite rising wages and the emergence of China as an attractive low-wage site. Although employment fell in 1997–98, value added and exports have continued to rise. However, government's effort to move firms into higher-value-added activities have yet to gain much success. A combination of human-capital deficits and ineffective promotional instruments to stimulate R&D has prevented firms from raising their R&D intensities to levels comparable to Taiwan and Korea.

What lessons does Malaysia's positive experience hold for poor economies? The answer depends on whether the electronics industry still follows the Babbage principles of division of labor. Put simply, is the industry still characterized by segments with varying skill requirements? Are assembly activities in some product lines still highly labor-intensive? With the exception of low-end *maquila*-type peripheral assemblies, where workers merely do simple assembly or packaging as in Mexico, much of today's electronics industry is too knowledge-intensive for economies (such as Cameroon, Chad, and Congo) wishing to attract labor-intensive operations. African economies, in particular, lack the non-labor-related advantages enjoyed by Malaysia and Mexico—notably regional rents from membership in regional trade agreements. Unless similar arrangements are formed, firms will tend to focus their labor-intensive segments in China, the Philippines, and Thailand. Countries seeking such manufacturing activities still require excellent basic infrastructure (export processing zones) and efficient airport service. Moreover, even the most labor-intensive segments of electronics assembly require a disciplined workforce prepared for occasional layoffs, as demand is volatile in these segments.

Nevertheless, the electronics industry can be an important source of economic dynamism in poor economies lacking a minimum threshold of human capital, but only if it diffuses technology to other industries. The diffusion and reengineering of electronics technology can produce synergies in many industries. Electronic control automation, for example, has transformed traditional industries such as garments, fishing, and even pottery, while the proliferating links between electronics and other state-of-the-art technologies (such as materials) have blurred the distinction between old and new economies.

Endnotes

1. Pioneers were defined as firms in industries classified by the government as new to the country.
2. Computed from Malaysia 2002: 5, using 1978 as the base year.
3. The British colonial administration had in place the rural industrial development authority (RIDA) to promote cottage industry (Rasiah, 1995: ch. 3).
4. This problem is also seen in public universities, where academic standards should have been the most important criterion for employment.
5. The managing director of Monolithic Memories Incorporated (MMI) reported to the author in July 1986 that several other firms had similar plans in 1985.
6. The double deduction tax incentive offered for training was less important than the technological dynamics of the industry in stimulating investment in training.
7. The author was a member of the Intensification of Research in Priority Areas (IRPA) review committee over the period 1996–2001.
8. Interviews conducted by the author on June 17, 2004, with officials in the Ministry of Human Resources, Kuala Lumpur.
9. See Best (2001) for a lucid account of the diversity that typifies dynamic clusters.
10. A cluster is considered to enjoy network cohesion when there is strong connectivity and coordination among the different economic agents present.

References

Alavi, R. 2000. "Growth of Malaysian Firms: The Case of Sapura." Unpublished paper. Kuala Lumpur.

Amsden. 1989. *Asia's Next Giant: South Korea and Late Industrialization.* New York: Oxford University Press.

Ariffin, Norlela, and M. Bell. 1999. "Firms, Politics, and Political Economy: Patterns of Subsidiary-Parent Linkages and Technological Capacity Building in Electronics TNC Subsidiaries in Malaysia." In Kwame Sundaram Jomo, Greg Felker, and Rajah Rasiah, eds., *Industrial Technology Development in Malaysia.* London: Routledge.

Ariffin, Norlela, M. Bell, and Paulo N. Figueiredo. 2003. "Internationalisation of Innovative Capabilities: Counter-Evidence from the Electronics Industry in Malaysia and Brazil." Paper presented at the DRUID Summer Conference on Creating, Sharing, and Transferring Knowledge: The Role of Geography, Institutions, and Organizations, June 12–14, Copenhagen.

Best, Michael. 2001. *The New Competitive Advantage: The Renewal of American Industry.* Oxford: Oxford University Press.

Best, Michael, and Rajah Rasiah. 2003. "Malaysian Electronics: At the Crossroads." Technical Working Paper No. 12. Small and Medium Enterprises Branch, United Nations Industrial Development Organization, Vienna. February.

Edwards, C. B. 1975. "Protection, Profits, and Policy: Industrialisation in Malaysia." Doctoral thesis submitted to University of East Anglia, Norwich.

Edwards, C. B., and Jomo, K. S. 1993. "Policy Options for Malaysian Industrialization." In K. S. Jomo (ed.), *Industrializing Malaysia: Policy, Performance and Prospects.* London: Routledge.

Freeman, C. 1989. "New Technology and Catching-Up." *European Journal of Development Research* 1(1): 85–99.

Hobday, Mike. 1996. "Innovation in South-East Asia: Lessons for Europe." *Management Decision* 34 (9): 71–81.

Hofmann, L., and T. N. Tan. 1980. *Industrial Growth, Employment, and Foreign Investment Growth in Malaysia.* Kuala Lumpur: Oxford University Press.

Ishak, Shari. 1995. "Poverty and Distribution in Malaysia." Unpublished paper, Kuala Lumpur.

Johnson, Chalmers A. 1982. *MITI and the Japanese Miracle: The Growth of Industrial Policy, 1925–1975.* Stanford, CA: Stanford University Press.

Jomo, Kwame S. 1986. *A Question of Class,* Singapore: Oxford University Press.

———. 1990a. *Undermining Tin.* Kuala Lumpur: Forum Publications.

———. 1990b. *Growth and Structural Change in the Malaysian Economy.* Basingstoke: Macmillan.

Lim, L. Y. C. 1978. "Multinational Firms and Manufacturing for Export in Less Developed Countries: The Case of the Electronics Industry in Malaysia and Singapore." Doctoral thesis submitted to University of Michigan, Ann Arbor.

———. 1998. "Profile of Electronics Firms in Penang." Unpublished paper, Penang.

Lin, Y. 2003. "Industrial Structure, Technical Change, and the Role of Government in the Development of the Electronics and Information Hardware Industry in Taiwan." Working Paper, Asian Development Bank, Manila.

Malaysia. 1976. *Third Malaysia Plan 1976–1980,* Kuala Lumpur: Government Printers.

———. 1988. "Malaysia: Incentives, Policies, and Procedures." Malaysian Industrial Development Authority, Kuala Lumpur.

———. 1996. *The Second Industrial Master Plan.* Kuala Lumpur: Ministry of International Trade and Industry.

———. 2001. *The Eighth Malaysia Plan 2001–2005,* Kuala Lumpur: Government Printers.

———. 2002. *The Economic Report.* Kuala Lumpur: Government Printers.

———. 2003. *Economic Report 2002–2003.* Kuala Lumpur: Bank Negara Malaysia.

Mathews, John A., and Tong-song Cho. 2000. *Tiger Technology: The Creation of a Semiconductor Industry in East Asia.* Cambridge, U.K.: Cambridge University Press.

Narayanan, Suresh, and Lai Yu Wah. 2000. "Technological Maturity and Development Without Research: The Challenge for Malaysian Manufacturing." *Development and Change* 31 (2): 435–458.

Nelson, R. 1993 (ed.). *National Innovation Systems.* New York: Oxford University Press.

Porter, Michael E. 1990. *The Competitive Advantage of Nations.* New York: Free Press.

Rasiah, Rajah. 1988. "The Semiconductor Industry in Penang: Implications for the NIDL Theories." *Journal of Contemporary Asia* 18(1).

———. 1993. "Pembahagian Kerja Antarabangsa: Industri Semikonduktor di Pulau Pinang." Master's thesis. Malaysian Social Science Association, Kuala Lumpur.

———. 1994. "Flexible Production Systems and Local Machine Tool Subcontracting: Electronics Component Transnationals in Malaysia." *Cambridge Journal of Economics* 18 (3): 279–298.

———. 1995. *Foreign Capital and Industrialization in Malaysia.* Basingstoke: Macmillan.

———. 1996a. "Institutions and Innovations: Moving Towards the Technology Frontier in the Electronics Industry in Malaysia. *Industry and Innovation* 3 (2).

———. 1996b. "Manufacturing as an Engine of Growth and Industrialization in Malaysia." *Managerial Finance* 22 (6).

———. 1996c. "Changing Organisation of Work in Malaysia's Electronics Industry." *Asia Pacific Viewpoint* 37 (1).

———. 1999. "Malaysia's National Innovation System." In K. S. Jomo and G. Felker (eds.), *Technology, Competitiveness, and the State.* London: Routledge.

———. 2002a. "Government–Business Coordination and Small Enterprise Performance in the Machine Tools Sector in Malaysia." *Small Business Economics* 18 (1–3).

———. 2002b. "Systemic Coordination and the Knowledge Economy: Human Capital Development in Malaysia's MNC-driven Electronics Clusters." *Transnational Corporations* 11 (3).

————. 2003. "How Important Is Process and Product Technology Capability in Malaysia and Thailand's FDI-driven Electronics Industry?" *Journal of Asian Economics* 14 (5).

————. 2004a. *Foreign Firms, Technological Capabilities, and Economic Performance: Evidence from Africa, Asia, and Latin America.* Cheltenham, U.K.: Edward Elgar.

————. 2004b. "Technological Capabilities in East and Southeast Asian Electronics Firms: Does Network Strength Matter?" *Oxford Development Studies* 32(3): 433–454.

Rasiah, Rajah, and Y. Lin. 2005. "Markets, Government, and Trust: Learning and Innovation in the Information Hardware Industry in Taiwan." *International Journal of Technology and Globalization* 1 (3/4): 400–432.

Rasiah, Rajah, and R. H. Osman. 1998. "Training and Productivity in Malaysian Manufacturing." IKMAS Working Paper 12, Bangi.

World Bank. 2002. *World Development Indicators.* CD-ROM, Washington, D.C.

————. 2004. *World Development Indicators.* CD-ROM, Washington, D.C.

CHAPTER 5

Explaining Malaysia's Export Expansion in Palm Oil and Related Products

Rajah Rasiah

MALAYSIA IS OFTEN VIEWED AS A country that evolved from dependence on tin and rubber to export-oriented manufacturing dominated by electronics assembly, but the commodity that took the country to the technological frontier is palm oil. Electronics firms either specialized in labor-intensive assembly based on technology imported from parent plants overseas or engaged in contract activities without extending their reach to higher-value-added segments. Palm oil firms, by contrast, are an integral part of value chains in which Malaysian companies play a significant role. Palm oil is now a major pillar of Malaysia's industrialization.

Malaysia overtook Nigeria as the world's leading exporter and producer of palm oil in 1966 and 1971, respectively (Gopal 2001: 122; Harcharan Singh Khera 1976; Malaysia 1975). By 1986, when the Industrial Master Plan (IMP) was launched, oil palm had become Malaysia's leading agricultural commodity and third-largest export earner (Malaysia 1986). Malaysia now accounts for about half of the world's production of palm oil; its plantations, processors, and manufacturers are generally regarded as operating at the industry's technological frontier. Malaysia evolved from simple cultivation and crude oil processing to become the industry's leading innovator, controlling the industry's value-added chain.

Our examination of the development and export expansion of the palm oil industry in Malaysia offers lessons for Sub-Saharan Africa, where cooking oil was first extracted from oil palms (in western and central Africa) and where the industry originated. Palm oil remains a major consumption item in those economies, but although climatic and soil conditions there are ideally suited for oil palm cultivation, Malaysia was able to displace Nigeria as the major producer of oil palm. Policy recommendations derived from Malaysia's success should help improve the cultivation, processing, diversification, and export of the commodity in Africa.

The framework adopted here examines the role of policy instruments and how those instruments, and the institutions they create, affect firms. We focus on government-led strategies to engender growth and structural change, how government policy can create or transform structural conditions to induce desired conduct in firms (Porter 1990), and how high-tech institutions help stimulate innovative activities in economies on a national scale (Freeman 1989; Lundvall 1992; Nelson 1993).

The chapter traces the critical drivers that helped make Malaysia the world's leading exporter of palm oil and related products. Three pillars support our analysis. The first consists of the policy instruments and institutions that were created to support the industry. The second involves network connections and coordination among economic agents directly related to the operations of palm oil firms. Those connections helped resolve information flow and allocation problems among firms. The third pillar concerns developments at the firm level (including plantations and smallholdings), where production is carried out.

We examine the sources of technical change in the industry and how Malaysian firms confronted change as latecomer learners and as innovators at the frontier. Owing to the process-based nature of the industry, much of the change involved efficiency-driven techniques designed to boost quality and speed delivery. Technology also contributed to the development of new uses for processed palm oil and by-products such as kernel oil and oil palm waste.

History

Oil palms were first introduced in Malaysia in 1870 but did not become an important commercial crop until 1920 (MPOB 2000). The Dura specie was brought to colonial Malaya from Indonesia, and between 1875 and 1917, oil palms were grown as ornamental plants (Gopal 2001: 123). Planting of high-yield varieties brought from West Africa began in 1917 (table 5.1). The wartime disruption in trade made experimentation easier. From 100 hectares in 1920, oil palm cultivation expanded to 20,000 hectares in 1930 (Gopal 2001: 124). However, until 1960 there was no special government-driven support to expand exports.

Export-oriented industrialization began in 1968 with the enactment of the Investment Incentives Act. Oil palm acreage expanded

Table 5.1 Phases of Oil Palm Cultivation in Malaysia, 1870–2002

Phase	Period	Critical emphasis
1	1870–1917	Ornamental plant
2	1917–60	Plantation crop
3	1960–79	Aggressive commercial cultivation to support exports
4	1979–86	Integration of palm oil processing
5	1986–96	External market expansion; establishment of oleochemical industry and product diversification
6	1996–	Product diversification and clustering

Sources: Compiled from Malaysia (1986; 1996).

dramatically, thanks in part to settler schemes under the Federal Land Development Authority (FELDA). The exemption of processed palm oil (PPO) from export duties after 1976 encouraged firms to switch from crude to processed palm oil. Between 1986 and 1996, under the industrial master plan (IMP), external markets expanded and firms made inroads into oleochemicals and other products. Since 1996, the focus expanded to new-product development, integration of value chains, and extensions into complementary products.

The government's promotion of oil palm was designed to diversify commercial crop production away from rubber, for which prices had been falling, and to contribute to the government's program of economic redistribution and poverty alleviation, expressed in the New Economic Policy (NEP) of 1971. Because that program aimed to eradicate poverty—absolute and relative—along ethnic lines, FELDA was targeted at the Bumiputeras (referred to as indigenous people but primarily Malays). In addition to promoting diversification through the agricultural ministry and land-development schemes, the government stepped up the displacement of rubber estates with oil palm when government-owned Bumiputera nominee companies began taking control of Western plantations in the late 1970s.[1] As a result of these policies, the participation of Bumiputeras in oil palm cultivation, as managers and laborers, expanded, along with the area planted in oil palm. However, whereas production of crude palm oil (CPO) expanded steadily from 1960 to 2000, production of PPO grew even more quickly from the mid-1970s. Export diversification efforts also caused the production of crude palm kernel oil (CPKO), processed palm kernel oil (PPKO), and palm kernel cake (PKC) to grow strongly after 1984 (Gopal 2001: 158). That the same equipment was used to refine CPKO and CPO facilitated diversification. The growth of CPKO also furthered differentiation into the oleochemical industry, including specialized and blended fats.

The Oil Palm Value Chain

Palm oil is the one industry in which Malaysia holds a considerable lead in global markets. As the main producer of a product consumed by hundreds of millions of people, Malaysia controls the value chain from raw materials to final consumer goods and is the engine for new-product development in the industry (figure 5.1).

The sector is an example of a resource-based industry in which Malaysian producers have learned to increase value added and pilot new-product development. To some extent, Malaysian firms have been able to participate in product-based, rather than just price-based, competition. With the benefit of extensive R&D in Malaysian institutions, Malaysian producers have led the world in refining existing products and defining new ones. As first movers in new products, they have no competition until rivals enter the market.

Oil palm has also emerged recently as an environmentally friendly crop (Basiron 2001); biodiesel is now used as a fuel in Europe. However, the product-related innovations are rarely firm-based, but rather emerge from an institutional framework and are shared by the participants in the processes. Although that tends to dampen profit levels, it increases diffusion through the value chain.

In addition to the incentives that palm oil firms enjoyed for R&D investments under two industrial master plans, firms enjoyed access to R&D carried out in the Palm Oil Research Institute of Malaysia (PORIM), the Malaysian Agricultural Research and Development Institute (MARDI), and universities. Personnel in those institutions, and in MARDI, the Malaysian Palm Oil Board (MPOB) and universities, strove for product diversification, new-product development, and higher value added in the oil palm chain. MARDI, MPOB, and the universities had access to RM 1 billion set aside by the government under the Intensification of Research in Priority Areas (IRPA) program, which was part of the Eighth Malaysia Plan (Malaysia 1991). The five-year program was renewed in 1996 and 2001 under the Ninth and Tenth Malaysia Plans (Malaysia 1996). Sustained financial support made it possible to introduce new products in markets (such as biodiesel, specialty fats, and vitamin A). The scope of R&D efforts expanded from oleochemical by-products to environment-friendly cultivation and manufacturing methods, productive recycling of waste, and raising value added in existing products.

Figure 5.1 The Palm Oil Value Chain

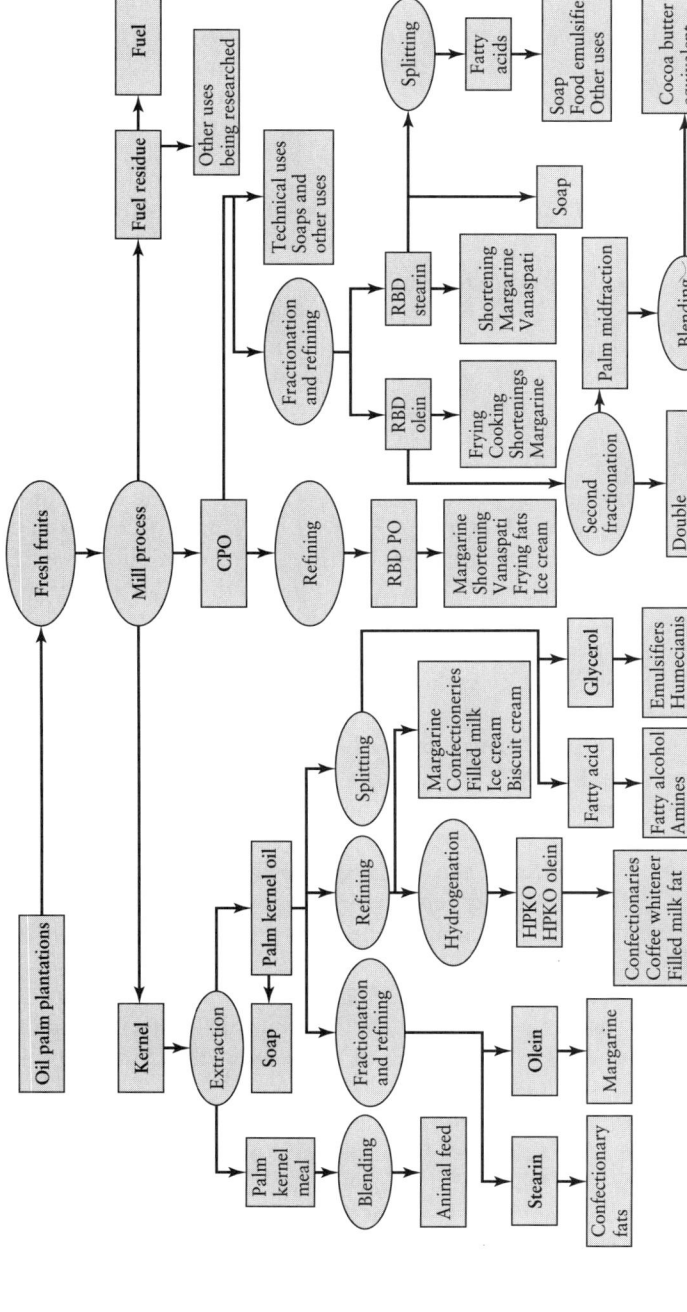

Source: Based on Gopal 2001 (figure 3.2).
Note: CPO = crude palm oil; RBD = refined, bleached, deodorized.

Thirty-five of the 50 firms interviewed believed that PORIM had played a critical role in training and market prospecting to encourage upgrading and new-product development in firms. Although PORIM is owned and operated by the government, its activities—including training—are strongly influenced by private member firms.

The Institutional Framework of the Oil Palm Industry

Cohesive relations among firms (made easier by the participation of medium-size and large firms), institutions, and policy instruments have created systemic efficiency and promoted the development of product technology in the palm oil industry (figure 5.2). Firms have developed ever-closer ties with universities where R&D is undertaken, with MARDI's specialized agricultural research institution, and with associations of planters and manufacturers.

Close coordination between the government, on the one hand, and the associations of planters, processors, and manufacturers has led to the formulation of contingency strategies to regulate supply in response to prices. Because Malaysia dominates the industry, accounting for more than half of all global exports of many oil palm products, the regulation of production has kept prices fairly stable. Unlike the situation in the early 1970s, when the government began to intervene in the industry, relations among the players are no longer asymmetrical. A smooth flow of information has led to the effective implementation of government policy.

The government strengthened the oil palm cluster by creating three vital institutions: the Palm Oil Regulatory and Licensing Association (PORLA), PORIM, and the Malaysian Palm Oil Promotion Council (MPOPC). The first two remain under government ownership and control, while the third is a privately registered company owned by the government. While PORLA played an administrative role, PORIM helped resolve collective action problems by deepening and broadening R&D in oil palm activities. MPOPC has also played a major role in promoting market expansion. PORLA and PORIM merged in the 1990s to form MPOB.

These institutions coordinated smoothly with university teaching and research units active in agriculture, and with MARDI, the Ministry of International Trade and Industry (MITI)—which controls the Malaysian Industrial Development Authority and the Malaysia External Trade Development Corporation—and industry associations (figure 5.2). MITI, in particular, promoted the international trade of manufactured goods, exhibiting new palm-oil products developed by

Figure 5.2 The Network of Institutions Involved in the Oil Palm Value Chain

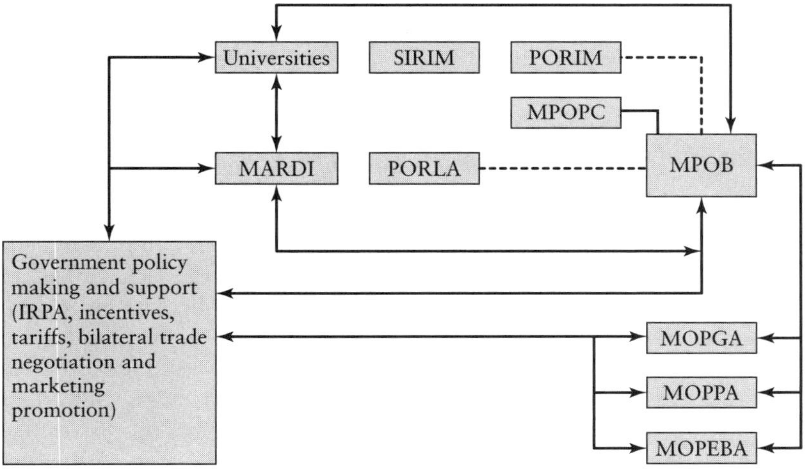

Source: Author.
Note: IRPA = Intensification of Research in Priority Areas, MARDI = Malaysian Agricultural Research and Development Institute, MATRADE = Malaysia External Trade Development Corporation, MIDA = Malaysia Industrial Development Authority, MITI = Ministry of International Trade and Industry, MOPGA = Malaysian Oil Palm Growers Association, MOPPA = Malaysian Oil Palm Processors Association, MOPEOMA = Malaysia Oil Palm Edible Oil Manufacturers Association, MPOB = Malaysian Palm Oil Board, MPOPC = Malaysian Palm Oil Promotion Council, PORIM = Palm Oil Research Institute of Malaysia, PORLA = Palm Oil Regulatory and Licensing Association, SIRIM = Standards and Industrial Research Institute of Malaysia.

Malaysian manufacturers. MPOPC and MPOB were dedicated to developing the oil palm value-added chain. A constant flow of information and discussion improved institutional support services to firms.

The private sector has conceived and advocated for government policies in the sector. Apart from initiating the first commercial cultivation of oil palms in 1917, private companies were also responsible for several important organizational innovations. Pooling, or bulking, was begun by four foreign-owned private plantations to achieve scale economies. That organizational technology was subsequently adopted throughout the country and later abroad. The private sector was also instrumental in lobbying the government to coordinate overseas promotional efforts.

Interviews show that the industry associations did a good job of representing the interests of private firms and avoiding potential problems. Discussions between the captains of industry and government officials, fostered by the associations, formed the basis of several policy directives that shaped the industry. For example, the government provided RM 12 per hectare to encourage growers to replant plantations with trees over 20 years old. The gesture was timed with a glut in world markets during 2000–1 to revive prices as well as encourage productive renewal. The replanting subsidy scheme was often used to reduce supply during times of glut so that mature trees gave way to new ones when prices were low. All 35 firms queried on this point considered this a major instrument that helped keep prices from falling sharply, while offering an incentive for firms to invest in new crops. Incentives to promote the manufacture of critical inputs, machinery, and equipment for the industry in the 1990s also attracted firms' participation. The participation of FELDA settlers (who obtained land titles) in these networks was handled through the FELDA management company. As late as September 2005, titles offered to individual settlers conferred ownership rights but not the right to sell their lots or change crops. If FELDA seeks to maintain parastatal governance structures to specialize in scale-intensive commodity cultivation, then such conditions may be necessary to preserve effective coordination.

Consistent with Porter's (1990) idea of high-tech clusters, following the IMP 2, the government encouraged strong connections among firms in the oil palm value chain and suppliers of human capital and R&D in universities, labs (both PORIM and MARDI), and the Standards and Industrial Research Institute of Malaysia (SIRIM). Government IRPA grants have also been extensively used by university academics to undertake R&D on palm oil products with joint support from the firms. The same degree of coordination among government, firms, and knowledge institutions has not been observed in many other sectors.

However, because the cluster selection criterion under IMP 2 used Porter's framework for selecting and supporting existing critical industries, the cluster examples drawn by the government defined clusters in a truncated way. The experts working on the oil palm cluster did not attempt to connect the industry with complementary industries, such as machine tools. Efforts were made to start new machinery and equipment firms rather than to connect the cluster to the existing machine tool cluster. Incentives were introduced for firms seeking to add value in the oil palm value chain—including critical inputs and machinery and equipment. What was not done was to attract such industries engaged in similar technologies domestically to extend their specialization to supply chemicals and fabricate machinery and equipment used by firms in the chain. The disjuncture between the approach adopted under IMP 2 and Porterian "dynamic clustering" arose because many things were imported at that time—there were few firms with the requisite technological capabilities operating domestically.

Policy Making for Sectoral Dominance

Under colonial rule the government emphasized plantation cultivation for CPO extraction primarily for export markets. After independence in 1957, government intervention focused on increasing value added, boosting exports, and alleviating poverty through land schemes. The colonial government had imposed an export tax on primary commodities, using much of the revenue to develop and maintain infrastructure (Lim 1968; Jomo 1986; Rasiah 1995). Most processing of palm oil was done in Europe. Unilever, which processed palm oil to produce cooking oil, margarine, soap, *vanaspati,* and detergents, established a plant in Kuala Lumpur in 1952 (Rasiah 1995: chapter 3; Gopal 2001: 149). Singapore-owned Lam Soon established a similar plant in 1962 (Lim 1968: 42).

Early Intervention

Under British rule, planters of oil palm specialized in primary production and received no subsidy or protection from the government. Specialization in primary production continued after independence. The government's first intervention came in the late 1960s, when foreign-owned estates were acquired by parastatals—among them the state economic development corporations, Permodalan Nasional (PERNAS) and, later, Permodalan Nasional Berhad (PNB). During the 1950s and 1960s the government extended the Rural Industry and Smallholders Development Authority (RISDA) to include oil palm cultivation and launched FELDA and the Federal Land and Crop Authority (FELCRA) to alleviate poverty.

When launched in 1957, FELDA applied to rubber cultivation; oil palm (375 hectares) was added in 1961 (Tunku and Thong 1988). Unlike the estate cultivation, which was motivated by market expansion and the search for profit, FELDA focused on alleviating poverty while improving efficiency. In line with the Second Malaysia Plan's objective of engendering restructuring along ethnic lines, only poor Bumiputeras, primarily those with experience in agriculture, were targeted (Malaysia 1971; Arif and Tengku Mohd Ariff 2001).

Government efforts to diversify Malaysia's exports to reduce the negative effects of poor terms of trade in rubber and tin[2] focused on oil palm (Malaysia 1971, 1981, 1984; Rasiah, Osman-Rani, and Rokiah 2000). As a consequence, rubber plantations gave way to oil palm plantations (Sekhar 2000). While agricultural land use has gradually expanded, rubber acreage has declined in absolute terms (table 5.2). Oil palm acreage grew from 320,000 hectares in 1970 to 3.3 million hectares in 2000.

Table 5.2 Agricultural Acreage, Malaysia, 1970–2000
(Thousands of hectares)

Crop	1970	1985	1990	1995	2000
Oil palm	320	1,482	2,030	2,540	3,338
Rubber	2,182	1,949	1837	1,679	1,590
Rice	533	655	681	673	692
Coconuts	349	334	316	249	116

Source: Arif and Tengku Mohd Ariff (2001: 2).

Instruments of Government Coordination

Industrial policy in Malaysia's oil palm sector was carefully coordinated with the private sector. Apart from FELDA,[3] which encouraged oil palm cultivation indirectly, the first government support for oil palm cultivation came with the launch of PORLA in 1974. Until then, commercial transactions of palm oil were recorded and governed from the United Kingdom, though pooling of output emerged in Singapore in 1953.

During World War II and until 1952, most palm oil was sold to the U.K. Ministry of Food, which determined prices on a long-term bulk contract. The Malayan Palm Oil Pool (MPOP) was started by the major palm oil producers in Singapore (then part of British Malaya) in 1953 (Gopal 2001: 233). Pooling locations grew as palm oil production expanded. The Joint Selling Committee (JSC), located in London, quoted prices and received supply commitments from shipping companies based on decisions made by brokers in Europe. The MPOP helped coordinate the pooling of scattered production units across British Malaya. The bulk pool was relatively easy to manage because large producers gathered the output of small producers (making use of their capacity for simple off-estate processing). Output to the pool therefore involved a small number of large producers, foreign and local.

The MPOP gave way to PORLA in 1974. By then, problems of coordination had become too big to handle, due to increases in the number of producers and their distance from pooling locations. However, pooling remained a key strategy, albeit between a collection of estates and smallholders, because it brought scale economies in transport, processing, and storage. Smallholders brought their crops to nearby estates equipped with crude processing plants and storage facilities under FELDA. Collaboration among planters with processing capabilities and those lacking such capabilities was established and strengthened through networks formed within the Oil Palm Growers

Association (OPGA). The government's opening of the large land tenure program under FELDA hastened MPOP's displacement. Resentment among FELDA's administrators over foreign plantations' control of pooling raised further calls for changes in the coordination role of MPOP. FELDA's participation in oil palm cultivation also led to the creation of the Malaysian Palm Oil Producers' Association (MPOPA), but this development changed pooling practices only by shifting the activity from London to Kuala Lumpur, where producers and a large segment of consumers were located. Transport costs alone made continued operations in London uneconomic.

Producers were assisted by the founding, in 1980, of the Kuala Lumpur Commodity Exchange (KLCE), which acts as an instrument for price setting, hedging, and dissemination of market information to reduce market risk (Mohd Arshad and Mohd Noh 1994).

In the 1960s, research and development (R&D) in oil palm breeding began to expand after the Malaysian Department of Agriculture established an exchange program with West African economies and four private plantations formed the Oil Palm Genetics Laboratory (OPGL) (Hartley 1988; Kajisa, Maredia, and Boughton 1997: 10). The government also established Kolej Serdang, which became the Universiti Pertanian Malaysia (UPM) in the 1970s (and was renamed the Universiti Putra Malaysia in 1997), to train agricultural and agro-industrial engineers and agro-business graduates to conduct research in the field.

Without a clearly defined quota structure, the government nevertheless promoted the appointment of Bumiputera managers on estates where they enjoyed influence and within the management of FELDA. Private plantations, too, bent to government pressure to hire Bumiputera managers, while the number of non-Bumiputera managers in the plantations gradually fell.

After strong lobbying from the OPGA and MPOPA, and with support from MARDI and UPM, the government set up PORIM in 1979.

Acting against the advice of international agencies, the government began in the late 1970s to encourage a shift from CPO to PPO. Before large-scale PPO production took off it was believed that Malaysia did not enjoy a comparative advantage in processing (Little and Tipping 1972). But the government went against this position. In 1976, Malaysia introduced an export tax on CPO to stimulate participation in PPO production (Malaysia 1986: 31). At the same time the government raised tariffs on bleaching earth, a key input used in PPO production. Bleaching earth accounted for 40 percent of processing costs in 2000. To prevent a domino effect on prices, the government tied the price of bleaching earth sold domestically with world prices. To encourage lumpy investment into refineries, it offered tax incentives on capital investment and regulated the number of refineries within the country to prevent overcapacity.

Incentives and Export Allowances

Early financial incentives for palm oil refining fell under the import-substitution initiatives of the Pioneer Industry Ordinance of 1958. This ordinance was not very successful owing to the small domestic market. But the export-oriented Investment Incentives Act of 1968 is generally regarded as the first major incentive upon which palm oil firms relied in making investment decisions (Gopal 2001: 251–252). The investment abatement allowance conferred a 40 percent abatement of corporate income tax for two years, which could be extended, and of excess profit and development taxes over eight years (Gopal 2001: 254). Palm oil refineries that obtained "pioneer status" enjoyed a tax holiday for seven years. Nine palm oil refineries obtained pioneer status between 1969 and 1974. After 1974 palm oil refineries were no longer eligible for the pioneer status. The investment tax credit (later the investment tax allowance) allowed firms to obtain tax exemptions through capital spending. In the period 1969–78, one firm obtained a 100 percent tax exemption, 22 firms a 50 percent exemption, 1 firm a 30 percent exemption, and 19 firms a 25 percent exemption under the investment tax credit instrument (Gopal 2001: table 6.2).

Compared to the generous incentives for most types of palm oil processing that it provided earlier, after 1978, the government tightened its hand. Palm oil refining and fractionation in developed areas and not owned, at least in part, by Bumiputeras were removed from the incentives list. Fractionated products and cooking oil continued to qualify for a 50 percent investment tax credit (Yusof 1979, cited in Gopal 2001: 258). Margarine, *vanapasti,* and shortening qualified for higher tax credits. The qualifying criteria included location in under-developed areas and Bumiputera ownership. Even here, incentives were removed by the early 1980s.

Although the Investment Incentives Act of 1968 was very generous, its application was governed along ethnic lines from the mid-1970s. The government launched the NEP with the Second Malaysia Plan in 1971 to carry out poverty alleviation and economic restructuring along ethnic lines. The Industrial Coordination Act of 1975 imposed ownership conditions on the basis of export-orientation and NEP criteria. Firms exporting all of their output did not have to meet the ethnic criteria of the NEP. The required share of Bumiputera equity varied with the percentage of sales going to the domestic market, but its application remained arbitrary until the promulgation of the Promotion of Investment Act of 1986. Export allowances were also offered, but the relief amounted to only 5 percent (FOB value) of gross income. Quoting Malaysia (1984: 304–305), Gopal (2001: 260) argued that this incentive was essentially redundant and was used by just 15 of 52 refineries in 1982. Malaysia (1984: 306) implied that the Export Credit Refinancing scheme that offered export-oriented firms loans with preferential interest rates was more widely taken up than the export allowance by PPO firms.

Three major explanations account for the success of some of the incentives offered to palm oil processors. First, the processors were big and coordinated easily with MITI, the authority offering the incentives. Second, because the major firms were also involved in oil palm cultivation and CPO, they were glad to expand into a higher-value-added segment because of the rent they enjoyed from processing downstream in relation to exporting CPO. Third, firms received strong support from MITI, SIRIM, MPOPC, and PORIM to expand into PPO activities. Support included technical and marketing know-how (including promotional exhibitions abroad) and privileged consideration for incentives.

Export Taxes

Gopal (2001: chapter 6) argues that the key factor behind Malaysia's export shift from CPO to PPO was the assessment of export duties on CPO simultaneously with export duty exemptions on PPO over the period 1968–84 (figure 5.3). Exemptions from export duties for PPO reduced government revenue to the extent that palm oil manufacturers shifted from CPO exports to PPO exports. The exemption represented a duty difference of 7.5 percent in 1968, which was sufficient to stimulate first-stage processing (Gopal 2001: 289). Figure 5.3 also shows that the period of subsidy on PPO was high during the period 1976–80 and highest during the period 1980–84.

The exemptions offered to PPO producers did not entail a transfer of government revenue from non-oil-related products. They meant only that the differential export duty increased the cost of producing CPO over PPO. In imposing the duties, the government had four goals: (i) to make PPO processing attractive, (ii) to avoid overburdening CPO producers, (iii) to protect duty revenue as much as possible, and (iv) to forgo financial support from other sources, even when the industry was not profitable (Gopal 2001: 290).

As with most measures of competitiveness, past studies do not establish a clear pattern. Ali and Osman's (1986) aggregate study at

Figure 5.3 Direct and Indirect Subsidies for Export-Oriented
PPO Products, Malaysia, 1968–84

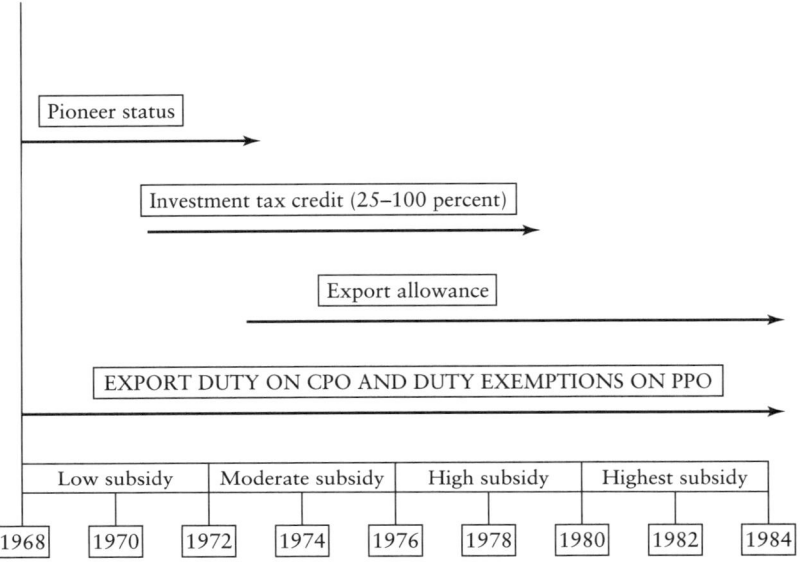

Source: Based on Gopal 2001 (figure 6.2).

the five-digit level showed that the palm oil industry was still profitable in 1981, even though its effective protection rate was negative
(−62 percent). Alavi's (1996) aggregate study using measures of domestic resource cost also found palm oil processing competing against
international prices in 1987. Although the differentials caused problems, as producers took advantage of a faulty customs mechanism that
led to leakage, the duties did have their intended effect of stimulating
a transition from CPO exports to PPO exports—so much so that the
learning involved in the transition helped Malaysian producers lower
costs below world prices by the mid-1980s. The new PPO producers
initially acquired knowledge from equipment suppliers or purchased
it through arm's-length transactions. Subsequently, the machinery and
equipment of foreign companies were bought by Malaysian companies. In fact, IOI and Golden Hope each reported acquiring a plant
from Unilever in Rotterdam in 2004.[4]

Industrial Master Plan

Palm oil was one of the industries flagged in the IMP of 1986 for sectoral support (Malaysia 1986). The task force appointed to implement

the plan emphasized the rationalization of palm oil refining and fractionation to increase efficiency and competitiveness in world markets. In addition, it called for the development of different segments of the industry in the value chain. The oleochemical industry was targeted for special support and hence promoted strongly with financial incentives. Incentives were also offered to promote downstream processing and production of oil palm products (Malaysia 1986: 28).

As with other sectors, the IMP set sectoral targets, which, given the low levels expected from the crisis of 1985, proved too modest and were achieved well before the target (Malaysia 1990). Only in oleochemicals was there a deficit.

At the launch of the IMP, the government decided that PPO production and the inputs used (including bleaching earth) had already become economic and of world-class quality. Hence it started scaling down tariffs on CPO exports and bleaching earth. The government's response to detariffication was also a consequence of complaints from oil palm growers and processors. But clearly the government's prior coordination with plantations and processors allowed for the unleashing of competition as capabilities were built. This example resembles the experience Taiwan had in the development of its machine tool industry (see Fransman 1985). The same framework was used to promote the manufacturing of acids, cartons, tins, drums, labels, adhesive tapes, plastics, and equipment used in the production of palm oil products (table 5.3).

Under the IMP, palm oil refineries enjoyed an abatement of taxes on corporate income amounting to 50 percent of export sales. Palm oil refineries also enjoyed the double-deduction tax benefit on export sales. Through a combination of these two provisions, many firms managed to avoid paying tax altogether, as long as their export sales remained high enough. The Export Credit Refinancing facility coordinated by the Bank Negara was continued under the IMP. Unlike CPO, where the large growers that enjoyed scale economies built oil palm processing facilities to reduce their dependence on others, PPO required substantially more sophisticated processing know-how and final markets at a time when European buyers were reluctant to lose their grip on PPO. In the end, however, European processors either declined in significance or gradually switched to other activities as a tariff-induced fall in exports of CPO reduced supplies available to European processors. The export tariff on CPO also raised PPO costs in Europe.

The oil palm industry was also one of the sectors promoted under the National Agricultural Policy (NAP). Launched in 1992, the NAP was designed to address concerns over increasing imports of agricultural products and their evolution following rapid expansion in

Table 5.3 Elements of the Palm Oil Value Chain Stressed in Malaysia's Industrial Master Plans

| | | Links in value chain | | |
Plantation	Crude output	Refining/fractionation	Downstream	Complementary industries
IMP (1985–95)				
Peninsular Malaysia Fresh fruit branch	CPO CPKO	Crude palm olein/palm Crude palm stearin/palm kernel stearin RBD palm oil/olein/stearin	Cooking oil, shortening, margarine, vanaspati, frying fat Cocoa butter substitute, dough fat, salad oil, confectionery fat, nondairy creamer Chocolate products	Bleaching earth, acids Cartons, tins, drums, labels, adhesive tapes, plastics Imported and locally made equipment Shipping, tankers, storage, bulk pumping stations, services at ports
IMP 2 (1996–present)				
East Malaysia Offshore Mass tissue culture Genetic engineering and biotechnology Cloning to get better pericarp breeding Mechanization	Increased supply of CPO/CPKO Specialty fats (e.g., high carotenes, high lauric, high olein)	Trans-fatty-acid-free POP Red POP Increased volume of current products	Microencapsulated POP, emulsifiers, food ingredients Powdered ice cream, salad dressing/oil Low-calorie products, palm oil–based cheese Biotechnologically modified oils/fats, vitamins E and B, carotenes Pharmaceuticals and other nutrient products	Competitively priced local products Specialized packaging materials to meet consumer and environmental requirements Locally made equipment for domestic use and export Adequate dedicated services and facilities

Source: Malaysia 1996: figure 6.3.
Note: CPO = crude palm oil; CPKO = crude palm kernel oil; POP = palm oil product; RDB = refined, deodorized, bleached.

manufacturing. It aimed to stimulate agricultural growth over the period 1992–2010 (Malaysia 1996: 203). However, the NAP did not have much impact on oil palm cultivation; its focus was on agricultural foodstuffs.

The IMP provided generous incentives to stimulate R&D activities in manufacturing. A tax allowance of 50 percent was offered on qualifying R&D expenditures over a period of 10 years. The allowance included expenses incurred on personnel, buildings, machinery and equipment, contract R&D, and materials.

Although a specialized training institution with national coordination did not emerge for many years, the IMP did extend incentives to stimulate training of staff. One incentive, in force between 1988 and 1992, allowed for the double deduction of approved training expenses. That program was replaced with the Human Resource Development Fund, created by law in 1992 and implemented in 1993 for firms with at least 50 employees. Smaller firms could still seek tax exemptions under the double-deduction scheme.

Since its founding in 1979, PORIM has been the key public (but privately coordinated) institution for advanced training in the sector, conducting training on chemistry, quality, analytical techniques, processing operations, transportation, and handling related to palm oil products (Gopal 2001: 266). From its original participation largely in training activities related to CPO and PPO, PORIM expanded its role into R&D after the IMP was launched in 1986. Under the IMP, PORIM's role in supporting the industry's marketing functions was expanded to include training and R&D in oleochemicals, specialty fats, and processed palm kernel oil.

Unlike the period after 1976, the government began to deregulate control over the oil refineries from 1986. The PPO industry was already regarded as a competitive player and hence deemed able to operate without subsidies and protection. This shift also explains partly why PORIM's activities also changed from PPO to new-product development following the IMP of 1986.

Industrial Master Plan 2

By the time the second industrial master plan (IMP 2) was launched in 1996, Malaysia's processing capacity exceeded the supply of CPO. CPO production reached 7.8 million metric tons in 1995, substantially less than the 10.1 metric tons that the 41 processing mills could handle (Malaysia 1996: 176). IMP 2 called for productivity gains and encouraged Malaysian firms to seek raw materials from abroad. Indonesia became a major supplier of CPO.

The exhaustion of labor and land reserves in Peninsular Malaysia led to the extension of IMP 2 to East Malaysia (table 5.3), where export processing zones were a relatively new development. In East Malaysia, IMP 2 offered incentives for labor-intensive and agro-processing industries. Among the effects of that extension was the opening of export-oriented processing and assembly plants. In addition to the development of basic infrastructure, IMP 2 also called for the expansion of bulking, onshore pumping, storage, and handling facilities in Sarawak and Sabah (table 5.4).

IMP 2 also stimulated participation in mass tissue culture, genetic engineering, cloning, and mechanization. In the crude-processing phase of the value chain, the focus was on stepping up production of CPO, CPKO, and specialty fats. IMP 2 also encouraged the production of complementary products, such as packaging, machinery and equipment, and related services (table 5.3). The human resource, technology, financing, physical infrastructure, and tax and regulatory requirements of achieving the IMP 2 are presented in table 5.4.

IMP 2 also called for the localization of machinery and equipment production, which had been largely imported (table 5.3). Special support was approved to manufacture machinery and equipment for the oil palm sector. Local firms have successfully used incentives to manufacture oil processing machinery and equipment: all six firms interviewed for this study asserted that domestic supply of machinery, equipment, and components—including repair and fabrication services—has helped reduce downtime and costs arising from freight charges and exchange-rate fluctuations.[5]

IMP 2 gave MPOPC the task of developing a comprehensive strategy to build Malaysia into an international leader in oils and fats and to market and distribute downstream products.[6] A privately registered but government-controlled institution, MPOPC has played a major role in promoting palm oil products, including raising consumer awareness of their content, benefits, and uses. Unlike other government-owned institutions, where accountability existed under the civil service domain, the MPOPC existed as a private entity, though it was owned by the government. The difference in the coordination was aimed at introducing private corporate management practices within a government-owned institution.

Downstream activities to increase value added were a major focus of IMP 2 (see table 5.3). The focus on biotechnology, in particular, has increased sharply since the plan began. Plugging the gaps in value chains and clusters required the efficient production of inputs and machinery and equipment. For that reason, as previously noted, IMP 2 emphasized the expansion of complementary industries for the oil palm cluster

Table 5.4 Institutional Support for the Oil Palm Industry Under Industrial Master Plans 1 and 2

Human resources	Technology	Financing	Physical infrastructure	Tax and regulatory agencies
IMP (1985–95)				
Training institutes, universities On-the-job training	Adapted process and R&D technology from PORIM Local fabrication	Equity, own fund, bank, access to offshore loan and venture capital	Cooking oil, shortening, margarine, vanaspati, frying fat Cocoa butter Substitute, dough fat, salad oil confectionery fat, nondairy creamer Chocolate products	Government incentives
IMP 2 (1996–present)				
PORIM—Institutions of higher learning to provide training, especially on downstream products Training of R&D personnel Overseas training	Adaptation, innovation, and development to enhance local technology for domestic use and export	Equities, own fund, bank, access to offshore loan and venture capital	Improved onshore pumping facilities, more onshore storage and handling facilities and utilities, particularly in Sabah and Sarawak, to meet growing demand.	Market-coordinated incentives

Source: Malaysia 1996 (figure 6.3).

(see table 5.4). Adding value also required the Malaysian oil palm processors to diversify into new products and to improve processing technology. To achieve that, PORIM was instructed to intensify R&D activities, including in downstream products. Joint-venture R&D activities were encouraged to facilitate early commercialization (Malaysia 1996: 178).

Because ties among the government, plantations, and downstream processing firms are strong, fluctuations in world prices are coordinated fairly smoothly. Oil palm firms have often used coordination meetings to lobby for support for their own initiatives. Also of significance is the support firms have received from the government to promote their products in developing economies (especially Africa, China, and India) and to negotiate bilateral trade agreements involving PPO exports to these economies. Given this level of cooperation, plantations tend to respond quickly to government initiatives, which help the firms in the long run. For example, a government plan launched in July 2001 to replant 200,000 hectares on which the trees were more than 20 years old was fulfilled by June 2002. In the interval, the reduction in supply helped improve prices. Relations between the government and firms are also kept fluid to allow for quick changes in direction. For example, the government managed to halt its plans, launched in 2001, to burn 500,000 tons of CPO as industrial fuel to push prices up. The quick reversal came with the emergence of new markets from barter-trade agreements with China, India, and Myanmar. Without resorting to price controls or caps, the government has played an active role to ensure acceptable and stable prices in the market.

Oil palm and its downstream and complementary products have figured prominently in government policies. The focus has shifted from diversification and processing in the 1970s to the expansion of exports through manufacturing (from 1986) and strengthening of sectoral clusters (from 1996). The government devised strategies to widen and deepen value chains—vertically and by involving complementary industries. Stronger institutional and systemic coordination achieved those ends. After 1996, training, R&D, and marketing promotion became important. Although MITI played the pivotal government role in the coordination of the oil palm value chain, MPOPC, MPOB, and producer associations also played pivotal roles.

Critical Drivers of Export Expansion

The critical drivers of oil palm export expansion were (i) efforts to adapt technology so as to move up the value chain; (ii) "network cohesion," or institutional arrangements between government agencies, firms, universities, and other organizations; (iii) governmental initiatives to diversify exports; and (iv) engagement of the poor. The first two drivers were discussed earlier in the chapter, under the headings "The Oil Palm Value Chain" and "The Institutional Framework of the Oil Palm Industry." The last two have been touched on throughout this chapter and are summarized below.

Government Initiatives to Diversify Exports

Realizing from historical experience with rubber and tin that dependence on narrow product lines can bring disastrous price downswings, Malaysia's government embraced diversification as a way to sustain production and exports. While rubber cultivation fell sharply beginning in the 1960s, the area under oil palm increased dramatically. Simultaneously, from specialization in CPO production, emphasis shifted to exports of PPO. As argued earlier, various instruments were used to encourage the shift from CPO to PPO exports.

Government support for the shift from the lower-value-added and price-volatile CPO to the more stable PPO caused a massive shift in exports from the former to the latter. Between 1960 and 1970 Malaysia exported no PPO at all. Thereafter, exports of PPO expanded sharply from 17,000 tons in 1970 to 9.7 million tons in 2002 (figure 5.4), whereas exports of CPO fell from 957,400 tons in 1975 to 13,100 tons in 1985, before rising again to a new peak of 1.3 million tons in 2001. CPO exports rose sharply in 2000/2001, offsetting a sharp dip in export revenue caused by falling prices. This rise was strongly influenced by bilateral barter-trading arrangements organized by the Malaysian government with China, India, and Myanmar.

One clear measure of the pervasiveness of the shift from CPO to PPO is the steady decline in exports of CPO as a share of CPO production. That share fell sharply from 76.1 percent in 1975 to a trough of 0.3 percent in 1985 as CPO was increasingly processed in Malaysia and exported as PPO (figure 5.4). Malaysia's share of world production of PPO rose from 2 percent in 1971 to a peak of 78 percent in 1982 (figure 5.5) and has not dipped below 62 percent since 1980. Meanwhile, the share of world PPO production held by the European Union and the rest of the world declined sharply from 53 percent and 45 percent respectively in 1971 to 9 percent and 27 percent in 1995 (figure 5.5). The massive expansion in export-oriented PPO production helped Malaysia raise its share of all processed oil production in the world from 0 percent in 1971–75 to 10 percent in 1995. A huge simultaneous expansion in acreage ensured that imports of CPO to satisfy domestic demand remained low.

Figure 5.4 Malaysia's Exports of Crude and Processed Palm Oil, 1960–2002

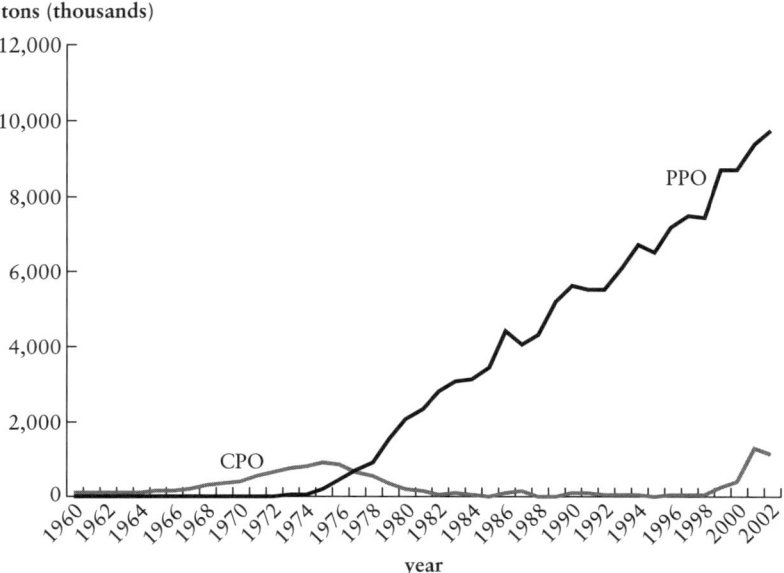

Source: Gopal 2001.

The 50 firms interviewed for this chapter foresaw no immediate threat to their exports of palm oil and related products. Indonesia's expanses of arable land and large labor force are obviously attractive to Malaysian firms eager to secure supplies of CPO. Some Malaysian companies have already established plantations. Interviewees reported that palm oil will continue to grow as a share of edible oils. In light of rapid expansion in the Chinese, Indian, and African markets, the U.S. soybean lobby, which mounted a campaign in the 1990s to persuade the public that palm oil was unhealthy, is not seen as a serious threat to check its growth.

Although the ease with which palm oil and its downstream products can be developed might be expected to encourage competitors to enter the market, climatic conditions would hamper cultivation in China and India, while infrastructure in Africa, Indonesia, Thailand, and the Philippines is still insufficiently developed to pose a serious threat to Malaysian firms in the medium term. Even if improvements in political stability and infrastructure allow those economies to expand cultivation, efficiency and new-product development are likely to improve in Malaysia as a result of the increased competition.

Figure 5.5 Malaysia's Share in World Production of Processed
Palm Oil and All Processed Oils, 1971–95

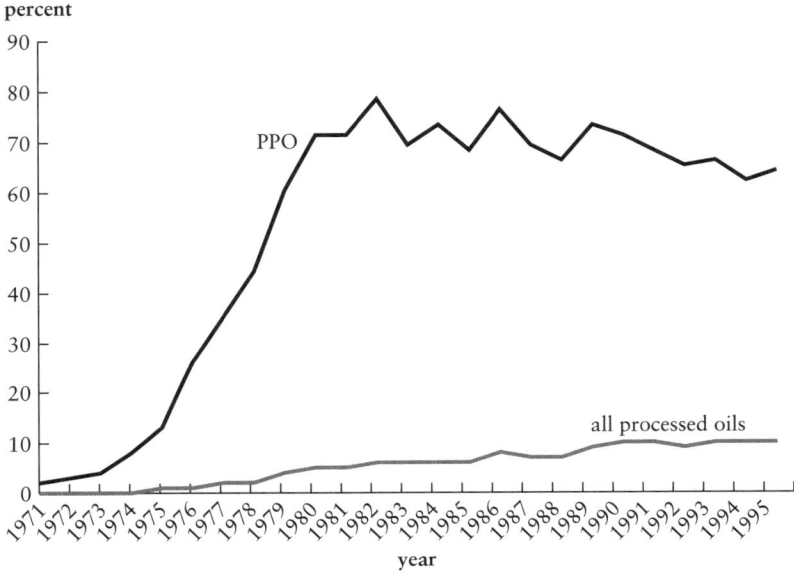

percent

Source: Drawn from Gopal 2001 (table 4.5).
Note: PPO = processed palm oil.

Engaging the Poor

Malaysia's efforts to expand palm oil exports were part of an economic restructuring effort that was intended to alleviate poverty and
inequality while making a domestic industry competitive in global
markets. The focus on reducing poverty and inequality has strengthened the country's position in the sector.

At independence, the Malays were not engaged in any significant
numbers in oil palm cultivation—a foreign crop brought to Malaysia
for commercial benefits. Most early workers on oil palm estates were
Indians who resettled from rubber estates. The picture changed
considerably in the 1980s, as workers migrated to factories. Post-
independence efforts by the government to build canals, drainage
systems, and other infrastructure to raise agricultural productivity
quickened urbanization. Plantations resorted to Indonesian labor to
slow down wage pressures.

With the exception of the royalty and a small middle class, most of
Malaysia's indigenous people were engaged in sedentary land tenure
without clearly defined ownership rights (see Jomo 1986). Some

enjoyed access to Malay reserve land that could not be sold to non-Malays. The FELDA land scheme was one of the initiatives the government devised to absorb Malay labor displaced from farming and other rural activities after independence.

FELDA land schemes became a major instrument for alleviating poverty and equalizing income distribution. Selection criteria gave priority to poor and landless settlers, as well as age, marital status, and physical fitness. The original group came from a background of agricultural workers (22 percent), estate workers (10 percent), rubber smallholders (14 percent), and paddy farmers (12 percent) (Arif and Tengku Mohd Ariff 2001: 11). Having begun with FELDA, Malay participation in the oil palm industry expanded to large estates after the government acted to acquire foreign estates on the London stock market in the late 1970s.

The concept of peasant participation in plantation crops emerged from a 1953 fact-finding mission headed by Francis Mudie that recommended, among other things, separate replanting funds for estates and smallholdings (Halim 1987). The World Bank then recommended land schemes in 1955 to raise the living standards of rural people. Rubber became the first crop planted under the FELDA program in 1957. Oil palm was added in 1961, with the incorporation of 375 hectares planted to the crop (Tunku and Thong 1988). By 2000 oil palm acreage under FELDA had expanded to 685,520 hectares—a fifth of the acreage under oil palm cultivation in Malaysia.

FELDA has operated as a resettlement scheme. Much of its new lands are currently opened in East Malaysia. Settlers were assigned to cultivate a parcel of 10 hectares in a large, cooperatively owned tract of land (Arif and Tengku Mohd Ariff 2001: 8; Mohd Arshad and Mohd Noh 1994). Ownership was bestowed once settlers had worked long enough to pay for their parcel. The system had three stages. In the first stage, settlers tended and harvested fields (Tunku and Thong 1988). In stage two, they managed small blocks to prepare them for greater independence. In stage three, settlers became landowners. Although land ownership shifted from cooperative to individual ownership, the nature of oil palm cultivation (with significant economies of scale) meant that FELDA managed the holdings. By 2000, 48,826 settlers held individual titles to 221,938 hectares of land. By 2004, 70 percent of settlers had become owners of their roughly 10 hectare farms (Nungsari 2005). In 2000, FELDA enrolled 102,750 settlers, two-thirds of whom were engaged in oil palm cultivation (Arif and Tengku Mohd Ariff 2001: table 8).

FELDA confronted enormous problems and complaints from the outset, causing the program to undergo considerable changes as problems were solved. Those problems included the wide disparity in skills, learning ability, and income of the settlers.

FELDA's success was not assured from the outset. The income levels of FELDA settlers fluctuated but remained higher than the poverty income level between 1980 and 2000 (Arif and Tengku Mohd Ariff 2001). Serious social and health problems afflicted the settlers in the 1960s, 1970s, and 1980s. Investments in infrastructure (especially schooling and health centers) gradually reduced those problems. Productivity levels were generally low in the early decades, compared with the private estates. A flexible plan characterized by continuous appraisal reduced the productivity gap—a remarkable accomplishment given that the land allocated to FELDA settlers was less fertile than the estates.

Once it was realized that oil palms are best managed on a large scale, FELDA has operated like a large estate made up of many individual owners. FELDA's management—including trading, finance, and marketing—is handled collectively. FELDA regulations require all settlers, including those with individual titles, to operate under its central management. Hence, the individual owner neither has the freedom to grow his or her own crops nor the opportunity to sell his or her own plot.

Despite problems and complaints, FELDA became a successful model of land tenure in which cooperative ownership gave way to individual titles, with management centrally coordinated to appropriate the benefits of scale economies. Centralized management not only facilitated pooling of clearing, plowing, weeding, application of fertilizers, harvesting, and refining, but it also strengthened financing, marketing (including promotion), and investment. Thus the centrally coordinated governance mechanism succeeded in Malaysia while similar initiatives failed in Africa. Success was due, in large part, to market discipline: FELDA settlers enjoyed the same sale prices of their produce as private estates.

While the FELDA model has been successful, serious challenges cloud its future. Relocating settler families in rural regions made socioeconomic sense when the farmers were poor, young, and uneducated. Attaching the settlers to cooperative farms offered them income and shelter, while the government's education policies gave settlers preferential access to higher education and scholarships. A direct corollary of these initiatives is that few of the children of the original settlers show interest in the collective farming system. Because the objective of alleviating poverty has been met, FELDA farmers should be given the option to sell their small plots—either to other poor settlers or to plantations. A dynamic policy must allow for such adaptations.

Conclusions

Oil palm cultivation in Malaysia and the subsequent expansion of downstream processing and manufacturing offer a unique example of natural resource–based manufacturing. Malaysia transplanted a nonnative plant and encouraged its use in higher-value-added manufacturing, eventually reaching the technological frontier and gaining first-mover advantages. From producing CPO in the 1960s, firms gradually upgraded into processing and eventually into products such as oleochemicals. Oil palm became a vehicle for economic diversification and a tool for ethnic-based economic restructuring.

Export diversification helped reduce the deleterious effects of over-production and falling prices associated with primary commodities (Rasiah, Osman-Rani, and Rokiah 2000). At the same time, ethnic restructuring helped an underprivileged community participate and eventually own agricultural land. Oil palm cultivation under the FELDA land-distribution program became a model for poverty alleviation and redistribution as thousands of indigenous Malays benefited from the program, while the program's centralized management helped coordinate the appropriation of scale economies in cultivation, transportation, processing, and exports.

Government instruments were also critical in the promotion of diversification and expansion in the oil palm value chain. From simple incentives under the Pioneer Industry Ordinance of 1958, which offered little help, tax incentives became important in stimulating firms to move downstream from CPO production. The initial export allowances were largely redundant, but subsequent incentives for exports, especially under the first Industrial Master Plan, had the desired effects, stimulating diversification, new-product development, and participation in complementary products under IMP 2. While the initiation for the movement up the value chain may have been the imposition of export duties on CPO and the exemption from duties of PPO (Gopal 2001), a broad-based strategy to support techno-diversification in the industry through new-product development, industry-government cooperation, and careful monitoring to prevent oversupply has been the prime driver of expansion in the industry. Nontax incentives—export-credit refinancing, R&D, and training grants—were instrumental in supporting short-term adjustments to raise efficiency or to enter new-product segments.

Policies to promote the widening and deepening of the oil palm sector also benefited from effective coordination among government and industry to resolve collective action problems, expand into international markets, and develop new products.

The specificity of palm oil, used for final consumption in developing economies and as an input into myriad products, makes comparison with other agricultural commodities in Malaysia difficult. Malaysia did not enjoy the same success in rubber as in palm oil, because final demand was controlled essentially by end users and cheaper plantations emerged in other countries. Rice farming is also dissimilar. Although the government attached great importance to rice as a staple cultivated by a major segment of the Bumiputera population, rice's position in world markets and its low potential for new-product development make it distinct from oil palm. Even cocoa and pepper—which some farmers in East Malaysia have cultivated to reduce their dependency on oil palm—have not produced the same level of success.

The oil palm industry offers substantial lessons for poor economies in equatorial and tropical countries with similar climatic conditions and soils—in Central and Western Africa, Central and South America, Southeast Asia (including Indonesia, East Timor, Cambodia, the Philippines, and Thailand), and Sri Lanka. With large-scale planting, the right policies, and effective coordination, oil palm cultivation can be made the center of a diversified economic cluster that includes products of high value added.[7] Diversified exports can finance the infrastructure needed to further develop the industry. Oil palms can also be planted by smallholders, though the management of scale economies is vital, to support poverty alleviation and redistribution.

Unlike Malaysia, however, many of the poor economies in equatorial and tropical areas lack the capital and infrastructure to encourage large-scale cultivation and processing of oil palm products.

Endnotes

1. The government kept control of the equity of these companies until Bumiputeras were able to acquire it.
2. Rubber faced competition from synthetic rubber and overproduction from new plantations in Indonesia and Thailand. Tin-mining was suffering from exhaustion of reserves and the entry of new producers, especially in China and Brazil.
3. In addition to FELDA, the government also promoted oil palm cultivation under the land tenure systems of FELCRA and RISDA, albeit on a small scale.
4. Interview by author on April 20–24, 2004.
5. Interviews carried out by the author from April 24 to May 2, 2004.

6. The launch of that initiative coincided with a campaign by the American Soybean Association to persuade the public that palm oil was nutritionally unhealthy. That campaign cut demand in developed economies. Nevertheless, the exponential growth in demand from the developing economies—especially India and China—has led to stable prices in the industry.

7. Oil palm grows well in areas that receive 80 inches of rainfall a year, with temperatures of 20–30°C, and in textured volcanic, alluvial, or marine clay soil (Gopal 2001: 133).

References

Ali, A., and R. H. Osman. 1986. "Malaysia's Industrial Strategies and Prospects with Special Reference to Resource-based Industrialisation." *Kajian Ekonomi Malaysia* 23(1): 27–44.

Alavi, R. 1996. *Industrialization in Malaysia: Import Substitution and Infant Industry Performance*. London: Routledge.

Arif, Simah, and Tengku Mohd Ariff Tengku Ahmad. 2001 "The Case Study on the Malaysian Palm Oil." Paper presented at the United Nations Conference on Trade and Development/Economic and Social Commission for Asia and the Pacific Regional Workshop on Commodity Export Diversification and Poverty Reduction in South and South-East Asia, April 3–5, Bangkok.

Basiron, Yusof (2001) "The Role of Palm Oil in the Global Supply and Demand Chain," Paper prepared for the 72nd World Congress of the International Association of Seed Clusters, "Industry Challenges of the 21st Century," September 17–20, Sydney.

Best, M. 1990. *The New Competition*. Cambridge: Harvard University Press.

Chamhuri, S. 1987. *Rural Development Policies and Programmes in ASEAN*. Singapore: Institute of Southeast Asian Studies.

FELDA (Federal Land Development Authority). [Retrieved in 2004 from http://www.felda.net.my]

Fransman, M. 1985. "International Competitiveness, Technical Change and the State: The Machine Tool Industries in Taiwan and Japan." *World Development* 14(12): 1375–1396.

Freeman, C. 1989. "New Technology and Catching-Up." *European Journal of Development Research* 1(1): 85–99.

Gopal, J. 1999. "Malaysia's Palm Oil Refining Industry: Policy, Growth, Technical Change and Competitiveness." In K. S. Jomo, G. Felker, and R. Rasiah, eds., *Industrial Technology Development in Malaysia: Industry and Firm Studies*. London: Routledge.

———. 2001. The Development of Malaysia's Palm Oil Refining Industry: Obstacles, Policy and Performance, PhD. thesis submitted to Imperial College, London.

Halim, Salleh. 1987. "Changing Forms of Labour Mobilisation in Malaysian Agriculture." PhD. thesis, University of Sussex, Brighton, United Kingdom.

Harcharan Singh Khera. 1976. *Oil Palm Industry: An Economic Study.* Kuala Lumpur: University of Malaya Press.

Hartley, C. W. S. 1988. *The Oil Palm.* Harlow, England: Longman Scientific and Technical.

Jomo, K. S. 1986. *A Question of Class.* Kuala Lumpur: Oxford University Press.

Kajisa, K., M. Maredia, and D. Boughton. 1997. "Transformation Versus Stagnation in the Oil Palm Industry: A Comparison Between Malaysia and Nigeria." Department of Agricultural Economics Staff Paper 97-5, Michigan State University, East Lansing.

Lim, C. Y. 1968. *Economic Development of Modern Malaya.* Kuala Lumpur: Oxford University Press.

Lim, B. A. 1979. "An Evaluation of the Palm Oil Refining Industry in Malaysia." Unpublished master's dissertation, submitted to Faculty of Economics and Administration, University of Malaya, Kuala Lumpur.

Little, I. M. D., and D. G. Tipping. 1972. *A Social Cost Benefit Analysis of Kuali Oil Palm Estate.* Paris: Organisation for Economic Co-operation and Development.

Lundvall, B. A. 1992. *National Systems of Innovation: Towards a Theory of Innovation and Interactive Learning.* London: Frances Pinter.

Malaysia. 1971. Second Malaysia Plan 1971–1975. Kuala Lumpur: Government Printers.

———. 1975. Third Malaysia Plan 1976–1980. Kuala Lumpur: Government Printers.

———. 1981. Fourth Malaysia Plan 1981–1985. Kuala Lumpur: Government Printers.

———. 1984. Report on the Palm Oil Industry in Malaysia: Industrial Master Plan, Volume 1. Kuala Lumpur: UNIDO/MIDA.

———. 1986. Industrial Master Plan. Kuala Lumpur: MIDA/UNIDO.

———. 1990. Action Plan for Industrial Technology Development, Kuala Lumpur: Ministry of Science, Technology and Environment.

———. 1991. Sixth Malaysia Plan 1991–1995. Kuala Lumpur: Government Printers.

———. 1996. Seventh Malaysia Plan 1996–2000. Kuala Lumpur: Government Printers.

———. 2001. Eighth Malaysia Plan 2001–2005. Kuala Lumpur: Government Printers.

Mohd Arshad, F., and K. Mohd Noh. 1994. "Agricultural Marketing Information for Selected Commodities in Malaysia," Food and Fertilizer Technology Center, Taipei. [Retrieved from www.agnet.org/library/article/eb392.html.]

MPOB (Malaysian Oil Palm Board). 2000. "History of Oil Palm in Malaysia." Unpublished paper. Kuala Lumpur.

—. 2003. *Malaysian Oil Palm Statistics*. Kuala Lumpur: Ministry of Primary Industries.

Nelson, R., ed. 1993. *National Innovation Systems*. New York: Oxford University Press.

Nungsari, A. 2005. "Critical Statistics on the FELDA Settlers." Unpublished paper.

Porter, Michael. 1990. *The Competitive Advantage of Nations*. New York: Free Press.

Rasiah, Rajah. 1995. *Foreign Capital and Industrialization in Malaysia*. Basingstoke: Macmillan.

—. 2002. *Systemic Coordination and Human Capital Development: Knowledge Flows in Malaysia's MNC-Driven Electronics Clusters*. Transnational Corporations, 11(3). United Nations University/Institute for New Technologies Discussion Paper.

Rasiah, Rajah, H. Osman-Rani, and Alavi Rokiah. 2000. "Changing Dimensions of Malaysian Trade." *International Journal of Business and Society* 1(1): 1–35.

Scherer, F. M. 1970. *Industrial Market Structure and Economic Performance*. Chicago: Rand McNally.

Sekhar, B. C. 2000. "State of the Rubber Industry in Malaysia." Unpublished paper. Kuala Lumpur.

Tunku, Shamsul B., and B. L. Thong. 1988. "The Development of the Oil Palm Industry in Malaysia." Unpublished paper. Kuala Lumpur.

CHAPTER 6

Salmon Farming in Chile

Jorge Katz

CHILEAN EXPORTS OF SALMON, ALMOST ENTIRELY cultivated, increased from less than $50 million in 1989 to around $1.2 billion in 2003 and $1.7 billion in 2005. Salmon now accounts for close to 5 percent of Chile's exports. From a negligible fraction of the world's salmon production—2 percent in 1987—Chile's share reached nearly one-quarter by the end of the 1990s (figure 6.1).[1] Chile is now one of the three major salmon-farming countries in the world, with Norway and Scotland.[2]

In the two decades during which the Chilean salmon-farming industry matured and attained international competitiveness, many new firms—Chilean and foreign—entered the market, sector-specific institutions and skills developed, and professional management came to dominate what had been a quasi-artisanal and family-based industry.[3]

Salmon farming is a major success story in which many forces—some scientific and technological, others institutional or socioeconomic—evolved together, influencing one another. The industry's growth path can be understood in the language of industrial organization, with reference to firms' learning processes, market entry and exit, and the gradual development of a sophisticated cluster of institutions, both public and private.

The Chilean salmon-farming case teaches valuable lessons about policy design and implementation. Reaching international competitiveness appears to be the result not just of adequate macroeconomic fundamentals, but also of an extensive set of sector-specific policies that trigger interactions among firms, government agencies, financial institutions, research and development (R&D) laboratories, universities, municipalities, and so forth.

Given the importance of sector-specific policies, one may well ask whether the salmon-farming story can be successfully replicated in other areas of the Chilean economy and exported to other countries wishing to become competitive in the processing of natural resources.

Figure 6.1 Chile's Exports of Farmed Salmon, 1991–2002

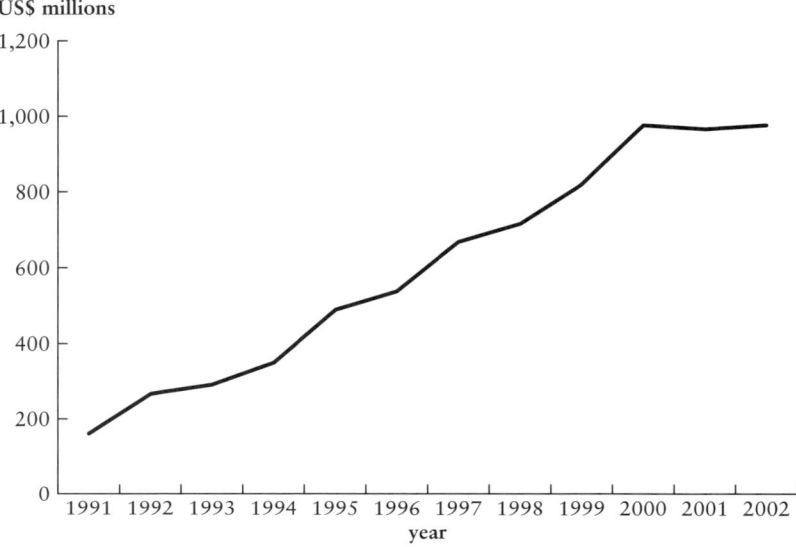

US$ millions

Source: Asociación de la Industria del Salmón de Chile A.G.

Are "virtuous trajectories" of the sort exhibited by salmon farming in Chile replicable in different institutional and economic settings?[4]

Those are the questions we explore in this chapter. In the next section, we elaborate on the three stages broached above. We then study the industry's technology and the dynamic changes it produced over time. The following section deals with industrial organization, individual firm learning, productivity growth, and the transition from a quasi-artisanal, family-owned industry to a mature oligopoly populated for the most part by large firms operating in a global setting. We then take up the question of institutional growth, public-private interaction, and the gradual consolidation of a sophisticated salmon-farming cluster. Although local R&D and knowledge-generation activities seem to be gradually expanding, they remain the weakest part of an otherwise successful cluster.

The last section summarizes our main findings and offers policy-oriented ideas about the industry's future, notably the need to build a stronger scientific and technological infrastructure to support marine technology and aquaculture in a country that seems ideally suited for these activities. Chile cannot expect to rely forever on foreign know-how and technologies. Having come as far as it has as a producer of marine products, it might now attempt to move from selling fish to selling related know-how and technology. Having understood this, the Corporación de Fomento (CORFO), a public development agency, and Fundación Chile, a public-private R&D organization, are moving in the right direction, but the level of investment is still small and needs to be expanded. Marine technologies and aquaculture clearly appear among the disciplines in which the expected social rate of return of further knowledge generation within Chile would pay off in the long run, thus justifying further concerted action between the public and private sectors.

Three Stages in the Evolution of Salmon Farming in Chile

Salmon farming in Chile can be seen as evolving through three quite different stages of growth. At each stage the problems faced by salmon-farming firms—and the structure, strategy, and core capabilities of individual companies—varied strongly, as did the complex fabric of interactions that salmon-farming firms developed among themselves and with suppliers of intermediate inputs, public sector regulatory agencies, and other organizations. These evolving structural links involved elements of learning and trust that help to explain why salmon farming developed into a sophisticated cluster of economic activity.

Salmon farming was brought into the Chilean environment almost entirely on the basis of imported genetic material and intermediate inputs. Chile's public sector set the wheels in motion, disseminating salmon-farming technology and designing and erecting new production facilities for salmon farming that were later transferred to the private sector. The Instituto de Fomento Pesquero, a publicly funded institute for the fishing industry, and the Secretaría de Agricultura y Ganaderia, the fishing and hunting division of the Chilean department of agriculture, established the foundations of a global regulatory framework, which strongly influenced the behavior of the industry.[5] Acting through CORFO and Fundación Chile,[6] the public sector set up the first commercial salmon-farming operations in the country and demonstrated that salmon farming could be successful in Chile. Reinforcing those important contributions were foreign firms and agencies— among them Nichiro Chile, the Japan International Cooperation Agency, and the University of Washington. Nichiro Chile started to cultivate salmon in Chile for the Japanese market, using genetic material imported from the United States. Only a handful of Chilean-owned businesses entered the industry at this stage.

The highly experimental foundation stage was characterized by trial and error within firms striving to identify and control the biological and environmental parameters specific to each farming location. Because no two salmon-farming operations in the world are identical, the learning path of each company is highly localized and specific.

In a second stage salmon farming expanded rapidly in size and complexity. Many new firms, mostly family enterprises, entered the market. Suppliers of intermediate inputs, service firms, producers of salmon food, pharmaceutical companies, and individual professionals also joined the industry. The impact was particularly large in Region IX and Region X of the country, remote and isolated territories in southern Chile, where more than 30,000 new jobs have been created by salmon farming. At this stage, the industry made strong inroads in international markets, particularly in Japan and the United States, to the point where Chilean exports provoked a charge of dumping from U.S. salmon farmers.[7]

In the 1990s the sector moved into its third and present stage, characterized by major transformations in industry structure and performance, world-class process technologies, and the reorganization of international marketing strategies. Large foreign salmon producers gained ground in the industry as it rapidly moved into global operations. Large international retailers—such as Wal-Mart and Carrefour—acquired a more significant role, affecting the production and marketing strategies of Chilean-based salmon farmers.

The dynamics described thus far can be seen through the lens of the classic product-cycle theory (Hirsch 1967; Posner 1961; Vernon 1966), in which an industry evolves from inception to maturity before reaching a plateau characterized by technological obsolescence and a decline in profit margins. Salmon farming in Chile teaches us that dynamic trajectories of this sort are not found exclusively in more developed economies and high-tech sectors but can shape the comparative advantages of developing nations as well.

Foundation

In 1974 Union Carbide's local subsidiary, Domsea Farms Chile, started producing salmon at Curaco de Velez using imported genetic material.[8] Around the same time, a second venture supported by CORFO began operating under the name Lago Llanquihue Ltd. Four years later it was exporting small quantities of trout and salmon to France. The firm was soon privatized.

In 1981, Fundación Chile bought Domsea Farms and created Salmones Antartica, which remains one of the industry's largest firms. Contemporaneously, Nishiro began Mares Australes in a joint venture with Pesquera Mytilus. Nutreco, a large Dutch company, acquired Mares Australes in 1988/89.

Thus public firms, foreign companies, and a few small and medium-size local enterprises participated in the industry's inception. Although there was active public involvement in the industry right from the beginning, it cannot be said that the government played a strong orchestrating role. Both CORFO and Fundación Chile had a proactive strategy of building up production capacity and disseminating know-how and technology, but from the outset Chilean entrepreneurs joined the industry and became the driving force behind its rapid expansion. Regulatory and sanitary functions—salmon-cultivation permits, environmental-impact studies, surveillance of imported salmon eggs, and so forth—were efficiently carried out by government agencies such as Sernapesca, Conama, and others. The legal infrastructure and authority required for those functions, already in place in the 1980s, was considerably improved to comply with standards emerging from the U.S. Food and Drug Administration, the United Nations Food and Agriculture Organization, and, later, the World Trade Organization (*AquaNoticias*, November 1997).

A large number of small and medium-size Chilean salmon-farming companies appeared in the 1980s. Simultaneously, cooperative mechanisms and institutions emerged as firms sensed the opportunity to capture economies of scale and lower production costs by jointly purchasing intermediate inputs or jointly selling their output to large international brokers.

Production practices in those early years were quasi-artisanal and strongly dependent on imported intermediate inputs and genetic material, salmon eggs, cultivation tanks, processing machinery, and so forth. Salmon food, a major component of salmon-farming costs, was made from fresh fish by each firm. At that time, three kilograms of food were needed to produce a kilogram of salmon, more than three times the conversion rate the industry exhibits today (*AquaNoticias*, July 1997: 24). The first rudimentary cultivation tanks, made of wood and, at 7 meters square, much smaller than those used today, were installed in 1981 by Nichiro Chile and Fundación Chile at Salmones Antartica, in Chiloe. Larger cultivation tanks, 12 meters square and made of galvanized steel, were introduced by salmon-farming companies throughout the 1980s (*AquaNoticias*, April/May 1998: 12). A similar process of gradually upgrading intermediate inputs—nets, processing machines, vaccines, veterinarian services, and transportation boats—can be identified, together with a gradual process of dissemination of best practices among firms.

Expansion and Transformation

New technologies, intermediate inputs, and production routines were introduced by local companies in the late 1980s and early 1990s as firms expanded their capacity and technological complexity. Many of the firms entering the industry in the late 1980s were small family undertakings that financed their expansion largely from family savings. Investment programs on the order of $4 to $5 million were common (*AquaNoticias*, February 1998).

Only in the second half of the 1990s, however, did Chile's farmed salmon industry mature. How the notion of a mature industry applies to salmon farming was made clear in a public conference by Torben Petersen, CEO of Fjord Seafood Chile, when he stated, "The real maturation process begins when we see that company actions are aimed at the markets and not at production, in other words, when salmon-farming growth is determined by its market and not by its production" (*AquaNoticias*, May 18, 2004).

World prices for salmon fell significantly in the second half of the 1990s, approaching the industry's long-term unit production costs, as competition increased and markets for salmon became more competitive. At the same time, the technological and competitive regime of the industry became more demanding as a result of mergers and acquisitions that left the average firm larger, more capital intensive, and technologically more sophisticated. Mergers also increased business concentration and turned the industry into an oligopoly. The process of structural transformation also involved greater vertical and horizontal integration of firms. Larger companies have proceeded to integrate under a single roof not just the three conventional stages of production—hatchery, cultivation, and processing—but also side industries such as salmon-food, salmon eggs, and other intermediate inputs and services. In other words, the search for economies of scale and cost reductions—stimulated by stronger international competition—induced many Chilean salmon-farming companies to introduce major changes in technology and production organization in the 1990s, driving the industry closer to world-class standards for process technology and product quality.

Many small and medium-size companies, in Chile as elsewhere, abandoned salmon farming during the 1990s in a worldwide trend toward oligopoly in the industry (table 6.1). Lack of adequate sources of finance, production know-how, and technological expertise induced many family-owned firms either to exit the industry or to sell their

Table 6.1 Worldwide Consolidation of the Salmon-Farming Industry

Country	Number of firms 1994	1999
Canada	40	7
Chile	65	35
Faeroe Islands	30	15
Ireland	15	4
Norway	360	180
United Kingdom	40	20
United States	22	5
Others	20	5
Total	592	271

Source: www.nutreco.com, cited in Montero 2004.

production facilities to larger companies. The abandonment reflects the fact that the industry has become much more capital intensive and technologically demanding than in the past. In many respects, it must now be considered a knowledge-intensive sector in which genetics and biology play an important role.

The following list of mergers and acquisitions expresses some recent trends in Chile (Montero 2004: 55):

- The Norwegian parliament authorized Statkorn to invest and take over Salmones Mainstream in 2000.
- Marine Harvest and Mares Australes, both owned by Nutreco, merged in 2000.
- Fjord Seafood of Norway took over Chile's Salmones Tecmar, Ltda. as well as Salmones Americanos, Ltd.
- Norway's Stolt Sea Farm, one of the industry's largest firms worldwide, purchased two Chilean firms, Eicosal and Ocean Horizons.
- Two Chilean firms, Salmones Unimarc and Pesquera Nacional, both owned by a local holding company (Inverraz), merged into Salmopesnac SA.
- Biomar, an enterprise of the Danish group KFK, purchased Ecofeed-Biomar, Ltd. and expanded into salmon.
- Provini, a world leader in salmon-food production, took over from its Chilean partners the production of intermediate products for salmon cultivation.

The world's top salmon-farming companies now control well over 50 percent of installed capacity in Chile's salmon-farming sector, including horizontally related activities such as salmon-food production and hatcheries, among others.

Vertical integration of the hatchery, cultivation, and final processing activities involved in salmon farming has further increased the capital intensity and technological complexity of the industry at the level of the individual firm.

Nutreco integrated horizontally into salmon-food production, a particularly profitable area of the industry, by taking over Biomaster (from Iansa), Marine Harvest, and Pesquera Mares Australes. Chilean-owned firms, such as Salmofood, Huillinco, and Alitec, have also attempted to develop their capacity to produce salmon food.

Although the largest firms in the industry have now also integrated (vertically) the production of salmon eggs and alevins (newly hatched salmon still attached to the yolk sac), only the most sophisticated companies have developed domestic experimental facilities in which to perform knowledge-generation activities of a certain caliber. Knowledge-generation efforts remain the weakest part of an otherwise successful production cluster.

The growth of salmon farming has been associated with the gradual expansion of related industries producing intermediate inputs such as vaccines, nets, packaging material, boats, transport services, software, professional services (legal and veterinary), maintenance, and so forth. In her study of the salmon sector, Montero (2004) indicates that in Chile's Region X, home to 60 percent of Chile's salmon-farming activity, some 22 firms produce nets, 13 provide painting and maintenance services, 18 build or maintain cultivation tanks, 10 provide ichthyopathological services, and so forth. New knowledge-intensive activities emerged in the late1990s—among them labor-training services, the production of software, and environmental care services. Employment creation from salmon farming in Chile's southern districts has not been marginal, as we shall see later on in the chapter, nor has the impact of the industry on regional development been negligible.

The characteristics of Chile's salmon-farming industry today are as follows:

- Large, capital-intensive, technologically complex and sophisticated plants
- Vertically and horizontally integrated firms
- Growing sales of salmon as a specialty in sophisticated market niches and through direct marketing arrangements with international retailers, such as Wal-Mart and Carrefour
- Forms of collective action by the industry

A summary of the growth of the industry is provided in table 6.2.

Table 6.2 Evolution of Salmon Farming in Chile, 1960–2000

	1960–73	1974–85	1986–89	1990–95	1996–2002
Exports (tons)	Negligible	1,000	11,000	100,000	500,000
Main products and markets		Fresh and frozen Coho salmon; trout	Coho salmon for Japanese market	Coho salmon for Japan; Atlantic salmon for United States	Diversification of markets: United States, Asia, Latin America
Key event in marketing		Brokers buy from producers	Brokers buy from producers and wholesalers	Collective export activities	Large foreign retailers buy directly
Issues to be resolved	Transition from catch and release to cultivation tanks	Established know-how for fresh-water and need to develop saltwater aquaculture	Rapid expansion in scale of production	Development of forward (egg and smolt) and backward linkages (food, vaccines)	Environmental control systems; salmon food; production of eggs, vaccines; traceability
Government policies	Technology transfer under government cooperation; support from CORFO, ministry of agriculture	Regulation and technology from CORFO, Fundación Chile, Sernapesca, JICA, others	Provision of basic road and ports infrastructure	Missions for market research, technology for supporting industries; regulation	Missions for environmental management; sources of productivity growth
Typical type of firm in industry	External cooperation; no industry yet	Family-owned; small firms; few foreign companies	Local SMEs grow very fast	Growing presence of foreign firms	Mergers and acquisitions by foreign firms

(Continued on the following page)

Table 6.2 (Continued)

	1960–73	1974–85	1986–89	1990–95	1996–2002
Intermediate suppliers	Very few	High degree of vertical integration; few domestic input suppliers	Hatchery, cultivation, and final processing begin to integrate	Outsourcing expands and many new suppliers enter the market	Cluster gets stronger and service industries develop
Expected externalities			Supporting industries develop	Clustering forces become stronger	International norms and standards diffuse; GMPs and traceability
Sources of competitiveness	Natural comparative advantage	Production	Rapid expansion of number of cultivation sites and scale of plant	Mostly local quality standards	Productivity, local and international standards; ISO 9000 and 14000; traceability
Relations among actors in industry	International cooperation; proactive state participation	Public-private cooperation; CORFO, Fundación Chile	Private sector cooperative activities expand	Initial forms of globalization emerge	Full-scale globalization after M&A

Source: Based on Iizuka 2004.
Note: GMP = good manufacturing practices, M&A = mergers and acquisitions.

Technology

Salmon farming involves three stages of production. The first is *hatchery* and pisciculture, in which salmon eggs, alevins, and smolts (young salmon) are produced. The activities involved at this stage are highly demanding, both in terms of natural resources—pure, uncontaminated waters, suitable climatic conditions—and skilled personnel capable of managing sophisticated routines of quality control and environmental protection. This stage in the salmon-farming process is followed by the *cultivation* stage, in which alevins and smolts grow in captivity into full-grown fish ready for market. The size and quality of cultivation tanks, feeding equipment, and food play a crucial role in this stage of the production process. The third stage is *processing,* when different products and packages are prepared—boneless, frozen fillets, smoked, sliced—for the market.

Although the three phases can be integrated under a single roof, they also can develop as separate industries, opening up opportunities for specialization, economies of scale, and intra-industry division of labor. Each employs its own specialized technology, capital goods, and intermediate inputs. Each has specialized suppliers. Salmon eggs and smolts, salmon food, and salmon-processing equipment were largely imported during the early stages of the industry. As the industry grew in size and complexity, demand for many intermediate inputs expanded and acquired economic significance, inducing domestic and foreign companies to create the capacity for local production. Many firms responded to an expanding demand for pesticides, chemicals, vaccines, nets, transport services, veterinarian services, production-organization logistics, legal services, and so forth.

In all three production stages—hatchery, cultivation, and final processing—an increasingly complex set of interactions has developed involving salmon-farming firms, universities, public sector agencies, suppliers of equipment and intermediate inputs, consulting firms, and financial institutions, creating a dense and sophisticated fabric of economic, institutional, and technological interdependencies among the various agents in the cluster. It is the economics of networking rather than the economics of independent production units that best represents the pattern of functioning of the industry today.

Salmon food accounts for roughly 45 percent of total production costs, while eggs and smolts add a further 18 to 20 percent. In other words, the first two stages in salmon-farming production—hatchery and cultivation—account for roughly two-thirds of production costs.

Table 6.3 Intermediate Inputs and Services in the Hatchery Phase

Inputs	Services
Salmon food	Transport
Cultivation tanks	Maintenance of cultivation tanks
Nets	Maintenance of nets
Salmon eggs (national and imported)	Veterinary services (vaccines)
Imported machinery	
Automatic food suppliers	
Computers	
Oxygen systems	
Machines to count salmon eggs	

Source: Montero 2004.

Hatchery

Alevins born from the artificial fertilization of salmon eggs grow in freshwater into smolts; under careful biological control they reach 80 to 100 grams, at which point they are moved into salted water for cultivation, a process that takes about six months. A strictly controlled environment—pure water, uncontaminated ecosystems—is the sine qua non for the successful production of smolts. Other requirements are detailed in table 6.3.

In 1991, some 51 million salmon eggs were imported into Chile. That figure increased to 62 million in 1992, 73 million in 1973, and 114 million in 1994. Salmon eggs were imported from Ireland, Scotland, the United States, and Norway for Atlantic salmon and from the United States, Finland, Sweden, Norway, and Denmark for the production of trout. The need to lower domestic production costs and reduce the incidence of infectious diseases—which increase with the importation of foreign eggs—induced some Chilean firms into local production of salmon eggs and alevins. Some 74 million units were locally produced in 1995, which would suggest that roughly 60 percent of the salmon eggs used locally came from imports, with the remaining 40 percent being domestically produced. The rate of local production vis-à-vis imports is somewhat higher for Atlantic salmon than for either Pacific salmon or trout (*AquaNoticias,* March–April 1977). After 1996 imports of salmon eggs fell somewhat, CORFO, the University of Chile, and the Instituto de Fomento Pesquero (IFOP) participated in the country's efforts to improve local reproductors, a project in which a small number of local firms also took interest (*AquaNoticias,* March–April 1997). Sernapesca also played a significant role in inducing the local production of salmon eggs as a way to limit the spread of contagious diseases.

Domestic learning processes and investments in the local production of salmon eggs and smolts have been quite significant. The need to reduce the ichthyopathologies that occur with imported genetic material induced local firms to undertake import-substitution efforts, and the entry of new firms to the hatchery stage has made that section of the industry more competitive.

However, local specialists in salmon farming indicate that the production of salmon eggs and smolts has not been associated with a concomitant expansion of more basic research activities in fields such as biotechnology and ichthyopathology, both of crucial importance in salmon farming. This failure could be taken as an indication of the fact that the more knowledge-intensive areas of the global salmon-farming process are relatively less developed in Chile than in more mature developed countries. Visits of local entrepreneurs and technologists—jointly organized and financed by Fundación Chile, CORFO, and local companies—to facilities for production, experimentation, and R&D in Norway and Scotland confirm that this is the case (*AquaNoticias,* February 1998: 78, and November 1997: 39).

Cultivation

Once smolts reach an adequate weight—between 60 and 80 grams—they are transported by specially equipped trucks or boats to cultivation centers, where they are kept for just over one year, gaining weight and getting ready for the market (box 6.1). Many forms of technological progress were incorporated into this stage of salmon farming in the 1990s, among them larger cultivation tanks, automatic feeding machines, and computerized sensors and control systems. Although the labor needed for the cultivation process is mostly unskilled, many

---

Box 6.1 Steps in the Cultivation of Farmed Salmon

Step 1   Transportation of smolts from lake or basin by truck equipped with oxygenated tank or by boat equipped with water circulation and oxygenation system.

Step 2   Smolts drained or pumped from transport tanks into cultivation tanks.

Step 3   Feeding by manual or automated system. Sensors and screening mechanisms may be used to monitor feeding process.

Step 4   Harvesting by manual or mechanical means.

Step 5   Transportation of harvested salmon to processing plant.

Source: Montero 2004.

of the technological innovations were designed to save on the high cost of labor in industrial countries.

Computer-based production technologies, in particular, have spread rapidly in cultivation operations worldwide. Some are gradually being adopted by Chilean firms, even though they will not necessarily confer a major competitive edge given the wide disparity in labor costs between Chile and the developed countries.

Salmon food and cultivation tanks are central to the cultivation stage. In both areas Chile has made major changes and improvements.

Because salmon food accounts for nearly half of the direct costs of cultivation, the industry closely monitors the rate of conversion of salmon food into fish weight. In Chile that rate was around 3 to 1 in the 1980s; since then it has fallen to 1.2 or 1.3 to 1. The improvement is linked to the use of richer diets, more digestible pellets and vitamins, and better food dispensers, among other factors. Experimental efforts have been essential in improving salmon food. New products enter the market each year, as salmon food becomes a knowledge-based industry in which product differentiation plays a major role (*AquaNoticias,* February 98: 33).

As in the global salmon-farming industry, the salmon-food sector has shown a significant transformation in market structure and company behavior over the last decade. From close to 20 firms producing salmon food in Chile in the early 1990s only 8 remained in the market at the end of the 1990s. Those firms are technologically sophisticated and close to world standards.

The size of the cultivation tanks and the material from which they are constructed changed significantly in the 1980s and 1990s. Newer tanks made from aluminum or plastic are larger (20 meters square) and more resistant than older wooden tanks. Local producers of state-of-the-art cultivation tanks have emerged in recent years, along with producers of automatic, computer-based food dispensers that permit consistent feeding during the lengthy cultivation stage. All of these improvements have lowered production costs and improved cultivation practices to close to world standards. However, Chile remains an early imitator rather than an originator of new process technologies.

Processing

At the end of the cultivation cycle, adult fish are transported to processing plants where they are weighed, eviscerated, and prepared for consumption in various forms (fresh, smoked, boned, canned, whole, steaks, fillets). Each product presentation requires specific machinery and intermediate inputs. Processing also varies significantly, depending on the destination of the product (Europe, Japan, North America).

Salmon-processing plants must be certified by EU and U.S. authorities before their products can be sold in European or U.S. markets. To be certified, plants must employ HACCP (Hazard Analysis and Critical Control Points) or PAC (Programa de Aseguramiento de Calidad) norms, under the supervision of Sernapesca. In 1997, only 26 Chilean processing plants had received such authorization (*AquaNoticias,* November 1997: 11), but the number has increased considerably since then. The treatment of solid and liquid residuals, the level of asepsis in various sections of the plant, how workers are trained, and the traceability of the product are all important in the certification process.

Fundación Chile plays a key role in this area, helping companies organize to meet ISO 9000 or 14000 standards. Likewise, the Asociación de Productores de Salmon y Truchas de Chile (APSTCH)—through INTESAL, the Salmon Technical Institute—plays a significant role in the upgrading of plants and in the certification process (*AquaNoticias,* November 1997).

The quality and sophistication of packaging—an important part of the processing of salmon—has increased significantly in recent years, with specialized firms, both foreign and local, actively competing in the markets for packaging materials.

Firm-Level Learning and Industry Evolution

In the third stage of the development of salmon farming in Chile, business concentration and foreign control increased as large firms, mainly Scandinavian, entered the market, often by acquiring Chilean firms. The industry came to resemble a mature oligopoly, with large international retailers becoming important actors in the industry.

That process of growth of the salmon-farming sector has important microeconomic underpinnings—associated with learning processes in individual firms.

The Microeconomic Dimension

Learning by individual firms and the evolution of in-house technological capabilities are crucial determinants of company performance. Most firms in Chile's salmon-farming industry began small and grew quickly between the late 1980s and the late 1990s. Rapid expansion demanded new machinery and equipment, new cultivation sites, and more qualified human resources, among other requirements.

Local salmon-farming companies started production with rudimentary production routines and an imperfect understanding of the ecological, environmental, and technological parameters underlying their

operational routines. More difficult salmon-farming locations—
involving more complex environmental conditions—were gradually
brought into production as the industry expanded.

Companies must be seen as moving along an idiosyncratic, firm-
specific learning curve. They acquire knowledge through trial and
error, in an experimental way, over time. No two salmon-farming operations in the world are identical; each must acquire know-how in-
crementally to overcome barriers to its growth. No firm began with
adequate knowledge of its production function; all had to discover
that function and work with it in an experimental way. *AquaNoticias,*
a publication of Fundación Chile, illuminates the process through a set
of company histories published over the last decade. It is clear from
those histories that what was true for Invertec Pesquera or Mar de
Chiloe is by no means true for Patagonia Salmon Farming or Salmofood Chile. Each confronted a unique set of questions resulting from
the constraints and opportunities that each faced.

Why is that? Do not all of these firms produce salmon and compete
in similar markets? The answer has several parts. For one thing, ecological and environmental parameters vary strongly across locations.
Water quality, temperature, salinity, and other ecological variables related to the microorganisms that populate each particular lake and
marine location suggest that each firm probably has only a partial, and
very imperfect, understanding of the parameters underlying its production process. Other factors are technological and organizational.
Understanding more about the role such variables play, and knowing
exactly what to do when production conditions, raw materials, and
consumer tastes change, demands experimentation and the ability to
generate incremental knowledge that firms must develop in-house if
they are to manage their specific circumstances. Those circumstances
are what make individual learning curves different and company
strategies so varied.

Some strategies turn out to be more successful than others in ad-
dressing the underlying uncertainties of salmon farming, which explains why some firms do better than others, open new markets, and
gain market share.

In Aggregate

The manufacturing data published by Chile's National Institute of Statistics (INE) aggregate the information of some 20 salmon-farming
companies and provide information on gross production value, investment in four types of assets (buildings, transport equipment, machinery
and equipment, and land), the ratio of skilled to unskilled personnel,

and the entry and exit of firms to and from the market for the period 1981–2000.

The data provide an aggregate account of the evolution of salmon farming in Chile, an account that we regard as highly consistent with our previous qualitative assessment of how the industry developed through time (figures 6.2a and 6.2b).

The number of salmon-farming companies increased until 1996, reached a plateau in the late 1990s, and then began to drop (figure 6.2a). Although there were fewer firms in the industry at the end of the period, the average salmon-farming company was larger and more capital- and technology-intensive, as evidenced by the higher ratio of skilled to unskilled personnel. Capital deepening occurred at two different moments over the last two decades: in 1986–89 and in the mid-1990s. With the benefit of hindsight, we know that the first of these moments occurred during the industry's inception in Chile—as many new, locally owned salmon-farming firms entered the industry. The second wave of investments accompanied the mergers and acquisitions that took place when large, international salmon-farming enterprises arrived in Chile in the 1990s. Many of these foreign companies entered the industry by purchasing small and medium-size Chilean firms. Concentration ratios—the share of output accounted for by the largest four plants in the sector—increased. The industry should now be thought of as a mature oligopoly, operating at a global scale.

The life cycle of the industry described here is consistent with the Schumpeterian notions of the entry and exit of firms to and from the market; innovation rents at the industry's inception period; their subsequent erosion following rapid new entry, mergers, and acquisitions in a later period (accompanied by foreign direct investment, FDI); and the eventual globalization of the activity.

Institutions and Institutional Evolution in Salmon Farming

Three different types of institutions—and associated forms of interaction between public sector organizations and salmon-farming firms—can be identified in Chile. All played a significant role in the inception of salmon farming in the country. None should be seen as static features of a particular market structure or model of industrial organization, but rather as dynamic forms of interaction undergoing change, adaptation, and learning over time. In other words, they should be seen as parts of an evolutionary picture rather than as a snapshot of a given system at a given time.

Figure 6.2 The Chilean Salmon-Farming Industry, 1981–2000

a. The number of firms and their gross output

b. Investment

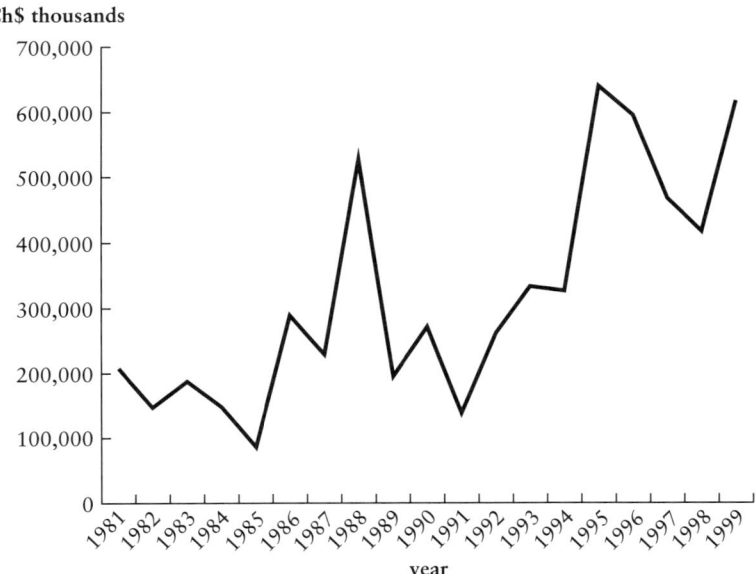

Source: National Institute of Statistics, Chile.

Some of the organizations deal with regulatory issues, others with industrial promotion, and yet others with knowledge generation and diffusion.

Regulation and Oversight

Starting up a salmon-farming operation and exporting fresh salmon to Europe, Japan, or the United States is not a simple undertaking. Environmental conditions have to be met; quality-control standards fulfilled; packaging, labeling, and traceability requirements observed—all of which demands a sophisticated legal and regulatory infrastructure as well as open and transparent production and monitoring practices. It also demands an independent set of public agencies with state-of-the-art knowledge and experimental capabilities. For the most part, those requirements are met in Chile, and the situation has gradually improved over time. The Comisión Nacional del Medio Ambiente (CONAMA, national environmental commission) normally requires environmental impact studies for the approval of new cultivation sites. Sernapesca, through its department of Sanidad Pesquera (fishery hygiene), closely monitors salmon-processing plants and is the local agency responsible for evaluating compliance with HACCP and U.S. Food and Drug Administration norms (*AquaNoticias,* November 1997). The Instituto Tecnológico de Salmon (INTESAL), a division of the salmon industry association, is the counterpart on the industry side, monitoring quality standards and awarding quality certificates to firms complying with standards for the disposal of industrial residuals and other norms that plants must meet if they are to be certified for exports. Fundación Chile provides technological assistance to individual firms seeking to upgrade their production processes and make the transition to a technologically more demanding export mix. Products are classified as super premium, premium, grade 1, or industrial. The different grades are intended for different uses and markets; price differentials are significant

In sum, the regulatory and oversight functions are well established in Chile, with transparent and dynamic public-private mechanisms of interaction in place.

Industrial and Technological Development

In spite of its rapid expansion in recent years and its growing technological sophistication, Chilean salmon farming continues to be based on imported machinery, equipment, and know-how, marginally supplemented by ad hoc knowledge-generation and adaptation efforts carried out locally. In other words, Chile has managed to develop a state-of-the-art salmon-farming industry, but it has not, so far, developed a similarly strong domestic scientific and technological infrastructure, nor a local capital-goods industry to serve the sector. Through regular study and business missions to Norway, Scotland, England, and the United States, firms, public agencies, and institutions responsible for knowledge generation and diffusion obtain systematic access to external scientific and technological information and production know-how (*AquaNoticias,* February–March 1998). Following such missions, reports are published in the specialized industry literature for the benefit of firms that could not attend. Production organization know-how is disseminated through regular seminars. Fundación Chile and CORFO play a major role in this respect, partly financing the exploratory missions that help the domestic industry keep in touch with the international frontier.

Both agencies were active in the design and start-up of salmon-farming operations in the early days of the industry. The industry no longer needs that type of coaching. On the other hand, country- and region-specific knowledge and technology are still needed. The domestic research and engineering capabilities for such knowledge generation and diffusion activities are still weak in the Chilean salmon-farming environment and are a major topic for policy discussion.

Knowledge Generation and Dissemination

A distinction must be made between agencies that exclusively finance research activities and those that perform (and finance) R&D efforts. In the first group are the Subsecretaría de Pesca (bureau of fisheries within the Ministry of Agriculture), CORFO, ProChile, and the Comisión Nacional de Investigación Científica y Tecnológica (CONICYT). These agencies finance research projects related to salmon farming through public tenders in which projects in other scientific and technological fields compete on equal terms. In the second group are the Instituto de Fomento Pesquero (IFOP, institute for the promotion of fisheries) and public universities, as well as public-private entities such as Fundación Chile, INTESAL, the Instituto de Investigaciones Pesqueras de la VIII Region (IIP, institute for fisheries research of Region VIII), and private universities.

Annual spending on R&D approaches US$10 million, with three-quarters of the total coming from public sources and the remainder from individual firms.[9] A sampling of research projects appears in table 6.4.

Table 6.4 Sample R&D Activities

FUND: Project	Executive agency and participating firms	Expected impact
FIP: Zoogeological studies between 41°50' and 48°49' latitude	IFOP and Gobernación Marítima, Region X	Information for the regulation of salmon-farming sites and concessions
FONDEF: Deformations in the mandibular structure of Atlantic salmon	University of Chile and EWOS Chile, Region X	Reduction of industry's production costs
FONDEF: Remote sensing systems for cleaning sea bottom	University of Concepción. Region VIII	Better oxygenation and renewal of waters Productivity improvements
FDI: Salmon vaccines	INTESAL and private firms	Reduction of infectious diseases in salmon farming
FDI: Value added products for aquaculture firms	Fundación Chile and private firms	To expand the industry's local value added and competitiveness
FONTEC: Salmon food and salmon-feeding technologies (CETECSAL)	Salmofood (private firm)	To expand the industry's technological capabilities; to improve productivity and competitiveness
FONTEC: Design and prototyping of floating salmon-feeding station	Simar (private firm)	Technological innovation and cost reductions
FONTEC: Development of integrated viral diagnostic system	Laboratorio Diagnóstico GAM (private firm)	Lowering industry losses from salmon pathologies
FONTEC: Technical mission—Risk assessment and prevention of introduction of disease stemming from importation of eggs	INTESAL and various private companies	Study missions abroad

Source: Montero 2004.
Note: FONDEF (Scientific and Technological Development Promotion Fund); FONTEC (National Technological Development and Production Fund); FIP (Fisheries Research Fund).

Lessons

Having examined the economic, technological, and institutional evolution of salmon farming in Chile, it is now possible to draw some generic conclusions. We have organized the discussion under seven headings:

- Natural resource–based industries and knowledge generation and dissemination
- Evolution of the industrial structure and technological regimes
- Environmental sustainability and standards
- University-industry relations
- Employment generation, regional impact, and clustering
- The role of FDI
- The role of government

On Natural Resource–Based Industries and Knowledge Generation and Dissemination

The restructuring of many Latin American economies in the aftermath of trade liberalization and market deregulation has favored natural resource–processing activities, notably mining; agricultural products and foodstuffs; pulp, paper, and wood products; fisheries; iron and steel; and gas and petroleum. The transformation opens up new questions about the unique physical, biological, genetic, and environmental features of these natural resources, their long-term sustainability, the source of the knowledge needed to exploit them in a sustainable way, and how firms should act and be regulated. Very little progress has been made in the region in this broad and complex territory that will become increasingly important in the years to come.

The sustainable exploitation of agricultural land, mineral resources, forests, and water and marine resources demands basic research and continuous supplies of new country-specific knowledge in biology, genetics, marine sciences, immunology, mineralogy, and other sciences. Applied research and technological development efforts are needed downstream to produce new equipment, vaccines, pharmaceutical products, and many other intermediate inputs needed for the rational and sustainable exploitation of resources. The crucial question is how to design and implement an incentive regime capable of avoiding market failures, dealing with negative externalities, and protecting and delivering public goods.

Additional questions—among them external vs. local know-how, of codified generic knowledge vs. location-specific knowledge and "localized" production functions—must be examined and answered in context. Appropriability of R&D efforts and intellectual property rights, university-industry relations in knowledge generation and dissemination, and controversial questions associated with WTO-related disciplines and initiatives (Trade-Related Intellectual Property Rights, the Convention on Biological Diversity, the Protection of Geographical Denominations, the Doha Convention) must also be taken into account. To master this complex institutional territory and implement long-term strategies for sustainable economic growth, developing nations need specialized human resources: economists, lawyers, biologists.

Our study shows that market forces are incapable of providing all the necessary answers. A well-conceived national strategy is needed alongside functioning markets. A strategy for the rational exploitation of natural resources demands an understanding of the links among the underlying economic, institutional, and technological forces, as well as the complex nature of the technology markets involved. Government responses must be pragmatic and flexible in order to maximize the potential benefits that countries can capture from WTO disciplines and agreements.

Salmon farming in Chile suggests specific roles for the state during the inception stage of a new production activity in the economy. But there are no easy-to-follow rules in this respect, and each country and sector of economic activity appears to demand localized treatment.

On the Evolution of the Industrial Structure and Technological Regimes

The present inquiry throws new light on the industry life cycle.

Many economic, institutional, and technological forces shaped the organization of Chile's salmon-farming industry and determined firms' profit rates, investment patterns, rate of expansion, and likelihood of survival. Chief among those forces were (i) the rate of entry of new firms and incorporation of new cultivation sites, (ii) the rate of exit from the market and of exhaustion—biological, ecological—of existing cultivation sites, (iii) access to new cultivation permits, (iv) economies of scale, (v) accumulated technological and financial capabilities within individual firms, (vi) flows of FDI, (vii) development of financial institutions to serve the sector, (viii) mergers and acquisitions, (ix) international market penetration by local firms, and (x) insertion of domestic companies in international commodity chains.

In the short period of just two decades, the Chilean salmon-farming industry went from a model of family-owned firms and quasi-artisanal organization of production to that of a concentrated oligopoly firmly inserted in global commodity chains. Gross margins have fallen significantly, and the rate of the industry's expansion appears now to depend more on the rate of growth of world demand for salmon than on the local potential to increase supply. That evolutionary process appears to be associated with growing market saturation, as technologically more difficult and complex cultivation sites were brought into operation over time, and as economies of scale were attained through mergers and acquisitions. Simultaneously, the industry as a whole has become technologically more demanding, as international quality control and environmental sustainability standards and traceability have become necessary conditions for successful competition within international commodity chains.

After a successful inception of the industry in a favorable local environment—in which exceptionally high rents prevailed—the industry today is much more susceptible to global competitive forces. That does not mean that the growth possibilities of Chilean aquaculture have diminished, but it does suggest that continued expansion may demand new markets, new (diversified) products, and other changes in industry structure and performance. Economies of scale and scope, capitalizing on companies' capacity to learn, may largely determine which firms succeed and which ones fail in the process of diversification and growth. Public sector agencies, such as CORFO and IFOP, and Fundación Chile may once again play a catalytic role—as in the early days of the industry—facilitating investment in the creation and dissemination of new products and process technologies among local producers.

On Environmental Sustainability and Standards

Natural-resource-processing activities pose new questions of environmental sustainability. Salmon farming and other forms of marine-based activities are no different. This is an area in which relying on imported knowledge of how to protect the environment is not entirely satisfactory, given the large number of country-specific problems and uncertainties that demand tailor-made solutions rather than generic, off-the-shelf knowledge. The biological and ecological conditions of salmon farming in southern Chile are not a carbon copy of the biological and ecological conditions found in Norway or Scotland, despite common elements. Dealing with local pathologies, managing local biological and ecological conditions, deciding which food best suits the morphological and biological features of local species—all are location-specific questions that must be dealt with, ideally through domestic knowledge and engineering.

A similar set of issues emerges in quality control and environment protection. In the industry's early years such standards were tacit and largely uncodified, with an artisanal cast to them. The transition to world-class production standards obviously demands a concomitant transition to internationally accepted and codified rules, notably those required by U.S., Japanese, and European buyers and regulatory agencies. Chile's firms have had to upgrade their in-house technological capabilities to accommodate the ISO 9000 and 14000 norms, online traceability, and other changes.

On University-Industry Relations

One of the weakest links of an otherwise strong domestic cluster is the one that connects salmon farmers to universities that perform agricultural R&D and train biologists, pharmacologists, marine geneticists, and other human resources. The topic of university-industry relations is in its infancy in Chile. Progress is being made, but the road ahead is long. Fortunately, it offers rewards for both parties.

The early establishment of Fundación Chile as a sophisticated knowledge generation and dissemination agency probably explains why local universities have not yet become a significant source of new knowledge and technology for industrial use. As previously shown, the University of Chile, the Catholic University, the University of Concepción, the University of Los Lagos, and a few others maintain R&D labs specializing in molecular and marine biology, pharmacology, genetics, and the like. Some operate cooperative programs with salmon-farming companies or producers of intermediate inputs—such as pharmaceutical companies. CORFO, CONICYT, and IFOP frequently act as funding agencies for such programs under risk-sharing contracts with individual firms. Questions of patents and intellectual property rights, of precompetitive cooperation between firms (to establish generic standards from which other actors can develop proprietary technology), and of the gradual buildup of a vibrant and dynamic domestic scientific and technological infrastructure to support the expansion of aquaculture have, so far, gone largely unexplored. There is certain overlap between what Fundación Chile currently does and what universities and other knowledge-generation agencies do. So far, the system has operated without much explicit coordination, but if the Chilean scientific and technological infrastructure for marine-based activities is to expand by an order of magnitude—as would seem desirable—more coordination and funding for basic and applied research from the public sector might be required. Salmon-farming entrepreneurs do not appear especially eager to embrace such an option, as many of them believe that the state should stay outside the field, but we note that significant differences of opinion prevail between large and small salmon-farming companies.

On Employment Generation, Regional Impact, and Clustering

Large areas of southern Chile (Regions X and XI, and cities such as Puerto Mont, Cohaique, and Castro) have been significantly affected in their production structure, patterns of social and institutional organization, and role in the Chilean economy and society as a result of the growth of salmon farming over the last two decades. Salmon farming has created more than 30,000 direct jobs in remote areas, reducing regional disparities in income per capita and consumption patterns, while a complex web of market and nonmarket forms of interaction has developed as the industry has grown. Many new sectors of production developed to meet the industry's varied needs—for pharmacology, legal services, net maintenance, boat construction, vaccines and related services, road construction, and telecommunications. The positive and negative externalities resulting from growth (such as higher rents for housing and office space, wear on roads, environmental degradation, and greater participation in municipal affairs) require careful monitoring from regional and municipal authorities.

On the Role of FDI

FDI did not play a major role in the Chilean salmon-farming sector until very recently. The industry began on the basis of small and medium-size enterprises and was strongly influenced by the initial artisan approach to production organization. This model persisted until the 1990s, when mergers and acquisitions and the insertion of Chilean production into global retail commodity chains made the influence of foreign agents more significant. Their role is likely to continue to grow.

   However, the technological catch-up that Chilean salmon-farming companies staged in the 1990s was not a consequence of the arrival of multinational corporations bearing superior technology. In fact, the converse is true, as can be seen in various cases in which foreign firms decided to return to Chilean-designed know-how that they had decided to phase out after taking over domestic production operations in the late 1990s. Technology brought from their headquarters performed worse than expected when transferred to the Chilean environment.[10]

On the Role of Government

Our study suggests that there might be many ways of inducing new and dynamic economic activities and coaching them to international competitiveness. The "Chilean way" in the case of salmon farming is certainly not the Norwegian or Scottish one, but, more surprising, it is not the one that has prevailed in the case of Chilean wine production or Chilean pulp and paper. Many things may explain the observed differences. On the one hand, the nature of the firms involved in each case surely plays an important part in setting the evolutionary path along which public-private interactions develop during the industry's expansion. Small, family-owned enterprises—as distinct from large domestic conglomerates—appear to call for a different public role in setting the wheels in motion, whether in the provision of finance for new investment projects, the promotion of design capabilities for building new production capacity, or the construction of associative channels for international marketing.

Similarly, the perception that large rents might be obtained through the skillful exploitation of available natural resources, and the lack of the domestic capabilities to do so, must surely have been a powerful incentive for the public sector to begin promoting salmon-farming activities in Chile.

On the other hand, stochastic factors—the role played by Fundación Chile, for example—also need to be considered among the forces that set the wheels in motion. Such factors were clearly obtained in the case of salmon farming but not in wine production. The PROFO program—a scheme designed to provide support for joint R&D efforts by small and medium-size enterprises—played a major role in wine production but failed to do so in salmon farming.

Finally, the design and enforcement of environmental protection legislation appears as a major source of public sector influence in setting up the industry's long-term technological learning path.

In summary, the diversity of roles that the state has played (in a context marked by good macroeconomic fundamentals) is the major lesson of the present inquiry.

Conclusion and Policy Recommendations

We wrap up with a policy-oriented discussion of our main findings. Salmon farming developed quickly in Chile over two decades, turning the country into a leading competitor in world markets for salmon. Its comparative advantages resulted from an adequate set of natural resources—pure waters and good ecological conditions—but also from the early alertness of local entrepreneurs and the readiness of public (CORFO) and public-private agencies (Fundación Chile) to play a catalytic role in getting the industry going. There was some external collaboration—from Japanese, U.S., and Canadian sources—but it was minor.

The industry's evolution proceeded along three quite different stages, as we have seen. The actors, the problems, the opportunities, and the underlying institutions varied significantly from one stage to the other. So, too, did the role of the state and the nature of public-private interactions.

In the early years of the industry—the late 1970s and early 1980s—the foremost concern of the average small firm was overcoming its fragmentary and highly imperfect understanding of the technological, organizational, and ecological/environmental conditions in which it was producing salmon. Each firm coped with a specific and localized set of constraints and opportunities. Trial and error and firm-specific learning processes were the major characteristics of firms' behavior during those years.

Initial forms of collective action and novel institutional developments emerged with the expansion of the industry. Government did not play a powerful role in forcing the industry's expansion, but it certainly played a crucial catalytic role in designing plants and other forms of knowledge generation and diffusion. That role complemented the alertness that entrepreneurs exhibited in the early 1980s.[11]

During the ensuing period of rapid growth, the name of the game was new entry, the expansion of capacity, and the introduction of process improvements embodied in new machinery and equipment—brought almost entirely from abroad. Firms grew in number and size and began to compete in world markets. Access to finance, ready supplies of intermediate inputs and services, good transport infrastructure, and predictable, transparent, and forward-thinking regulation also favored growth. The Chilean public sector contributed by building roads, modernizing docks and shipping facilities, and inducing firms—through regulatory agencies—to adopt international quality norms and standards. In other words, a public-private system grew up around salmon farming throughout this period.

As the expansion process acquired momentum and a large number of firms—domestic and foreign—entered the market, the catalytic role of government became less important, but it remained significant in environmental regulation and in the generation and diffusion of knowledge. The public-good nature of some of the assets thereby involved clearly justify the government's active involvement.

In the third stage, profits diminished, growth slowed down, and large international firms entered the market through a series of takeovers and acquisitions that gradually brought the industry to its present market structure and *modus operandi*. The industry is now much more capital intensive, firms are larger and technologically more complex. The more dynamic ones are proceeding into products with a higher domestic value added, competing globally by selling under proprietary trademarks to large international retailers.

But there is a blot on the otherwise bright picture of the Chilean salmon-farming industry. In contrast to the industry in Scandinavia and Scotland, Chile's salmon-farming sector, so far, has not developed a strong capacity to generate and export knowledge and technology, nor has it induced the expansion of the capital goods industry catering to salmon farming. There is no reason why Chile should not be able, eventually, to proceed into more domestic value-added-intensive activities, complementing its exploitation of static comparative advantages with more dynamic, knowledge-intensive services and technologies. But to make that happen, a proactive government policy seems to be needed, given the many imperfections that underlie the markets for knowledge and the public-good nature of some of the intangible assets involved.

Once again—as in the early years of salmon farming—the catalytic role of government will be required if the Chilean salmon-farming cluster is to move to the next knowledge-intensive stage of development. Orchestration would be overkill. What is needed are clear signs that the next step is not only attainable, but also within reach. Government can provide those signs by coordinating collective action among salmon-farming firms, public-private knowledge-generation institutions, and financing agencies with an eye to overcoming market failures in the generation and dissemination of knowledge.

Endnotes

1. Additional information can be found in Chile Innova (2004).
2. Other natural resource–processing industries—mining, pulp and paper, wine production—as well as services, such as telecommunications and energy, grew rapidly in Chile during this period. Between the mid-1980s and 1998 Chile's GDP grew at an average annual rate of 7.5 percent, while exports expanded at double-digit rates to reach nearly one-third of GDP. Although the Chilean economy became open and deregulated after the 1973 military coup d'état, trade liberalization was reversed during the Latin American debt crisis of 1982–84. The country resumed its movement toward market deregulation and trade liberalization in the late 1980s and pursued them steadily throughout the 1990s (Moguillansky 2002).
3. Professional management accompanied the expansion of salmon farming and wine production in Chile, both emblematic of outward-oriented success during the 1990s. The professionalization of management has been identified in the business organization and innovation literature as a distinct stage in the evolution of production organization, a stage at which

institutions and technological absorption capabilities develop significantly (Nelson and Wright 1992; Kogut 1992).

4. At the end of the 1980s, the Chilean fishing industry held a place close to that of Peru in the international market. It was basically involved in the production of fishmeal, a highly standardized commodity. Today, the Chilean fishing industry is a leading supplier of high-value-added specialties, such as smoked salmon and trout fillets. Peru's industry remains largely unchanged, as do those of Ecuador and Argentina.

5. In 1978, the Chilean government created a bureau of fisheries (Subsecretaría de Pesca) and a national fisheries service (Servicio Nacional de Pesca, Sernapesca) in the Ministry for the Economy (TechnoPress 2003).

6. Fundación Chile is a privately owned, nonprofit corporation created in 1976 by the Chilean government and ITT Corporation of the United States. Its mission is to introduce innovation and develop human capital within the key clusters of the Chilean economy through technology management and in alliance with local and global knowledge networks.

7. *AquaNoticias,* January 1997. *AquaNoticias,* cited throughout this chapter, is a bimonthy news bulletin of the Fundación Chile. Fighting the dumping charge, at a cost of more than $10 million, strengthened networking and collective action within the sector.

8. This section draws heavily on Montero 2004.

9. It is our impression that this figure understates the amount of (informal) knowledge generation that individual firms undertake. Although no studies of R&D efforts in the salmon-farming sector have been carried out so far, private communications from company officials suggest that a great deal is going on at the plant level that is never measured or accounted for.

10. Private communication, obtained during the course of company interviews.

11. It is interesting to note that CORFO and Fundación Chile also tried to act as a catalyst of growth in the forestry sector, extending to the production of furniture. In fact, Fundación Chile designed and brought on stream a modern plant—Centec—that was intended to act as a demonstrator of Chile's potential in the field. The experience cannot be considered a success, however. The differential role played by local entrepreneurs in the two cases seems to be the major explanatory variable. In a private communication on this topic, a CORFO representative pointed out that the opportunities and constraints underlying the two cases might account for the differences in observed entrepreneurial attitudes.

References

AquaNoticias. Santiago: Technopress. Various years.

Chile Innova. 2004. "Chile 2010: Prospectiva de la industria de la acuicultura." Unpublished paper. Programa de Prospectiva Tecnológica, Santiago.

Hirsch, Seev. 1967. *Location of Industry and International Competitiveness.* Oxford: Clarendon.

Iizuka, Michiko. 2004. "Organizational Capability and Export Performance: The Salmon Industry in Chile." Paper presented at the 2004 Druid Winter Conference, January 22–24, University of Aalborg, Denmark.

Kogut, Bruce. 1992. "National Organizing Principles of Work and the Erstwhile Dominance of the American Multinational Corporation." *Industrial and Corporate Change* 1: 285–326.

Moguillansky, Graciela. 2002. "La inversión en Chile. Fin de un ciclo de inversión?" U.N. Economic Commission for Latin America and the Caribbean, Santiago.

Montero, Cecilia. 2004. "Formación y desarrollo de un cluster globalizado: el caso de la industria del salmon en Chile." *Desarrollo Productivo* 145 (January): 1–75. United Nations Economic Commission for Latin America and the Caribbean, Division of Production, Productivity, and Management, Santiago.

Nelson, Richard R., and Gavin Wright. 1992. "The Rise and Fall of American Technological Leadership: The Postwar Era in Historical Perspective." *Journal of Economic Literature* 30 (December): 1931–1964.

Posner, M. V. 1961. "International Trade and Technical Change." *Oxford Economic Papers* 13 (3): 323–341.

TechnoPress S. A. and SalmonChile. 2003. *Aquaculture in Chile.* Santiago.

Vernon, R. 1966. "International Investment and International Trade in the Product Cycle." *Quarterly Journal of Economics* 80 (2): 190–207.

CHAPTER 7

Wine Production in Chile

José Miguel Benavente

CHILE IS THE WORLD'S TENTH-LARGEST WINE producer and fifth-largest wine exporter.[1] But only 15 years ago Chile's role in international wine markets was negligible. Today, the country accounts for almost 4 percent of world wine exports (figure 7.1).

Chile's soil and climate give it natural advantages in wine growing. In the vineyard zones, the climate is practically devoid of rain from December, when the grape bunches appear, until after harvest, allowing the grapes to mature without the risk of *botrytis cinerea* infection. It also provides a high level of homogeneity in wine quality over time, because differences in the amount of water absorbed by the grape are consistent from one year to the next. Daily temperature variations of up to 20 degrees Celsius result in a high concentration of aromatic components in Chilean wine. Because of its isolated geography, Chile is the only country free of *phylloxera*, a root louse that destroyed Europe's vineyards in the nineteenth century. Therefore, Chile possesses some of the oldest vineyard plantings in the world.

Despite these unique characteristics, it was only during the last decade that Chile became an important player in the international grape and wine industry. What accounts for the change? We argue that Chile's current position is the result of decades of failures and successes shaped by the country's institutional framework. After a brief historical review, we examine technology and the role of the government in shaping the evolution of the industry.

First Years

The Spanish conquistadors brought new foods to the Americas: new grains, new meats, and wine. Wine quickly became part of the diet of the colonized territories. Domestic consumption was initially met by imports that over time were replaced by local production. Imported

Figure 7.1 Chile's Share of the World Wine Market,
1988–2000

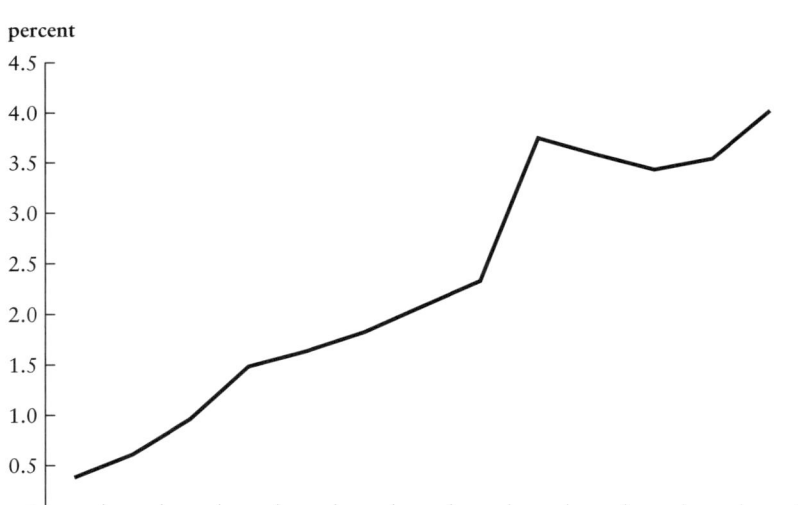

Source: Costa Barros 2001.

wine was costly and of very poor quality due to the absence of containers capable of preserving the product adequately.

The first grape variety planted in Chile was Pais, a relative of the
Mission grape, the first *vinifera* variety planted in North America
(Robinson 1986). Pais was grown by small colonial producers whose
plantations stretched from the region of Coquimbo in the near north
down to the region of Bio-Bio. Until the middle of the eighteenth century, the wine produced in Concepción was considered the best in the
country. However, the wine sector was characterized by informal domestic production. As the colony prospered, imports of French wine
increased.

Vineyards of larger scale began to appear in the nineteenth century.
These were established by businessmen, most of whom did not specialize in agricultural activities. Vineyard development was based on
four factors: investment in human capital, investment in irrigation, the
growth of agricultural loans, and the introduction of French varieties,
such as Cabernet, Merlot, Pinot Noir, and Sauvignon Blanc.

By 1880, Chile's main grape and wine producers were well-known
families that owned some 23 vineyards producing wine mainly for

domestic consumption. Production techniques were rudimentary and adapted from French production practices. Foreigners appeared in the industry in the early twentieth century, mainly as *négociants* and *éléveurs,* middlemen who bought wine from producers and aged it in their cellars for later sale in the domestic market.

Chilean agriculture began to take off during the middle of the twentieth century, with unprecedented growth rates. Grapes soon became the country's second most important crop. The export of wine was promoted, but with little success because of poor quality, distance, and tariff barriers. Chilean wine remained a low-price product destined primarily for local consumption.

In 1938 the Alcohol Law limited the number of hectares that could be planted in vines. The law led to price increases, making the business profitable, but it limited the potential for expansion. No major changes were observed in production and consumption during these years.

In 1969, most vineyards changed their legal structure, becoming publicly held corporations. The expropriations carried out under the agrarian reform of the 1960s did not affect the vineyards, which were considered to be agro-industrial companies. Even under the socialist government of Salvador Allende, wine growers were considered "integrated wine companies," rather than strictly agricultural entities, because they had facilities for production, storage, and sales.

The influence of the Alcohol Law of 1938 and the land reform came to an end with the military coup of 1973, giving way to an era of economic and regulatory liberalization in the agricultural sector. The new arrangements principally benefited fruit production and the forestry sector.

Wine production increased slightly at the end of the 1970s and then dropped sharply in the early 1980s with the currency crisis of 1982. Exports continued to be modest, increasing slightly at the end of 1990.

The 1990s

Wine exports began to increase sharply after 1990 due to several factors, some of which pertained to trade liberalization. Those factors brought dramatic changes to the old-fashioned, family-dominated wine industry, turning it decisively toward the international market.

In 1979 the government dropped prohibitions against making wine from table grapes. This made possible the production of new, more sophisticated types of wine, which were well received in international

Table 7.1 Production and Export of Chilean Wine
(Thousands of liters)

Year	Production	Exports	Percentage exported
1989	390,000	28,610	7
1990	350,000	43,050	12
1991	282,239	64,673	23
1992	316,534	74,029	23
1993	330,245	86,630	26
1994	359,838	107,904	30
1995	327,548	125,351	38
1996	387,765	184,084	47
1997	454,919	216,208	48
1998	547,477	229,802	42
1999	428,014	229,843	54
2000	641,937	264,749	41
2001	545,178	308,941	57
2002	562,000	355,300	63
Change (%)	44	1,142	

Source: Viñas de Chile.

markets. At the same time, especially toward the end of the 1980s, domestic consumers began to substitute other products for wine. Morel-Astorga (2001) suggests that this process was directly linked to the relative prices of wine and its substitutes, mainly beer. By 1990, Chilean wine consumption had fallen to half the level of the 1960s, while beer consumption nearly tripled.

Meanwhile, per capita wine consumption in the United States, United Kingdom, and Canada increased sharply. Together, these developments represented an export opportunity. Between 1989 and 2001, the exported share of Chile's wine production exploded from 7 to 63 percent (table 7.1).

The 1990s also saw growing cooperation between Chilean wine companies and international investors. Knowing the positive characteristics of Chilean soils and vines, international producers such as Spain's Miguel Torres, France's Baron de Rothschild and Chateau Lafite, and America's Robert Mondavi established themselves in Chile, investing in the country's natural advantages.[2] These investments came in response to general investment-promotion policies implemented in Chile starting in the mid-1980s.

The new companies invested heavily in modern technology and revitalized and replanted vineyards. Small producers responded to the change by forming collective corporations to develop and promote the production, elaboration, and sale of fine wines using tools provided by the Corporacion de Fomento (CORFO), an arm of the government responsible for promoting industrial development. We analyze this in more detail later in the chapter.

During the late 1990s new vineyards financed by foreign capital appeared in Chile's central valley between the Valle de Aconcagua and the region of Curicó due to its unique climate and land conditions. Plantings of Chardonnay increased in response to growing world demand for the variety. At the same time, producers from the northern regions of Coquimbo and Atacama began to replace wine grapes with table grapes and varieties suited for the production of pisco, a form of brandy. The zone that underwent the greatest change, however, was the southern Maule Valley, where many small producers maintained vineyards in which they planted the Mission grape for production for the now-depressed national market, contrary to the dominant export strategy.

Decree 464 of 1994 established standards for denominations of origin, indication of varieties, and year of harvest, as well as other matters related to the sale and labeling of the wines (table 7.2).

The second half of the 1990s was characterized by a sustained growth in plantings of French grape varieties—chief among them Cabernet Sauvignon, Merlot, Carmenère, Pinot Noir, and Syrah (figure 7.2). Today in Chile, 170,728 hectares are planted in vines. The largest areas are in Region VII (47,552 hectares), Region VI (40,644 hectares), the Metropolitan Region (22,223 hectares), and Region IV (21,408 hectares).

The main importers of Chilean wine changed during the decade. For reasons of (low) quality, exports were concentrated in Latin America in the 1980s; in the 1990s Europe became Chile's largest market, buying more than 50 percent of Chilean wine exports. Britain is the largest consumer (figure 7.3). The number of firms exporting wine from Chile increased from 37 in 1997 to 110 in 2003.

How did an industry that had been domestically oriented suddenly begin looking abroad? It is true that an improved macroeconomic environment and favorable tax policies encouraged investment in the sector. Although a necessary condition, however, these are not sufficient to account for the change, as they did not produce similar change in others sectors. To account for the growth of Chile's wine sector, we must examine its institutional framework.

Table 7.2 Chile's Wine Zones

Region	Subregion	Zone	Area
Atacama (II)	Valle de Copiapó Valle del Huasco		
Coquimbo (IV)	Valle del Elqui Valle del Limarí Valle del Choapa		Vicuña, Paiguano, Ovalle, Monte Patria, Punitaqui, Río Hurtado, Salamanca, Illapel
Aconcagua (V)	Valle del Aconcagua Valle de Casablanca		Panquehue
Central Valley (RM, VI, VII)	Valle del Maipo Valle de Rapel Valle de Curicó Valle del Maule	Valle de Cachapoal Valle de Colchagua Valle del Teno Valle de Lontué Valle del Claro Valle Loncomilla Valle Tutuven	Santiago, Pirque, Puente Alto, Buin, Isla de Maipo, Talagante, Melipilla, Rancagua, Requinoa, Rengo, Peumo, San Fernando, Chimbarongo, Nancagua, Santa Cruz, Palmilla, Peralillo, Rauco, Romeral, Molina, Sagrada Familia, Talca Pencahue, San Clemente, San Javier, Villa Alegre, Parral, Cauquenes
South (VIII)	Valle del Itata Valle del Biobío		Chillán, Quillón, Portezuelo, Coelemu, Yumbel, Mulchén

Source: Viñas de Chile.
Note: In Chile, "region," "area," and "zone" refer to geographical conditions, whereas "subregion" and "zone" refer to administrative designations, whereas "subregion" and "zone" refer to geographical conditions.
RM = metropolitan region (the region around Santiago).

Figure 7.2 Plantings by Variety, 1985–2001

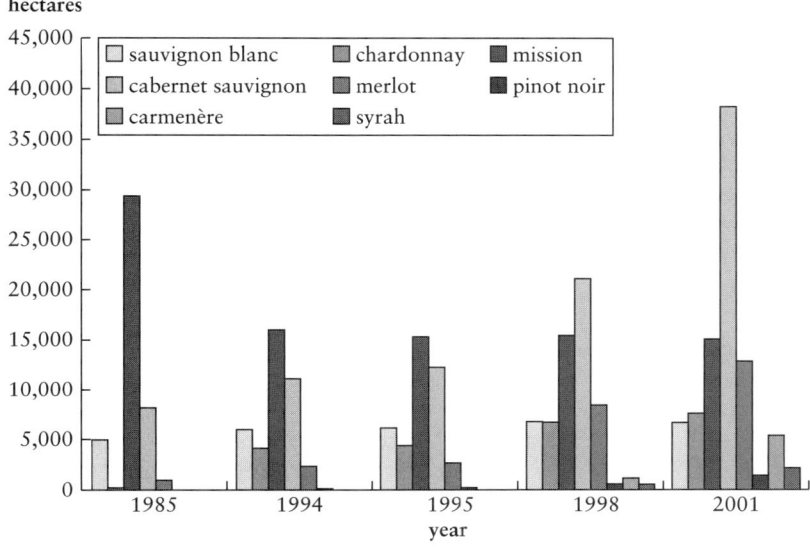

Source: Viñas de Chile.

Figure 7.3 Chilean Wine Exports by Destination, 1980–2002

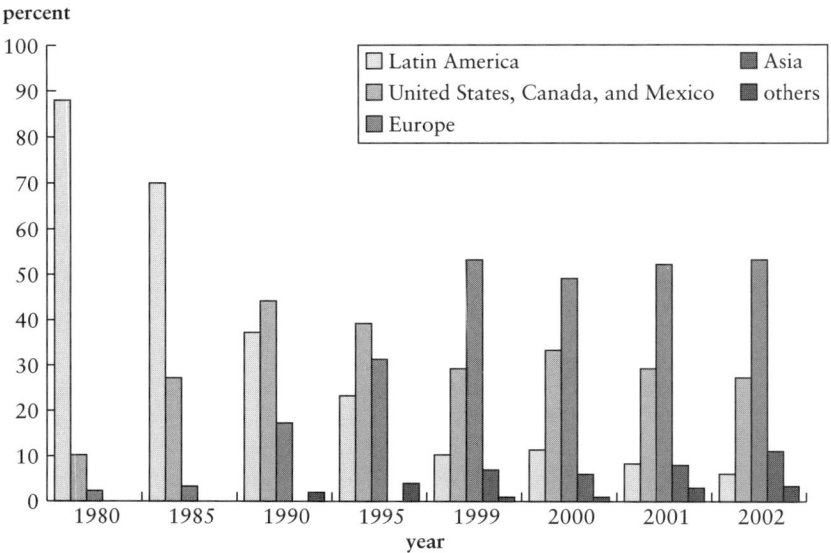

Source: Costa Barros 2001.

Technology in the Chilean Wine Industry

Industry participants interviewed for this chapter suggested that the adoption of technology in the Chilean wine sector came in response to market developments. After the arrival of the Spanish conquistadors, few changes were made until French experts arrived in the nineteenth century, driven from Europe by the *phylloxera* epidemic. French oenologists brought with them their varieties and technologies—notably production techniques to heighten quality, special grape-growing procedures, proper distances between plants, controlling yields to improve concentration, and so on.

The French arrivals found a ready welcome among wealthy and status-conscious Chileans, most of whom had made their wealth in other sectors. Like the Cousiño family, which owed its wealth to mining, rich Chilean families invested in vineyards, not to make more money, but to acquire cachet. The fact that Chile's new wine industry was based on the quality-conscious French model, particularly the model of Bordeaux, distinguished Chile from other Latin American countries.

From the late nineteenth century until the production quotas of 1938, little changed. The 1938 law turned the focus of wine growing to volume defined by quotas instead of quality. With profits in direct relation to the number of liters produced, cheap production became the goal. A black market flourished, supplied by wine from vineyards on properties devoted mainly to other uses.

After 1973, new economic policies were applied, the market was opened, and limits on planting were lifted. These changes paved the way for the development of exports and allowed imports of equipment and inputs needed for the technical development of the industry. The crisis of overproduction in the early 1980s convinced wineries that the only way to survive was to find export markets, which in turn required innovations in technology. The first major innovation during this period, introduced by the Canepa firm, was the use of stainless steel in place of wooden vats in the fermentation process.

A more significant and immediately influential change, which came just before the export boom of the late 1980s, can be attributed in part to the arrival of Miguel Torres. This major Spanish producer introduced small (220-liter) oak barrels, already long used in quality production nearly everywhere else, to replace the 4,000-liter *fudres* in which all Chilean reds had been fermented. The large vats encouraged oxidation, giving Chilean wines a prematurely old taste.

The use of smaller barrels, despite being more expensive, improved quality markedly. The rest of Chile's wineries soon followed Torres's example. With this radical change, Chilean red wine, which always had been made from excellent fruit, suddenly had potential for the world market. From that point forward, a massive program of improvement, especially in aging and storage, brought Chilean wine production to the frontier of world standards.

Until the 1990s, most of the industry's innovations arose from transfers rather than from domestic research and development (R&D). Since then, however, the picture has changed a bit. According to the industry participants interviewed for this chapter, there are at least two main channels by which foreign technology is absorbed at the country and firm level.

The first channel is "learning by looking." Chile's viticulturists and oenologists travel to other countries (Australia, Europe, and the United States) at harvest time. These "technology-capture tours" are typically financed by the government. The second channel, less important than the first, is international consultants. Foreign oenologists are invited to Chile at harvest time to work with their Chilean counterparts and share their experiences. This oenological exchange system, practiced worldwide, has the advantage of exposing practitioners to different practices with different varieties, allowing all participants to learn and apply what is most relevant to their own situation.

There are other ways that domestic producers update their knowledge of recent developments in the grape and wine sector—among them visits to international wine fairs, publications, congresses, courses, and seminars. Notably absent has been cooperation between Chilean universities and firms, a failing by no means limited to the wine sector. Linkages between Chilean firms and universities are quite weak compared to those in Australia and the United States, where governments invest the equivalent of 1 percent of total sales in applied research executed by the university.

Some changes can be seen in this area. Viñas de Chile, an organization of 45 wineries representing some 90 percent of Chile's domestic and export sales, is cooperating with the Universidad Católica de Chile on a three-year project with a budget of $1.5 million. The project focuses the resources of the university's engineering, agronomy, and biology departments on recognized wine-industry problems. In this pioneering R&D program, Viñas de Chile has agreed to provide annual funds to enable researchers to apply for matching public resources. The initiative promises economies of scale and scope from which all members of the association stand to benefit, while avoiding duplication of effort.

To improve fruit quality, wine producers have transferred newly acquired knowledge to grape growers. Business contracts typically specify handling procedures, irrigation systems, and performance indicators (such as maximum yield per hectare). Wine producers often provide free in-the-field technical assistance.

Small-scale producers remain vulnerable to the price consequences of overproduction. They could be protected by the establishment of cooperatives to produce, store, and commercialize wine from members' grapes. Such cooperatives are common in France and Italy, for example. Those in Chile have not thrived, however, in part because of low quality.

Institutional Framework

Legislation passed in 1938 limited the production of wine in Chile to 60 liters per person per year. Surpluses had to be exported or converted to industrial alcohol. This law, along with protectionism in Europe and other trade restrictions accompanying the Great Depression, discouraged investment in the industry (Morel-Astorga 2001).

For most of the twentieth century, the policy of the Chilean government toward the country's wine industry had two broad directions. First, wine was heavily taxed, as were other inelastic goods. Second, several government agencies, notably CORFO, promoted agriculture in general, and grape production in particular. In fact, CORFO created an agricultural directorate that administered two programs, one of which, the Fruit Development Program (Plan de Desarrollo Fruticola), had significant implications for wineries, because grapes benefited from public support for applied research.

Between 1958 and 1973, government policies were dominated by the implementation of land reforms. The first reforms were moderate but they deepened in the 1960s. For the wine industry, the reforms were a threat to property rights, but because most vineyards were already producing domestic goods, few were confiscated.

Government Policy After 1973

The military regime that took power in the 1973 coup implemented great changes in economic and institutional conditions. Limits on vineyard plantings were eliminated, and restrictions on imports of equipment (previously permitted only from within the Andean Zone) were relaxed. Property rights were now well defined and secure, just as wine consumption began a steep decline—from 58 liters per person per year in 1972 to 11 liters in 2000.

The economic crisis of 1982–83, brought on by external shocks and a badly regulated domestic banking sector, posed severe problems for agriculture and the wine sector. Economic hardship accentuated the steady decline in wine consumption (Morel-Astorga 2001) in favor of lower-priced substitutes, notably beer and pisco.

At the same time, liberalization of the wine market led to overproduction, which peaked in 1982. As a marketing strategy to cope with the glut, the wine industry consciously lowered the social status of its product, while producers of substitutes did the opposite (Morel-Astorga 2001).

After the crisis the government initiated a major technology-transfer program directed at the agricultural sector. The program, known as Grupos de Transferencia Tecnologica (GTT, technology transfer groups), was designed to create formal links between organizations, such as the Agricultural and Farming Research Institute (INIA, Instituto Nacional de Investigacion Agricola) and agro-producers. The goal was to facilitate the dissemination of foreign knowledge and promote the use of new technologies by local farmers. In the wine sector wineries encouraged the transfer of knowledge to small vineyard owners by raising their quality requirements (Poblete 1999).

After 1990, the new democratic government focused on the creation of Centers of Business Development (CDE). The centers, financed and supported by CORFO, were designed to create critical masses of firms focused on export development. ChileVid, a major consortium of small and medium-size wine producers, was initially financed by this program. A similar group of firms, the Chilean Wine Corporation (CCV), was created with support from CORFO.

The government worked in other ways to promote exports. The Export Promotion Office, or ProChile, was founded in 1974 to support and coordinate export activities by small and medium-size firms from all productive sectors. In the case of the wine industry, ProChile's support proved very important. It financed nearly half of the costs of activities to promote Chilean wine abroad, including fairs, travel costs, and marketing materials (Morel-Astorga 2001). A peculiar characteristic of this agency is its continuity; although created during the military regime, its importance has increased during the democratic years.

Wine exports did not take off on their own. By setting a friendly regulatory framework and supporting programs that reduced information and transaction costs for entrepreneurs, the government played a major role. However, it is also true that Chile's wine sector was not the beneficiary of specific government policies. Under the military regime, industrial and agricultural policy flowed from, and was subordinate to, macroeconomic fundamentals. The loosening of exchange rates and interest rates created conditions in which the wine sector could grow, but it does not fully account for the observed growth. The creation of institutions specifically designed to reduce the learning costs associated with export development must be taken into account. Those institutions focused not on predefined sectors but on all sectors where the country enjoyed a comparative advantage. Foreign investors, too, were major players in the expansion of the wine sector, notably through their influence on improving technology in viticulture and wine production.

Was Government Support Helpful?

Wine producers have come to Chile from Europe and the United States primarily because of the country's excellent natural conditions: weather, soil, absence of diseases, and low costs. But the government has an indirect role in attracting producers, who, like entrepreneurs everywhere, appreciate a healthy and nonbureaucratic economy. Compared to other countries of the region, Chile meets those requirements, allowing investors to build their businesses with confidence.

During the last two decades, the government's chief contributions to the sector have been the creation of a stable economic system, healthy macroeconomic indicators, well-established property rights, and clear rules. The national export promotion office, ProChile, played an important role in supporting producers in their first experiences in exporting grapes and later wine. Modest financial support through CORFO, too, was valuable in supporting cooperative efforts among small producers, producing information and maps, and guiding small-scale producers to foreign markets.

The Servicio Agricola y Ganadero (SAG, a branch of the Ministry of Agriculture devoted to agricultural and phytosanitary controls) has a fundamental role in the industry as a whole. It enforces national laws pertaining to wineries and vineyards and compiles statistics on planting and production, allowing growers and producers to make decisions in light of national trends and so avoid oversupply of one or another variety of grapes or wine. The service's staff is small, however, which hinders the dissemination of fundamental information related to crops, production, and exports.

In general, the opinion of industry participants is that the government fulfills its role as legislator and regulator, allowing the industry to function in a flexible and unrestrictive environment, but that the government should improve its role as a generator and diffuser of information.

One of the criticisms leveled at government is related to the law on taxation of alcohol. Wines face a special tax surcharge of 15 percent (in addition to the 19 percent VAT). Firms believe that the reason for this tax (to limit consumption for health reasons), is no longer valid for wine because of the demonstrated importance of wine in health. An interesting proposal for the use of the tax is to finance investments in the grape and wine sector, especially in R&D.

Through a CORFO subsidiary, the government encourages grape growers and wine producers to take individual and group actions to maximize the Chilean wine industry's long-term competitiveness. The Chilean Wine Corporation (CCV) promotes collaborative-development grant projects known as PROFOs (*proyectos de fomento*), provides technical-assistance funds, and offers support programs for suppliers and management (CORFO y CCV 2001).

PROFO grants are aimed at small and medium-size companies that wish to improve their competitiveness. PROFO grants fund projects that, because of their nature or scale, are better tackled by a group than by a single firm. Technical-assistance funds finance specialized consulting and support in areas such as finance, marketing, quality, and irrigation. The supplier-development program aims to improve competitiveness in the production chain and to support the creation of long-term relations, thus encouraging mutually beneficial specialization and complementarities. The management-support program provides cofunding to allow large wine companies to engage specialized consultants to increase productivity and quality.

CCV has 21 projects under development, 16 of which are PROFO projects. Most are located in the Maule region, where small producers suffered from technological lags.

The CORFO financing system has encouraged interest among research institutions in providing research services and transferring knowledge to the productive sector. In the Valle de Cochagua, evidence suggests that the process of accumulating knowledge was not entirely sustained by private firms (Giuliani 2003).

Very little R&D has been supported by the government to date, and that has not been well directed, according to the industry participants interviewed for this chapter. It would appear that the industry and the government have yet to agree on a set of priorities and a long-term vision for the sector. Such an agreement would include an appraisal of factors critical for the future of the sector. With such a framework in place, both public and private R&D could be better targeted.

Some recent progress has been made. In 2004, the government supported a study of the grape and wine sector in which academics, producers, and researchers were asked about the future of the sector, potential threats, major changes in technology, and likely future demand. The finished study (Programa Chile Innova 2004) gives interested individuals and firms a clearer idea of what should be done in terms of production routines, research, and marketing. Although most producers do not see the study as a guide for action, it appears that the government will use it to plan future public support for the sector.

Although the study is a promising first step toward public-private cooperation in the sector, future linkages are far from assured because the wine industry fears that cooperation might increase bureaucratic control of the industry. This reaction to the prospect of public involvement suggests that many in the private sector would prefer that the government concentrate on generating information and supporting research activities without participating actively in planning and production.

Some interviewees have suggested that a joint committee system organized by the Ministry of Agriculture could create a useful forum in which representatives from the private sector, universities, and government could come together to discuss important industry issues such as support for sectoral R&D from existing funding sources (CORFO, the Fondo de Desarrollo Científico-Tecnologico, and the Fondo de Innovacion Agraria). Such committees could help allocate funds in the most productive directions, make long-term plans, and facilitate communication among industry players. One may expect that the private sector will try to lead the committees and thus avoid government encroachments on what it sees as its prerogatives.

With regard to the legal framework of the sector, those interviewed favored more flexibility, believing that the current legal status of the grape and wine sector is too complicated. Producers prefer fewer but clearer rules and more predictable enforcement. It was agreed that rules in Chile are not strictly followed, as they are in Australia. Of particular concern is the issue of denomination of origin. Chile has a geographic denomination system, but it is not enforced. A bill before the Chilean congress at the time of this writing is designed to address this issue.

What More Should the Government Do?

The industry appears to believe that the government should play a facilitative role. Because the government has no *a priori* technical or commercial capacity to establish what should be investigated and how, it should capture and distribute resources only in accordance with clear, fair rules and guidelines. Beyond that, the government must, through regulation, facilitate market functioning and promote exports—without intervening directly in the industry. An important area of regulation that could bear strengthening is enforcement of the law on denomination of origin, a measure used to enhance quality and image in nearly all other important wine-producing states.

The industry does believe that the government is well positioned to consolidate Chile's image overseas. ProChile should provide direct support to promote Chilean wine as a brand in international markets.

Another government role, through the Phytosanitary and Agriculture Service, is to provide the industry with relevant information about national and international trends in cultivation, production, and supplies. Such information helps industry participants anticipate national and foreign levels of supply and demand and so make informed decisions.

Marketing

International competition in wines is intense—and growing more so. Chilean producers have responded with two marketing strategies, one collaborative and one competitive.

All matters related to Chile's image as a wine producer and foreign consumers' knowledge of Chilean wine fall under the umbrella of a national export strategy. Industry participants agree that Chile has serious problems of international marketing, image, and positioning. One problem is that Chile is not well known worldwide, not only for wine, but in general. Despite the extraordinary price-quality ratio of its wine, Chile is seen as a small and distant country, isolated between mountains and sea; selling top-quality wines at low prices only confirms failures in the national image.

Overcoming that image and penetrating important markets is the industry's goal. To meet international competition, Viñas de Chile is investing in positioning Chilean wine in foreign markets. But penetration is only half the battle. What is most important is to gain market share.

Those interviewed agreed that aiming at better market niches should be part of the marketing strategy, following the Australian model of focusing on geographical markets rather than competing on price. The industry is trying to reduce sales of bulk wine, increase sales of bottled wine, and increase the percentage of reserve and ultrapremium wines. Chile also needs to increase the range of wine varieties it offers for export sale and strive to give a personality to each region of the country.

Perspectives and Prospects

During the 1950s, a major technological transformation began in the international wine industry, centered in the United States and France. New developments appeared in harvesting machinery, stainless steel fermenting tanks with automatic cooling systems, and other techniques and equipment for the mass production of wine. There was commensurate progress in genetics and vineyard maintenance.

None of these new advances were seen in Chile until 1979, when Spain's Miguel Torres arrived in the country. Before the arrival of the Torres firm, Chile's wine sector was almost totally oriented toward the domestic market. Quality was low and production inefficient. A few family-owned firms controlled the market. Public support for the industry was nonexistent. The threat of land reform loomed throughout the 1960s. Production was limited by law. To say that the government did very little to develop the sector would be an understatement.

After 1979, the industry changed dramatically. Foreign ideas (notably stainless steel tanks for fermentation and small oak barrels for aging) revolutionized the way wine was produced and helped make wine fashionable, rather than just a staple good. Newly prosperous consumers enjoying Chile's improving economy came to value the remarkable improvements in the quality of their country's wines.

Technological innovation and the accumulation of knowledge have been fundamental in the development of Chile's wine sector, taking advantage of the country's favorable geography and climate. Torres's success in selling its Chilean product in international markets led other Chilean firms to turn their attention abroad—for knowledge as well as markets. In the space of a few years all domestic producers had updated their production practices. Technological improvements were diffused through study tours and "learning by looking." Large investments have been made in recent years by large, medium, and small producers in winemaking plants, machinery, equipment, and leading-edge technology. Those investments, together with the skill and diligence of technicians and professionals, have led to the creation of products that are winning market share in the United States, Europe, and Asia.

Those changes coincided with an explicit effort by the government to support agro-industrial exports. A stable macroeconomic environment, along with exchange rates and wages well below their long-term equilibrium levels, also favored the sector's international initiatives. But at the end of the 1980s the industry was still characterized by a few family-owned firms producing wine from their own vineyards or from grapes purchased from small growers.

The government played a major role in promoting collaborative association among these small producers to exploit economies of scale and scope.

Today the government's role in the grape and wine sector is chiefly to improve coordination (in viticulture and production, for example) and deal with market failures (such as lapses in phytosanitary practices, violations of laws on origin, and coordination between state agencies and associations of small producers). Particularly notable is the role played by the phytosanitary office of the Ministry of Agriculture in disseminating information about diseases and pests and in compiling statistics on vineyard plantings.

The government's chief contributions to the Chilean grape and wine sector have been to coordinate market players, generate and disseminate information, promote research, support export efforts, and engage in forecasting—all roles that no single private actor would or could provide. Prospective studies and the implementation of rules of origin are some results of the coordination role.

Having long been satisfied with the government's relatively passive role, private producers faced with new competitive challenges in an oversupplied market are urging a more active role for the government, notably in providing greater funding for the exploration of new production techniques and help in recruiting foreign oenologists and other qualified workers until the small program in viticulture and oenology recently begun at the Catholic University of Chile can be geared up.

Chile's success in wine is explained by a mixture of extremely favorable natural conditions, a long history of producing wine, a readiness to adopt up-to-date technology, and a government that in recent years has been active without being intrusive. That role will soon grow, with calls for the government to improve enforcement of the system of rules of origin and to participate more closely in R&D and innovation activities and the promotion of exports. The government's ability to assume new roles could spell the difference between the consolidation of Chile's position in the international market or its relegation to a minor role.

There are threats on the horizon, however. According to Viños de Chile, the trade association, the potential for increasing the price of Chilean wine is limited by the persistent oversupply of wine worldwide. The international wine market is currently saturated, due to the aggressive increase in supply by various producers, while new exporters (from Eastern Europe, for example) have entered the low-price market in recent years. Chile has been exporting much longer than the new entrants, experience that creates a special advantage with regard to distribution channels, but Chile must improve its image. In most countries, it is known as a producer of cheap wine. Exporters of high-end Chilean wine spend little on marketing—less than $2 million annually, about half as much, proportionally, as other countries. That will have to change. The increasingly competitive international market suggests that the Chilean wine industry will have to intensify its efforts in the next few years in order to consolidate its position in its major markets and secure promising new niches in higher-value segments of the industry.

Endnotes

1. The top 10 producers, in order, are France, Italy, Spain, the United States, Argentina, Germany, Australia, South Africa, Portugal, and Chile.
2. Torres's investment could be seen as a greenfield investment, whereas those of Rothschild and Mondavi were takeovers of local incumbents.

References

"CORFO y CCV: Ayudando a crecer vino chileno." 2001. *Revista Vendimia* 3(19). May/June.

Costa Barros, V. 2001. "La Vinicultura Mundial y La Situación Chilena en 2001." Servicio Agricola y Ganadero, Santiago, Chile. December.

Del Pozo, José. 1999. "Historia del Vino Chileno." Santiago de Chile: Editorial Universitaria.

Giuliani, Elisa. 2003. "How Clusters Learn: Evidence from a Wine Cluster in Chile." Paper prepared for the EADI Workshop on Clusters and Global Value Chains in the North and the Third World, Universitá del Piemonte Orientale, October 30–31, Novara, Italy.

Morel-Astorga, Paulina. 2001. "The Chilean Wine Industry—Its Technological Transformation and New Export Orientation." *Iberoamericana, Nordic Journal of Latin American and Caribbean Studies* 31(2): 85–101.

Programa Chile Innova. 2004. "Prospectiva Chile 2010."

Robinson, Jancis. 1986. *Vines, Grapes, and Wines*. New York: Knopf.

CHAPTER 8

Bridging the Knowledge Gap in Competitive Agriculture: Grapes in India

Gopal Naik

A COMPETITIVE AGRICULTURE SECTOR IS CHARACTERIZED by suitable crops supported by appropriate technologies, markets, institutions, and policies. This is especially true with perishable foods like fruits and vegetables, which are vulnerable to substandard postharvest management regimes—regimes that in developing countries are themselves vulnerable to poor infrastructure. These countries therefore face several challenges. As trade barriers for agricultural commodities fall, most developing countries should be building on their favorable climates and low-cost labor forces to create a competitive fruits and vegetables agrosector. But these advantages can be offset by rising quality standards in developed-country markets, which require improvements in technology and practices up and down the value chain. In countries like India, where the value chain is very diverse and fragmented, these improvements are hard won. Unless developing countries address these difficulties, they may not be able to build competitive agricultural sectors.

In this chapter, we examine the case of grapes in India, a crop grown in a few pockets of the country at yields that are among the highest in the world. Indian farmers have grown table grapes for local consumption for centuries. But since the 1960s, as grape production grew, local prices fell, creating the need for growers to explore farther-flung markets. Efforts here were constrained, however, by postharvest regimes, infrastructural constraints, and simple lack of know-how. In the state of Maharashtra, producers responded to production constraints by forming an association in the 1960s. Together with growers' associations and cooperatives, the state government helped to expand the market for grapes and to develop institutions for research and commercial handling.

In this case study we examine the history of modern grape production in India in an effort to show how technology, institutions, markets, policies, and, most important, the knowledge base were transformed to make the grape industry a competitive sector in India.

The Fruit Sector

India's diverse climate allows the production of a variety of fruits—from tropical guava and mango to the temperate pear and strawberry—and India is one of the world's largest fruit producers, with annual production standing at around 40 million tons, or about 10 percent of the total world production.[1] India processes only 2 percent of its fruit. Compare that figure with those of Brazil at 70 percent, the United States at 60–70 percent, Malaysia at 83 percent, and Israel at 50 percent. In India nearly 30 percent of fruit production spoils before sale, much higher than the 10 percent spoilage in other Asian countries, such as Japan, Republic of Korea, and Taiwan (China).[2]

After steady growth between 1950 and 1989, fruit production in India increased dramatically—between 1990 and 1999, for example, it grew at an average annual rate of 5.5 percent. Harvested area and yields rose at 3 percent and 2 percent, respectively (FAO 2001). Fruit production in India contributes 10 percent on average to the gross value of total agricultural output, and around 13 percent of the export earnings from major agricultural products. India's major fruit crops include mango, apple, banana, grapes, guava, pineapple, and citrus. The major fruit-producing states are Maharashtra, Tamil Nadu, Karnataka, Andhra Pradesh, Bihar, Uttar Pradesh, and Gujarat.

Reliable data on horticultural crops were unavailable until recently (Economic Survey 2002/3). Trends in the 1990s suggest that production increased by about 2 million tons annually and that from 1999 to 2002 the shares of different fruits in the total production show that banana (34 percent), mango (23 percent), and citrus (10 percent) together contributed 67 percent. Grapes account for 3 percent of fruit production. The average yield of grapes from 1999/2000 to 2001/2 was 24.5 tons/hectare, third-highest among fruits. Although India has a very high share in the world production of mango (46 percent), papaya (24 percent), and banana (20 percent), its share in their exports is significant only for mango.

India exports less than 1 percent of its domestic production. The major export markets for Indian fruit include the Near East, Far East, and Western Europe. Considering that fruit exports typically earn up to

30 times more foreign exchange per unit of cultivated area than cereals (FAO 2001), increasing fruit exports can substantially improve a country's farm income and trade balance.[3] India's export earnings from fresh fruits doubled from 1991 to reach $74.5 million in 1998.

In India, 64 percent of rural households reported fresh fruit consumption in 1999, compared with 84 percent of urban households.[4] The income elasticity of fruit demand for the highest income group (using an overall weighted average) was estimated at 0.283, compared with 0.826 for the lowest income groups for rural areas and 0.293 and 0.782, respectively, for urban areas (FAO 2001).[5] Therefore, as income and population increase, fruit consumption will increase significantly.[6] The government is promoting fruit consumption with national and regional educational programs that create more demand for higher-quality fruit and for more varied diets (including greater fruit consumption). Large agro-processing industries have initiated promotional programs utilizing mass media and in-store promotions for processed fruit and juice products in recent years (FAO 2001).

Prior to 1997, India banned or restricted imports of fresh and dried fruits, except dates and figs. As it stands, imports of all fresh fruit (except citrus, grapes, and litchi) are included in the free category, and imports are allowed under an open general license. For imports of citrus, grapes, and litchi a special import license is required. There are no export duties on fruit from India.

Domestic Indian fruit markets are not very efficient, displaying a substantial price differential between wholesale and retail markets, primarily because of the high margins retained by intermediate buyers and sellers, product perishability, and the distances between wholesale and retail markets (FAO 2001).

Grapes showed the greatest increases in area and production between 1990/91 and 2000/2001 (table 8.1). The other fruits with substantial increases in area and production were mango, banana, guava, and citrus. Production changes are much higher than the area changes for mango, apple, grapes, and citrus, indicating substantial improvement in the yield for these crops. For grapes, the production increase is double that of area changes.

In an effort to improve fruit and vegetable production, the government established a National Horticultural Board and identified horticultural crops as a way for producers to diversify and become more profitable through (i) the better use of natural resources and (ii) the creation of higher-skilled jobs for rural workers (Ministry of Finance 2003). The government is also encouraging private investment in high-tech horticulture, including micropropagation, protected cultivation, drip irrigation, integrated nutrient and pest management, and up-to-date

Table 8.1 Production of Major Fruits in India, 1990/91 and 2000/2001
(Area in thousands of hectares; production in thousands of tons per hectare)

Fruit	1990/91		2000/2001		Increase during 2000/2001 over 1990/91 (%)	
	Area	*Production*	*Area*	*Production*	*Area*	*Production*
Mango[a]	660.12	3,659.53	918.87	7443.56	39.2	103.4
Apple[b]	33.3	342.07	31.36	421.5	−5.83	23.22
Banana[c]	152.02	5,898.41	245.22	9,746.7	61.31	65.24
Grapes[d]	22.98	383.64	40.62	963.77	76.76	151.22
Guava[e]	27.63	186.53	42.94	296.53	55.41	58.97
Pineapple[f]	0.87	27.5	0.44	16.97	−49.43	−38.29
Citrus[g]	156.37	964.77	218.15	2,019.58	39.51	109.33
Total	1,053.29	11,462.45	1,497.6	20,908.61	42.18	82.41

Source: Agricultural Statistics at a Glance 2002 and 2004, Ministry of Agriculture, Government of India.

Note: Estimates are based on Pilot Study on Crop Estimation Survey on Fruits and Vegetables in 11 states: Andhra Pradesh, Gujarat, Haryana, Himachal Pradesh, Karnataka, Maharashtra, Orissa, Punjab, Rajasthan, Tamil Nadu, and Uttar Pradesh.

a. The estimates are for Andhra Pradesh, Gujarat, Haryana, Karnataka, Maharashtra, Orissa, Punjab, Tamil Nadu, and Uttar Pradesh.

b. The estimates are for Himachal Pradesh.

c. The estimates are for Andhra Pradesh, Gujarat, Karnataka, Maharashtra, Orissa, and Tamil Nadu.

d. The estimates are for Haryana, Maharashtra, Karnataka, and Tamil Nadu.

e. The estimates are for Gujarat, Haryana, Karnataka, Rajasthan, Tamil Nadu, and Uttar Pradesh.

f. The estimates are for Tamil Nadu.

g. The estimates are for Andhra Pradesh, Himachal Pradesh, Karnataka, Maharashtra, Punjab, Rajasthan, and Tamil Nadu.

postharvest technology for perishable commodities. The government budget allocation for horticulture rose from Rs 10 billion in the Eighth Five-Year Plan (1992–97) to Rs 14.5 billion in the Ninth Plan (1997–2002). That allocation was further increased to Rs 21 billion in the Tenth Plan (2002–7). In 2003/4 new initiatives began on technology missions for integrated horticulture development, greater cold storage facilities, high-tech horticulture, and precision farming. A National Horticulture Mission has been announced. Its objective is to double horticulture production by 2010.

The Grape Sector: An Overview

Consumption

In India grapes are consumed mainly as fresh fruit, especially in high-income urban households. The per capita consumption has risen to about 1 kg per year from very low levels in 1961 (table 8.2). Recent growth in consumption has been strong, and, with rising population and incomes, increasing nutritional awareness, and greater consumption of different products such as raisins and wine, consumption is likely to continue to rise.

The per capita consumption of wine in India is only 4.5 ml per year, compared with 375 ml in China, 8 liters in the United States, and 60 liters in France. By regulating wine and liquor similarly (an administrative decision), India constrains demand. Maharashtra, a major grape-growing state, has responded by declaring wine a food item and issuing licenses for wine bars at concessional fees.[7] Growing disposable incomes, changing lifestyles, availability, wine tastings, wine clubs, and greater media attention have all boosted wine consumption in the state, which is growing at 20 percent per year. Nearly 80 percent of premium wine is sold in four cities—Delhi, Mumbai, Goa, and Bangalore, but throughout the subcontinent, wine drinking is becoming *de rigueur* at social gatherings.

Production

Grapes, believed to have originated in Armenia, can be cultivated in various climates, from tropical to arid, and in various ecological settings, from mountainsides to seacoasts. The Mediterranean is thought to have an ideal climate for grape cultivation. The vine goes dormant during cold periods and produces fruit during hot and dry seasons (Singh 2000). Europe leads the world in grape production, with the

Table 8.2 Consumption of Grapes in India

Year	Per capita consumption (kg/yr)	Growth rate in previous decade (%)
1961	0.172	n.a.
1971	0.273	4.77
1981	0.305	1.12
1991	0.478	4.59
2001	1.026	7.93

Source: Computed based on FAOSTAT data on production, imports, and exports.
Note: n.a. = not applicable.

Table 8.3 Share of Major Producing Countries in World
Production, 1961–2003
(Percent)

Countries	1961	1971	1981	1991	2003
Argentina	5.06	5.31	4.75	3.71	3.89
Australia	1.24	1.01	1.21	1.52	2.91
Chile	1.98	1.68	1.79	2.12	2.87
China	0.17	0.23	0.35	1.84	6.46
India	0.16	0.25	0.34	0.74	1.97
Iran, Islamic Rep. of	0.98	1.20	2.23	2.90	4.15
Romania	1.75	2.05	2.85	1.51	1.77
South Africa	1.20	1.54	1.67	2.35	2.30
Turkey	7.42	7.09	6.00	6.42	6.00
United States	6.87	6.67	6.56	8.99	9.65
EU-15	52.18	51.11	47.18	43.84	39.48
World production (thousands of tons)	42,988	54,306	61,618	56,061	60,883

Source: Computed based on data from FAOSTAT.

EU-15 countries accounting for 46 percent of the grape-growing area
in 2003; their share has, however, been declining. Since 1961 India increased its area share in the world from 0.05 percent to 0.54 percent
in 2003. Worldwide, grape-growing areas have risen by 7 percent over
the past two decades. Australia, Chile, China, the Islamic Republic of
Iran, and the United States have increased their shares substantially.

World grape production increased from 43 million tons in 1961
to 61 million tons in 2003 (table 8.3). As noted above, the European
Union has a large share in grape production, but as this is less than
its area share, its yields are lower. In contrast, China increased its
production share from 0.17 percent in 1961 to 6.5 percent in 2003.
Australia, Chile, India, Islamic Republic of Iran, South Africa, and
the United States have increased their shares over the past four
decades. India's production share has increased from 0.16 percent in
1961 to 2 percent in 2003.

India has the distinction of having led the world in grape yields
since the 1960s (figure 8.1). It is followed by the United States and
Australia. Yields in the European Union are almost half those of India.
Yields are largely a function of the variety cultivated. Grape varieties
suitable for table use have higher yields than the varieties used to make
wine, and Indians consume mostly table grapes. In the United States,
fresh and dried grapes (raisins) account for nearly 60 percent of the
total production. These consumption patterns explain why yields in

Figure 8.1 Grape Yields in Major Producing Countries, 1962–2001

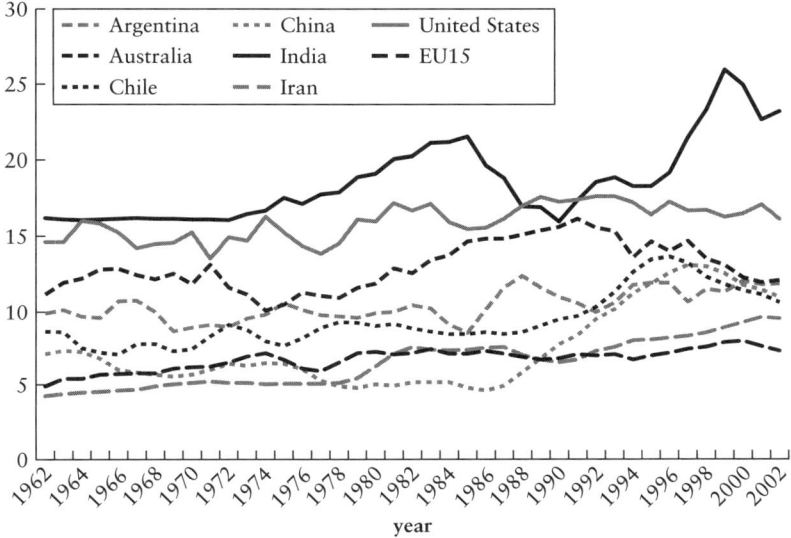

t/ha (moving average)

Source: FAOSTAT.
Note: t/ha = tons per hectare.

India and the United States are higher. China's yields, too, have increased substantially.

Exports

Grapes are one of the most traded fruits in the world, with the United States and Europe constituting the largest markets; the former accounts for approximately 1.1 million tons per year. The EU share in worldwide grape exports, however, declined from 48.6 percent in 1961 to 32 percent in 2003. Within Europe, Italy is the leading exporter, followed by Spain and Greece. The world's biggest exporters are the European Union, Chile, and the United States (table 8.4), with Chile having seen its share in total grape exports rise from 1 percent in 1961 to 25 percent in 2003.

The unit value of world exports of grapes increased from $0.16/kg of exports in 1961 to $0.92 in 2001 (table 8.5). Australia, the United States, and South Africa have fetched higher unit values relative to those of Turkey and Romania. The unit value for Indian grapes was high through 1981. It dropped a bit in the next decade, but rose significantly by 2002 to reach the world average.

Table 8.4 Grape Exports Among the Major Producing
Countries, 1961–2003
(% world total)

Countries	1961	1971	1981	1991	2003
Argentina	0.80	0.32	0.06	0.52	1.40
Australia	0.16	0.10	0.15	0.00	2.16
Chile	1.05	1.76	8.57	25.35	24.57
China	0.02	0.25	0.12	0.07	0.22
India	0.00	0.00	0.05	0.67	0.96
Romania	3.89	6.40	5.01	0.00	0.03
South Africa	3.60	3.08	2.67	3.38	7.78
Turkey	0.67	1.58	0.91	0.74	2.88
United States	12.38	13.22	10.50	14.90	13.92
EU-15	48.58	48.00	55.92	46.15	32.16
World (thousands of tons)	629	989	1,093	1,654	2,704

Source: Computed based on data from FAOSTAT.

Table 8.5 Prices Paid for Grape Exports by Major Exporting
Countries, 1961–2001
(US$/kg)

Countries	1961	1971	1981	1991	2001
Argentina	0.24	0.41	1.29	0.73	1.36
Australia	0.49	0.55	1.60	0.00	1.21
Chile	0.20	0.26	0.74	0.72	0.73
China	0.42	0.44	0.76	2.40	1.03
India	0.50	0.50	0.98	0.68	0.92
Romania	0.10	0.13	0.39	0.34	0.29
South Africa	0.34	0.52	1.45	1.12	0.74
Turkey	0.13	0.11	0.39	0.57	0.41
United States	0.21	0.28	0.85	1.20	1.38
EU-15	0.49	0.48	0.56	0.46	0.40
World	0.16	0.20	0.61	0.97	0.92

Source: Computed based on data from FAOSTAT.

Grape Sector Performance in India

Cultivation and Varieties

Grape cultivation is one of the most remunerative farming enterprises
in India (Shikhamany 2001). The fruit is grown in three distinct agroclimatic zones—the subtropics, the hot tropics, and the mild tropics.

The subtropical region comprises the northwestern plains, including Delhi and parts of Uttar Pradesh, Haryana, and Punjab (corresponding to 28° and 32° N latitude). Because vines go dormant in this region, there is only a short growing season from March to June (90–95 days), so the early ripening Perlette variety is cultivated here.

The hot tropics comprise parts of Maharashtra, Andhra Pradesh, and northern Karnataka. This area accounts for 70 percent of India's grape production (Shikhamany 2001). Because vines do not go dormant in the tropics, they need two prunings to produce a single harvest. Poor soil, water salinity, and drought are problems. Thompson Seedless and its clones (Tas-A-Ganesh, Sonaka), Anab-e-Shahi, Sharad Seedless, and Flame Seedless are the main varieties in this region.

The mild tropical region (10° and 15° N latitude) includes southern Karnataka, Andhra Pradesh, and parts of Tamil Nadu. Bangalore Blue (Isabella), Anab-e-Shahi, Gulabi, and Bokhri are the varieties cultivated here. They are harvested twice a year.

Thompson Seedless has a 64 percent share in terms of area and 55 percent of production in India. The other varieties with major production shares are Anab-e-Shahi (13.5 percent), Bangalore Blue (18 percent), and Perlette (6 percent).

Area, Production, and Yield

India had 49,400 hectares under grape cultivation in 2001/2, up from 15,400 hectares in the late 1980s (table 8.6)—an annual growth rate of 8.7 percent. Production increased from 251,000 tons in 1987/88 to 1,209,000 tons in 2001/2, a rate of 11.9 percent. Yield increased at a 3 percent rate from 16.3 tons/hectare in 1987/88 to 24.9 tons/hectare in 2001/2. The share of grapes in the total area under fruit cultivation increased from 0.5 percent in 1987/88 to 1.2 percent in 2001/2. The production share increased from 0.9 percent to 2.8 percent during the same period.

From 1985 to 1995, the number of hectares under grape production grew substantially, although a recent drought has depressed these numbers. Although India has the highest yields in the world, its potential yields for almost all the varieties grown are much higher than the actual yields (Shikhamany 2001).[8]

Although Indian grapes are harvested nearly year round (and at least one variety is available throughout the year), most are harvested in the hot tropics in March and April—mainly Anab-e-Shahi and Thompson Seedless and its clones. About 12 percent of the production, mostly of Thompson Seedless and its mutants, is dried for raisins, 2 percent is crushed for juice, and the remaining 1 percent is used to

Table 8.6 Area, Production, and Yield of Grapes in India, 1987–2002

Year	Area (thousand of hectares)	Share in total fruit area (percent)	Production (thousands of tons)	Share in total fruit production (percent)	Yield (tons per hectare)
1987/88	15.4	0.5	251.0	0.9	16.3
1991/92	32.4	1.1	668.2	2.3	20.6
1995/96	35.6	1.1	603.6	1.5	17.0
2001/2	49.4	1.2	1,209.7	2.8	24.9
Growth rate (percent)	8.68	6.45	11.89	8.44	3.07

Source: Indian Horticulture Database 2003, National Horticulture Board.

make wine (Shikhamany 2001). The most widely planted wine grapes are Bangalore Blue, Cabernet Sauvignon, Chenin Blanc, Chardonnay, Merlot, Pinot Noir, and Ugni Blanc.

Performance of the States

Maharashtra, Karnataka, Tamil Nadu, Punjab, Andhra Pradesh, and Haryana are the major grape-producing states; Madhya Pradesh, Mizoram, and Jammu and Kashmir are minor producers. Maharashtra predominates both in area and production. Its area share has increased from 29 percent in the late 1980s to 66 percent in 2002. In production, the increase in the state's share has been even more dramatic, rising from 19.6 percent to 75 percent in the same period. The annual growth rate in area was 15 percent, in production 23 percent, and in yield 7 percent during this same 15-year period. Maharashtra's yields are now the highest in India (figure 8.4). Although Karnataka, a neighboring state, had the largest share in area and production in 1987/88, it did not match Maharashtra's progress in later years (figures 8.2 through 8.4). The share of all other major states in area and production has declined since 1987/88.

Exports of Indian Grapes

India has made substantial progress in the grape trade. A net importer in the 1970s (by about 15,000 tons), it is now a net exporter (figure 8.5). As domestic production increased, imports declined and are now negligible. Exports rise and fall as market requirements and domestic

Figure 8.2 Grape Plantings in the Major Producing States of
India, 1987–2002

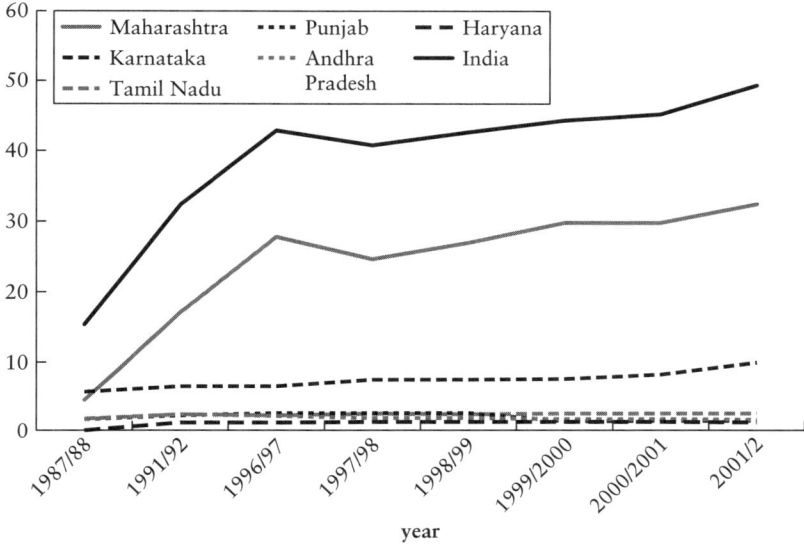

Source: National Horticulture Board.

Figure 8.3 Grape Production in the Major Producing States
of India, 1987–2002

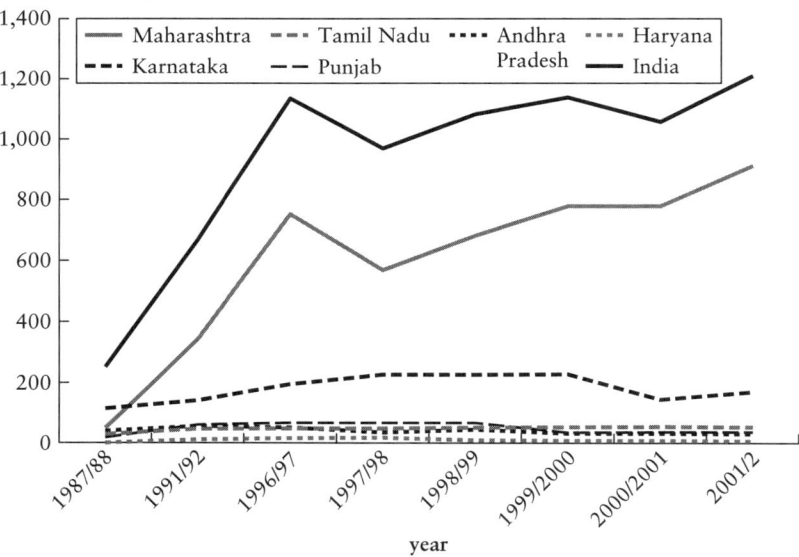

Source: National Horticulture Board.

Figure 8.4 Grape Yields in the Major Producing States of India, 1987–2002

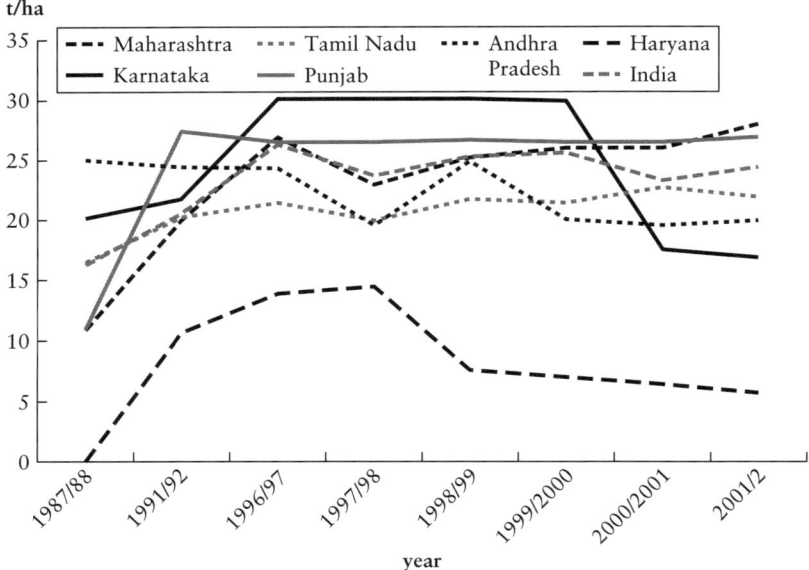

Source: National Horticulture Board.

Figure 8.5 India's Grape Exports and Imports, 1970–2002

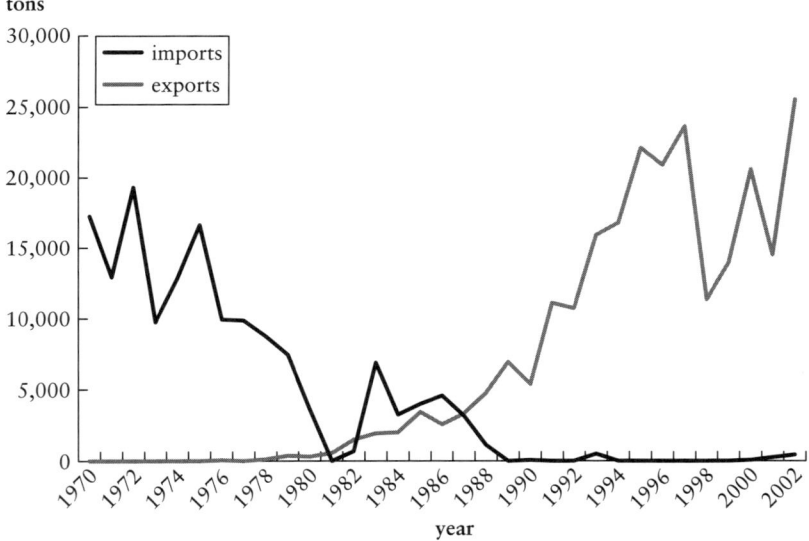

Source: FAOSTAT.

254

Table 8.7 Exports of Grapes and Grape Products from India, 1990–2003

Year	Fresh grapes (tons)	Grape marc[a] (thousands of liters)	Raisins (tons)	Sultanas and other dried grapes (tons)
1990/91	5,347.7	790.2	0.0	0.1
1991/92	11,139.8	861.6	0.0	4.5
1992/93	10,770.1	569.8	1.8	29.1
1993/94	15,928.7	1,366.7	0.7	2.2
1994/95	16,813.4	490.5	5.3	58.2
1995/96	22,150.9	422.9	29.2	234.0
1996/97	20,957.7	1,295.3	1.2	41.6
1997/98	23,679.8	114.1	180.2	128.3
1998/99	11,382.2	27.3	0.1	143.2
1999/2000	14,005.6	328.6	0.0	77.0
2000/2001	20,646.1	429.7	132.8	1.5
2001/2	14,571.0	1,063.9	44.0	35.0
2002/3	25,568.0	543.2	54.8	112.6
Growth rates of India's exports				
1991–2001	14.5	−5.9	71.3[b]	37.97
1991–2003	13.9	−3.1	40.8[c]	87.41

Source: India Trades.
a. Spirit obtained by distilling grape wine.
b. For the period 1992/93 to 2000/2001.
c. For the period 1992/93 to 2002/3.

production change, and as infrastructure and know-how rise to meet world standards, notably those of the European Union. Raisins also made some progress in the export market, but the quantities are very small (table 8.7).

In the early 1990s, before India undertook improvements across the board to comply with export market requirements, the unit prices it obtained were considerably below world average (figure 8.6). The improvements brought India's prices up to par with world trade.

Export Markets

India exports mostly Thompson Seedless grapes to markets in the European Union and the Middle East (table 8.8). Given Europe's stringent quality requirements, India's success with this market is commendable. The market for Indian grapes in Europe gives the country a toehold for exports of other perishables, but because these products are differentiated, they require sustained efforts in brand building.

Figure 8.6 Prices Paid for Indian and World Grape Exports, 1991–2002

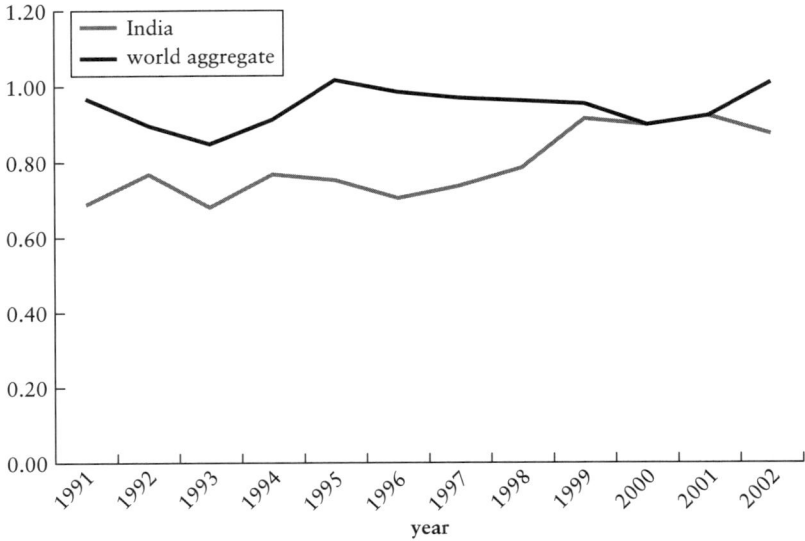

unit value (US$/kg)

Source: FAOSTAT.

Table 8.8 India's Exports of Fresh Grapes, by Destination, 1990–2003
(% total exports)

| | Destination | | | Total exports (tons) |
	Middle East	Europe	Asia	
1990/91	84	0	15	5,348
1991/92	87	2	11	11,140
1992/93	74	9	17	10,770
1993/94	64	19	17	15,929
1994/95	35	46	18	16,813
1995/96	29	48	23	22,151
1996/97	25	58	17	20,958
1997/98	15	60	25	23,680
1998/99	23	48	25	11,382
1999/2000	32	58	8	14,006
2000/2001	32	56	11	20,646
2001/2	42	50	7	14,571
2002/3	30	60	9	25,568

Source: Monthly Statistics of Foreign Trade of India.

Typically, Indian grapes can be sold in the European market during a one-month window from mid-April to mid-May. Before then, Chilean grapes dominate; afterwards, grapes from Brazil and Spain meet the demand. In price and quality, Indian grapes are not terribly competitive. Europe imports 90,000 metric tons of grapes each winter month from five to six suppliers (APEDA 2000); in summer the figure falls to 40,000 metric tons per month. The summer suppliers are Chile, South Africa, Israel, and India.

For export sale, produce goes through grading according to the importer's specifications, after which it is precooled, kept in cold storage, and finally dispatched in refrigerated containers. For the domestic market, produce is packed in 8–15 kg cartons, loaded in unrefrigerated trucks, and transported. Normally, precooling operations are not used for the domestic market.

Changing Market Requirements

Exports of grapes from India have been largely stagnant over the past decade. While a mix of internal and external factors is responsible for this trend, nontariff barriers are an increasingly critical factor (Chaturvedi and Nagpal 2003). To export to the European Union, grape-producing farms have to meet the standards issued by EurepGAP, an initiative of retailers belonging to the Euro-Retailer Produce Working Group. EurepGAP's aim is to develop widely accepted standards and procedures for the global certification of good agricultural practices (GAP). EurepGAP's July 2003 standards require that exporters meet a set of conditions pertaining to worker training, planning and production, pesticide record-keeping and disposal, testing for pesticide residues at the farm, and postharvest operations. One estimate contends that compliance with EurepGAP increases production costs by 40 percent (Chaturvedi and Nagpal 2003).

Physical and chemical characteristics are used to determine grape quality. Standards vary by country.

- *United Kingdom.* The United Kingdom has set rigid standards. The minimum acceptable diameter of an individual grape is 18 mm, and the berry must be a "light, milky green." Grape bunches must be a uniform 350 to 700 grams and packages are to hold 9 kilograms. British grocers sell grapes by the bunch, not by weight, the practice in India.
- *European Union.* The minimum acceptable diameter of the berry is 16 mm. The specified color is a light, milky green. The packaging has to be a uniform 4.5 kg.
- *Middle East.* India's main export market is an extension of the local market and is lax.

Given India's small-scale production, individual farmers are hard-pressed to meet demand for quality grapes. Moreover, the effort increases costs by an additional Rs 50,000 to Rs 75,000 per hectare.

Pruning is the most important practice in quality grape production. To achieve market-specified sizes for bunches and berries, yields must be sacrificed. Growers must thin shoots and clusters and apply gibberellic acid, a growth hormone that promotes elongated clusters. Berries themselves must be thinned and girdled. These practices reduce the number of bunches per vine, giving the fruit ample space to attain the required size. Thus yields are much lower in vineyards that cater to the export market. The average yield for a vineyard catering to the local market is about 35 tons per hectare; from vineyards that cater to the export market, the yield drops to about 25 tons per hectare.

Besides lower yields, the quality of the export and the exportable volumes also matters. Even in for-export vineyards, only the very well maintained ones are able to ensure that the entire available yield is exportable. In other vineyards, on average, only about 80 percent are of export quality—i.e., about 20 tons/hectare. It has been estimated that from most export-oriented vineyards, about 8 tons would meet the United Kingdom's quality standards; the remaining 12 tons could be exported to other EU countries.

The color specifications ("milky green") of the U.K. and EU markets require certain changes in preharvest practices—namely, the use of shade or light-cutting nets. Exposure to sunlight turns the berries golden, an acceptable color for the domestic market but deemed undesirable abroad. The costs of these shade nets must be borne by the farmer.

To achieve the desired size of berries and bunches, grape farmers must employ techniques that lead to crop suppression, namely the application of synthetic chemicals and plant-growth regulators, the residues from which may be toxic. All pesticides have minimum residue limits prescribed by the importing countries. The U.K. standards are more stringent than those of other countries and in fact are stricter than the international Codex Standards. To keep pesticide residues within the limits, there is a waiting period prior to harvesting. Because this waiting period is different for each pesticide, spraying must be done according to a special, sequenced schedule. Grape growers normally require technical assistance.

The application of chemicals also adds to the costs by way of testing and certification, practices required for the European market. Grape exports require two kinds of certification.

- *The EurepGAP certification.* This is a mandatory certificate needed to market produce in Europe. The certificate is based on three safety criteria: labor, environmental, and consumer.
- *Certified pesticide-residue tests.* The cost of testing and certification for pesticide residue is currently Rs 7,000/sample/pesticide. The number of tests undertaken depends on the export destination. Countries with more stringent measures, like the United Kingdom, require more tests, incurring higher costs.

In addition to certification, other postharvest processes for export grapes include grading, packing, precooling, cold storage, and transport. Grapes are immediately transported to the packhouse after harvesting and then sorted and graded. Bunches are put in polypacks, which are placed in properly cushioned boxes treated with grape guard,[9] according to market requirements. At each stage, the proper temperature and humidity levels have to be maintained, which constitutes a major bottleneck to growth in exports because of infrastructure limitations. The investments are substantial and normally beyond the means of individual farmers. Facilities are currently provided either by cooperative societies or private enterprises.

Economics of Production and Export

The start-up costs for a vineyard range from Rs 500,000 to 600,000 per hectare. After the second year, annual maintenance for export production ranges from Rs 300,000 to 350,000 per hectare, which includes labor costs to prune, train, weed, fertilize, spray, thin, harvest, pack, and load the grapes. Then there are the costs of the supplies themselves—manure, fertilizer, and pesticides—and of transportation. All told, these costs exceed by Rs 50,000–75,000 those required for the domestic market. The cost components of the export-market grapes, which are over and above those for the domestic market, include packing, precooling, cold storage, slow- and quick-release guards (SO_2), and shipping refrigeration (1–2°C, 95 percent humidity).

When cultivating grapes for export, yields can drop by as much as 30 percent. The premium on export-quality grapes, however, is sufficient to cover the loss, because the return on exported grapes is twice that for domestic grapes (table 8.9). On average, reducing yield from 35 tons/hectare for domestic consumption to 25 tons/hectare for export still produces only about 8 tons that meet the United Kingdom's exacting standards; 12 tons would qualify for the rest of the EU market. The remaining 5 tons would be sold in the domestic market. Although returns on the local market compare well with competing

Table 8.9 Economics of Grape Production for Local and
Export Market

Type of garden	Yield (tons/hectare)	Cost of production (Rs/hectare)	Prices (Rs/kg)	Returns over cost (Rs/hectare)
Local market	35	250,000	12	170,000
Export market	25	325,000		465,000
of which:				
U.K. standard	8		45	
EU standard	12		30	
Domestic standard	5		14	

Source: Computed based on field data.

crops, grape cultivation can be daunting for the small farmer. On top
of the costly initial investment and maintenance, growers need considerable know-how about production practices and techniques.

Institutional Development

Maharashtra's phenomenal growth in grape production and its growing access to the high-value export market have depended on the efforts of several institutions, mainly those created and developed by
producers.

The grape producers' association and its network. The Maharashtra
State Grape Growers' Association (Maharashtra State Drakhsa
Bagaithdar Sangh) is a prominent institution in this regard. Formed in
1961 by a group of 25 grape producers across the state, the association's sole objective is to improve cultivation practices appropriate for
conditions in the state. As cultivation problems mounted, the association began to involve scientists in extension work, and their field
research found answers to many of the problems encountered by association members. Initial research focused on canopy management and
hormones. The importation of gibberellic acid began when researchers
realized that it could substantially increase yields but was not available
in the local market. The association succeeded in getting the import
duty reduced and supplied the chemical to members at cost. As the association became more active, more farmers joined, both smallholders
and those running large operations. The association now has some
22,000 members.

Producers and the national research centers/university networks.
The association established contact with agricultural universities and
other Indian Council of Agricultural Research (ICAR) centers in an
effort to understand and improve cultivation practices; it also organized

meetings of scientists and producers. It carried out research trials on members' fields and at its own research farm, established in 1990. This research focused on training, pruning, crop physiology, soil and crop nutrition, pathology, entomology, agro-techniques, and postharvest technology.

Research institutions and the creation of a knowledge base. ICAR has a vast network of research stations that cater to the fruit sector's research needs. The Indian Institute of Horticulture Research (IIHR), in Bangalore, was established in 1968 to meet the need for research into the production and postharvest treatment of horticultural crops. One of the fruit research stations is located in Pune, where the grape producers' association is headquartered. Knowledge about the technologies required to produce export-quality grapes was available in other countries. IIHR adapted that knowledge to local conditions by conducting field trials.

ICAR institutes and the state agricultural university developed production technology pertaining to multiplication (rootstocks), pruning, fertilizing, irrigation scheduling, canopy management, insect pests and diseases, chemical application to encourage uniform bud formation, bunch thinning, and extending shelf life. They also developed suitable packages for harvested produce. Indigenous cold-storage facilities were upgraded with precooling and temperature-management technologies; humidity and airflow were optimized to extend the storage life of grapes. In 1997 ICAR established the National Research Center for Grapes to collect and maintain grape germplasm, to breed grape varieties resistant to various biotic and abiotic stresses, to increase production of export-quality grapes, and to improve raisin quality. In collaboration with exporters, the center conducts research on both the vineyards managed by the grape growers' association and those owned by individual farmers.

Postharvest technology and the infrastructure gap. Until the 1970s, owing mostly to their perishability, the production and consumption of grapes were highly localized, so India lacked a postharvest infrastructure. During the peak season, the fruit may have traveled a few hundred kilometers to reach another market. The government was more concerned with the production and distribution of staple food grains and important cash crops such as cotton and sugarcane. With improvements in road infrastructure (national and state highways) and transport systems, however, the government began to explore more distant markets for other agricultural products, including fruits and vegetables. Even then, because grapes had a restricted travel time, transport costs were high and brought spoilage, loss of quality, and a risky sales environment, since demand was unknown. The costs of transporting the produce long distances were prohibitive. Further, given India's low per capita income and its high price elasticity for fruits, demand was very low for costly produce.

Government-led development of the domestic market. During the early 1980s the government undertook systematic efforts to improve postharvest technologies. Because apples were the major crop sustaining the livelihoods of millions of farmers in Himachal Pradesh, the state began with apples. A state government agency called the Himachal Pradesh Horticultural Produce Marketing and Processing Corporation (HPMC) initiated several market-development activities for apple growers. Under a World Bank project, HPMC set up packing, grading, and cold-storage units for apples. HPMC opened branches and began to reach far-flung markets inside India. To facilitate marketing, a separate fruits and vegetables market was set up in New Delhi with better methods of transaction. This encouraged distant producers, cooperatives, and traders to sell their produce there.

The discovery of foreign markets. The experience with apples spilled over to grapes. Continued improvement in production technology created excess supply in the 1980s. Producers and traders began to investigate markets in other parts of India. Producers from Maharashtra used ice packs and fans to keep their produce cool in trucks headed for markets in Ahmedabad, Delhi, and Kolkata. This helped to widen the market for grapes, but transport bottlenecks and middlemen in the terminal markets cut into profits.

During this time, the Middle East had absorbed many workers from India and neighboring countries in various skilled and unskilled jobs. There was a large market for Indian food, including Indian fruits and vegetables. It took three days to ship grapes to the Middle East—a manageable timeframe, given the fruit's shelf life of seven days.

The Indian government responded in helpful ways. It created cold-storage facilities for export grapes and encouraged infrastructure development. It established a National Horticulture Board (NHB) in 1984 to develop integrated horticulture practices. The Agricultural and Processed Food Products Export Development Authority (APEDA) promoted agricultural and processed food exports.

Exports to the United Kingdom. The Maharashtra State Grape Growers' Association next began to focus on marketing constraints. In 1986, one enterprising Maharashtra farmer took a few cartons of grapes to the United Kingdom by air. There he found export potential. He continued to export small quantities each year by air. While the price he obtained was high, freight costs were staggering.

In 1988 the Maharashtra state government sent a seven-person delegation to Europe (five members were farmers). The delegation recommended that the state develop infrastructure to support exports. Next, APEDA invited businessmen from Singapore and France to India. They expressed interest in grapes, but noted that India had no precooling infrastructure in place, explaining that their markets demanded precooled grapes. This information inspired an association-sponsored study tour of California; 30 members went. In California, the growers learned that with the proper technology grapes can be stored for up to 12 months.

Storage technology is one thing, but meeting Europe's very stringent quality requirements was quite another. Higher-quality grapes demanded different equipment and technology and practices at all levels in the value chain. Government institutions like APEDA and NHB came forward to help in this regard.

Mahagrapes—a public-private partnership. The members of the producers' association saw exports as a way to dispose of excess supply. The market window (mid-April to mid-May) for the EU market was very attractive to India but required infrastructure that the grape growers did not have. Producers therefore formed cooperatives to help create the needed improvements. Together with a state government agency called the Maharashtra State Agricultural Marketing Board (MSAMB), the association helped to form 19 cooperatives in the state's major grape-growing regions. Precooling and cold-storage facilities were built for each cooperative at a cost of Rs 5.2 million; India imported the precooling machines from the United States. The National Cooperative Development Council provided 50 percent of the loan, and state government contributed 45 percent as share capital; the remaining 5 percent was borne by members. The National Horticulture Board provided a subsidy of Rs 5 million. By 1991, the growers' cooperatives began to export their grapes under one brand name, Mahagrapes, in an effort to help farmers to realize higher and more stable prices.

Study tours. Association members are routinely sent abroad to study foreign markets, their requirements, and export procedures. Mahagrapes found that airfreight costs were very high—more than the value of the produce. Sea freight cost only about 10–15 percent of the airfreight charges but required longer transport times. The solution? Better packing and storage technologies.

International experts. At the outset of India's foray into grape exports, the rate of rejection of consignments was as high as 50 percent. It took the growers a few years to perfect their understanding of market requirements—from seed quality, to harvest and postharvest care, to packaging practices. India began to invite international experts from major grape-producing countries such as Australia, Chile, and South

Africa to advise on production and postharvest issues. Experts at the Indian Institute of Horticultural Research and the Mahatma Phule Agricultural University helped to improve packaging technology.

A new focus on quality. By 1995 Mahagrapes and the cooperatives had the infrastructure and knowledge they needed to produce and export quality grapes. Rejected consignments fell to less than 10 percent by 1995. But while Mahagrapes was able to build brand recognition, exports fell from 1,800 tons in 1991/92 to 357 tons in 1994/95 (figure 8.7). The drop grew out of a conscious decision to export only quality produce. Since 1998/99, exports of Mahagrapes have increased to about 800 tons, about 2 percent of the EU market during the period. The prices commanded for Mahagrapes have steadily improved over time.

Mahagrapes, in short, succeeded in establishing a system for the export of high-value perishables from India. Mahagrapes is extending its product line to mango, pomegranate, strawberries, and other fruits and vegetables.

Ten new cooperatives in Maharashtra have recently emerged to export grapes, and it is estimated that cooperatives now account for

Figure 8.7 Grape Exports by Mahagrapes, 1990–2003

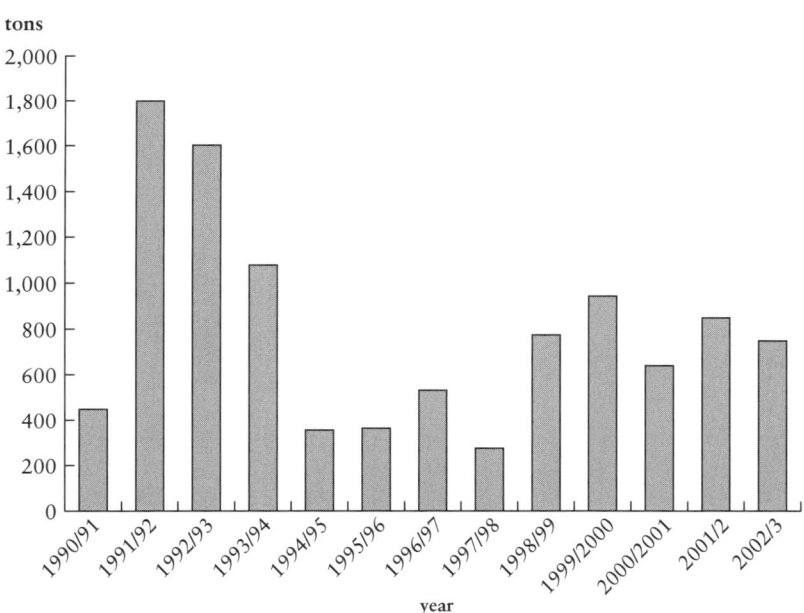

Source: Mahagrapes.

30 to 35 percent of total grape exports. Many producers are also exporting grapes individually. These and other exporters, a total of 150, have gained from association efforts to building a grape-cultivation and -export knowledge base. India now is able to send about 60 percent of its export produce to the high-value EU market and capture about 35 percent of that market during the export season.

New challenges are emerging for fruit and vegetable exports, however. The now-mandatory EurepGAP certification is costly, with a registration charge of Rs 35,000 per farm, making exports prohibitively expensive for the small grape grower. Pesticide-residue limits (set by importing countries) are growing more stringent, and satisfying them creates additional costs and delays. The growers' association and other organizations are seeking to meet these challenges in order to keep the grape sector competitive and profitable. APEDA, for example, has developed a pesticides-monitoring program, while the Maharashtra state government is arranging to issue pesticide-residue certificates.

Policies and Technological Progress in Grapes

India is a constitutional republic governed by an elected parliament. Both the central and the state governments are involved in policy making on agriculture, health, and education. Soon after independence in 1947, India instituted protectionist policies, directing investments so that the country would become self-sufficient in food and clothing. To generate employment, it reserved labor-intensive and low-tech products for small-scale enterprises. Protectionist policies, which continued through the 1970s, produced large inefficiencies, hampering competitiveness (Rakshit 2004). India's share in world trade shrank from 2.5 percent at the time of independence to a low of 0.4 percent. Together with higher oil prices and a sizeable debt, these policies created balance-of-payment difficulties into the early 1990s.

In the early 1980s India's exports were able to finance only 60 percent of its imports. As trade and investment declined and tax rates fell, a policy shift in the mid-1980s allowed India to finance its deficits through commercial borrowing. In 1991 India implemented an economic reform package proposed by the International Monetary Fund. The fiscal deficit was lowered through spending cuts, tax reform, and privatization. Public sector entities were encouraged to operate on commercial principles. As reforms in these sectors continued, WTO commitments became a major policy focus after 1995. The government removed quantitative restrictions on many consumer and agricultural product imports. Old trade-related laws were modified and new ones passed to comply with the WTO agreements.

Agricultural Development and Policies

The Bengal Famine of 1943 had a profound influence on Indian agricultural policy. Independent India set food self-sufficiency as its principal goal, followed by meeting raw material requirements. Toward this goal the government formulated various policies, set up new institutions and strengthened existing ones, and built infrastructure. Intermediaries in the land market were abolished through land reforms in the first three decades after independence.

In the 1960s, new technology in the form of high-yielding plant varieties (the Green Revolution) brought burgeoning growth in the agricultural sector. Budget allocations to agriculture under the first four Five-Year Plans were substantial, ranging from 11 percent to 17 percent. In subsequent plans, the allocation fell to 4–6 percent. In the late 1980s policies began to shift from land reform to technology adoption, with new subsidies for inputs such as fertilizers, water, and power. Emphasis on government procurement of food grains was meant to encourage food production.

Infrastructure development was a primary goal of the Five-Year Plans. The First Five-Year Plan allocated 19.7 percent of the total budget for irrigation and flood control. In subsequent plans, at least up to 1989/90, the share remained a substantial 10 percent. As a result, today, more than 40 percent of India's gross cropped area is irrigated. Rural electrification and road building were emphasized as well.

In the period after the Uruguay Round, Indian policies sought to liberalize the agricultural sector. Quantitative restrictions on imports have been removed, and applied tariffs have been much lower than the bound levels. In addition, India has encouraged market mechanisms, such as the futures and insurance markets, and eliminated government restrictions on various market activities, such as transportation and storage. Finally, the government is encouraging private sector participation in both production and marketing activities.

With increasing pressure on land for nonagricultural uses, and more Indians dependent on agriculture, farms are becoming smaller. The share of marginal holdings (farms of less than 1 hectare) increased from 51 percent in 1970/71 to 62 percent in 1995/96, while the share of holdings larger than 10 hectares decreased from 3.9 percent to 1.2 percent. Despite significant improvements in irrigation, fertilizer use, and mechanized farming, the overall growth rate for agriculture and allied sectors was significantly lower (1.3 percent) during the

Ninth Plan (1997/98 to 2001/2) than during the Eighth Plan (1992/93 to 1996/97), when annual growth averaged 4.7 percent.

Yields slowed in the 1990s. For food grains, the growth rate declined from 3.1 percent in the 1981–91 decade to 1.2 percent for the 1991–2001 period (table 8.10). Overall growth in yields was 2.3 percent in wheat, 1.3 percent in rice, and 2.9 percent in maize. With the exception of maize, yields for cereals and other commercial agricultural commodities fell substantially during 1990s. Moreover, the growth pattern for agricultural production is not uniform across the country. The northwestern and southern states have progressed admirably, while the central and eastern states have not. Another disturbing trend in Indian agriculture is investments. These have declined from 1.6 percent of GDP in 1993/94 to 1.3 percent in 2001/2. The performance of the agricultural sector in total trade is another area of concern. From 1990 to 2002, the share of agriculture imports in India's total imports increased from 2.8 percent to 6.7 percent, whereas the share of exports for the same time period decreased to 14.2 percent from 18.5 percent.

State government has a greater role to play in developing the agrosector than does the central government. Different states have different priorities for agricultural development. As a result, states have not aligned their agricultural policies. For example, in some states, electricity is supplied to farmers free of cost, whereas in others farmers pay for it. Some states are more receptive to central government policy suggestions. The federal structure is proving problematic for India as it adapts to a liberalized trade regime.

Government Policies

Political support. One prominent politician from Maharashtra served both as a minister in the central government and as chief minister for the state government. He was keenly interested in developing the grape sector as an alternative to sugarcane. The cooperatives have been the backbone of a strong sugar sector for many years. State politicians supported extending this model to the grape sector, and supplied the backup for various central and state government initiatives.

Having attained comfortable levels of food grain production by the 1980s, the Indian government turned to horticulture, establishing the National Horticulture Board in 1984. The board has a mandate to integrate the country's horticulture, develop proper production and postharvest management, offer financial assistance to develop cold storage facilities, develop technology, and transfer and develop market

Table 8.10 Growth Rates of Area (A), Production (P), and Yield (Y) of Major Crops in India (*Percent*)

	1953–61			1961–71			1971–81			1981–91			1991–2001		
	A	P	Y	A	P	Y	A	P	Y	A	P	Y	A	P	Y
Coarse cereals	0.72	2.40	1.68	0.22	2.59	2.33	−0.95	−0.54	0.44	−1.40	1.21	2.62	−1.79	−0.50	1.31
Cotton	2.17	6.83	4.34	0.00	−1.53	−1.64	0.26	3.85	3.67	−0.53	3.42	4.00	1.40	−0.31	−1.68
Food grains	1.56	4.16	2.56	0.73	2.83	2.08	0.19	1.80	1.61	0.09	3.13	3.04	−0.55	1.24	1.79
Gram	3.07	5.20	1.90	−1.74	−1.90	−0.16	−1.66	−1.88	−0.09	1.29	2.30	0.81	−3.60	−3.20	0.44
Groundnut	3.86	6.50	2.51	1.17	2.43	1.13	−0.71	−1.97	−1.24	2.01	4.14	2.08	−2.27	−1.57	0.78
Maize	2.54	4.42	1.91	2.80	6.23	3.28	0.34	−0.69	−0.98	−0.17	2.54	2.74	1.13	2.92	1.84
Nine oilseeds	2.64	5.11	2.26	1.86	3.21	1.34	0.59	−0.21	−0.84	3.19	7.06	3.78	−0.55	−0.11	0.49
Pearl millet	0.79	0.35	−0.43	1.19	9.37	8.08	−1.01	−4.00	−3.01	−1.06	2.58	3.69	−1.93	−2.32	−0.39
Potato	4.86	3.98	−0.93	2.36	5.87	3.24	4.28	7.23	2.89	2.56	4.63	2.06	4.02	5.68	1.54
Rapeseed & mustard	4.12	5.68	1.70	1.30	3.63	2.43	2.19	1.41	−0.59	3.53	8.50	4.91	−2.51	−2.11	0.34
Red gram	0.00	2.68	2.28	1.18	−1.00	−1.79	0.36	0.51	−0.29	2.54	1.84	−0.23	0.00	−0.42	−0.85
Rice	1.61	5.29	3.59	0.98	2.01	1.04	0.65	2.42	1.75	0.63	3.32	2.68	0.46	1.35	0.89
Sorghum	0.63	3.57	3.02	−0.56	−1.89	−1.33	−0.96	2.53	3.54	−0.92	1.18	2.12	−3.68	−4.35	−0.63
Tobacco	0.00	0.00	1.59	2.26	2.92	0.56	0.00	2.26	2.77	−2.21	1.84	2.42	−2.84	−6.70	−0.26
Total pulses	2.22	4.11	1.92	−0.43	−0.73	−0.28	−0.04	−1.07	−1.02	0.94	3.04	2.03	−1.94	−2.59	−0.60
Wheat	3.50	4.90	1.37	3.50	8.02	4.38	2.05	4.31	2.23	0.82	4.26	3.42	0.60	2.38	1.73

Source: Computed based on data from Indiastat.com.

information for horticultural products. The board has put together a database on horticulture products.

In 1986 the central government also established APEDA to develop agricultural commodities and processed foods and to promote their export. APEDA develops and implements schemes for the market, for infrastructure and quality improvements, for research and development, and for transportation assistance. Market schemes focused on packaging, feasibility studies, surveys, consultants, database upgrades, and export and international market development. Infrastructure schemes provided for common projects and the purchase of specialized transport units, in addition to helping exporter-growers move toward greater mechanization of their harvests, with more sheds for intermediate storage and grading. Other operations include cleaning, mechanized handling facilities—namely, sorting, grading, washing, waxing, ripening, package pelletization, and so forth—and precooling facilities.

Other schemes include quality-control assistance by providing (or strengthening) laboratories and improving management, quality assurance, and quality control, including consultations and certification. Training, seminars, and study tours help growers with technical and managerial matters. Research and development is supported through public, cooperative, and private organizations. APEDA's residue monitoring plan addresses certification and other issues related to pesticide-residue levels.

The Indian Institute of Horticultural Research provided rootstocks to establish vineyards in areas affected by salinity and drought. It also offered research help on production and postharvest technology until 1997, when ICAR established a separate National Research Center for Grapes in Pune, Maharashtra. The center maintains germplasm, develops resistant varieties, sustains productivity, and improves grape quality. It identifies research needs by consulting with grape growers, officials of the state departments of horticulture, and scientists from the agricultural schools in the major producing states. The center also conducts field surveys to understand producers' research needs. Working with the Maharashtra State Grape Growers' Association and Mahagrapes, it also conducts research on growers' fields.

National and state government partnership. In order to encourage value addition and export, the Indian government has announced Agri-Export Zones and Grape Wine Parks in grape-growing areas of Maharashtra. The government will develop infrastructure facilities and encourage the private sector to set up manufacturing and export facilities.

The state governments help in implementing various schemes of the central government and also may have their own schemes. In

Maharashtra the state government has been active in promoting the grape sector. The Maharashtra State Agricultural Marketing Board (MSAMB) played a major role in organizing Mahagrapes and other co-operative societies. The export market is a major focus of MSAMB, which collects market and technical information for farmers. The board also provides information on raisins and wine. The Maharashtra Industrial Development Corporation (MIDC), a state agency, is now in charge of promoting the grape sector and is playing a critical role in promoting the wine industry as well. In Maharashtra's Department of Horticulture implements APEDA's residue-monitoring plan. Under this plan, exporting farmers pay Rs 100 (US$2) to register with the state horticulture department. The department organizes inspections of farms and documents, draws and tests samples, and issues certificates. APEDA provides a subsidy of Rs 3,000 per sample tested (out of the total cost of Rs 7,000 per sample). As of 2004/5, 5,979 vineyards were registered for certification. Samples are tested in accredited public and private laboratories.

Building Forward Linkages from Grapes to Wine Production

Under the Maharashtra Grape-Processing Industrial Policy of 2001, the grape-processing and wine industry receives preferential investment. The policy offers tax and excise duty relief and simplified licensing procedures for new wine businesses and sales. The new businesses are grouped under the food processing industry. The same policy established the Wine Institute to train farmers and the Maharashtra Grape Board to promote industry and export. The MIDC is charged with establishing wine parks in Maharashtra; there are parks in Nasik and Sangli, and Sangli is also headquarters for the Grape Processing and Research Institute, which offers various diploma and certificate courses for farmers.

A "mother winery" offers high-quality infrastructure and services and also sells land at reasonable rates to wine-making entrepreneurs. High-quality viticulture equipment (pneumatic presses, crushers, destemmers, filtration systems, etc.) is housed there. Most of the equipment is imported. Entrepreneurs can rent the facilities. To promote wine consumption, the Maharashtra government is issuing licenses to wine bars at concessional fees. These policies have encouraged private investment in wineries; in 2004 nine new enterprises joined the existing 14 businesses. Ten new wineries are expected to begin operations in 2005. Out of the total of 3.6 million liters in India, 2.8 million liters are produced in Maharashtra. The burgeoning wine industry increased land under wine grape cultivation from 1,000 ha in 2003 to

1,500 ha in 2004. Wine grape production helps to stabilize the income of farmers in two ways: contract prices and more stable yields associated with wine grapes.

Although the Maharashtra state government is encouraging wine production, public research facilities are scarce at both central and state government facilities. Strengthening public sector research and training will improve the knowledge base and thus the industry.

Spillover from the Maharashtra Experience: National Learning

Karnataka, a major grape producer neighboring Maharashtra, has instituted its own state agency, the Karnataka State Agricultural Produce Processing and Export Corporation Limited (KAPPEC). Established in 1996 under the state's agricultural policy, KAPPEC promotes, develops, and encourages the export of agricultural, horticultural, and floricultural produce, including grapes.

In 1999–2000, KAPPEC initiated a joint venture project, in which 1.5 metric tons of grapes worth Rs 0.13 million were exported. The effort revealed that the entire value chain of grape exports required streamlining. So in 2002 KAPPEC launched a pilot project in two Bijapur villages known for their excellent, but unexploited, climatic and soil conditions. Collaborating agencies included SAM Agritech, the Indian Institute of Horticulture Research (IIHR), and the Karnataka Grape Growers' Association (Bijapur office). KAPPEC financed the project, IIHR provided technical assistance, SAM Agritech used its vast linkages to implement the project, and the Grape Growers' Association was the coordinating agency. KAPPEC hopes to form trade linkages with similar organizations in other states. It has long-term plans to create postharvest management infrastructure to fulfill the state's potential for procurement, processing, and export of agricultural products. It is also committed to importing advanced technologies and inputs in an effort to increase the quality and productivity of its food products.

Conclusions

Increasing demand in the domestic market, export opportunities, and remunerative returns make crops such as grapes an obvious candidate for development. The crop's high value creates many opportunities for rural farmers and entrepreneurs. Developing such crop sectors under a liberalized trade regime, however, must account for competition in the international market. In the case of grapes, the disparities in local versus export quality made it impossible to compete in high-value markets without first making substantial changes along the entire value chain.

In developing countries like India, such an overhaul requires support across the board: government, research systems, and technology suppliers must all develop appropriate facilities and practices along the value chain. The National Agricultural Research System created a knowledge base and then quickly adapted it to the field. Both central and state governments established appropriate institutions for technology and market development in partnership with producers' associations.

The Maharashtra State Grape Growers' Association played a key role in linking government agencies, research institutions, and the NHB and APEDA. It focused first on improving grape production, but as supply exceeded demand in the local market and prices fell, the organization began to investigate alternative markets. This led to the creation of Mahagrapes, an export-supporting organization of producers, which along with MSAMB successfully introduced technology and practices that support grape exports. The association then disseminated the technology among its producer members.

Governments became involved only after an individual producer demonstrated, with a simple flight to the United Kingdom, that an Indian could export grapes to the European Union. Afterwards, both central and state governments played cooperative, strong supporting roles, helping producers gain access to knowledge and learning about technologies, markets, and finance. To maintain India's image in foreign markets in an effort to regulate exports, they financed access to and communication with foreign agencies, local universities, and institute-based research. The reinforcing roles of producers' associations and government agencies not only helped Indian growers to overcome both technology and market barriers (and provide a knowledge base that would keep the grape sector competitive), but also they provided a model for other states and crop sectors.

The lessons are clear. In the presence of clear market opportunity, agro-sector underdogs like the Indian grape industry can adapt and compete in the world market. The government can exploit market opportunities more easily when the knowledge base is strong. Building know-how in the public sector enables countries to take advantage of opportunities when they arise.

Endnotes

1. Per capita consumption of fruits is very low, at 83 g per day compared to the requirement of 200 g for a balanced diet for an adult male who works moderately (Swaminathan 1985).

2. www.fftc.agnet.org/library/article/ac1993d.html.
3. Average farm sizes are already very small (1.57 ha in 1990/1) and declining in India, as opposed to large and increasing farm sizes in many developed countries. Agriculture accounts for 65 percent of employment and 24 percent of GDP. After 23 years, the current account of India's balance of payment recorded a surplus equivalent to 0.3 percent of GDP in 2001/2.
4. National Sample Survey Organization (1999–2000). Of the 1.02 billion people in India (2001), 72 percent live in rural areas, but the proportion is slowly declining. See Ministry of Agriculture, "Agricultural Statistics at a Glance 2002."
5. The incidence of poverty in India in 1999–2000 was 27 percent in rural areas and 24 percent in urban areas.
6. In the 1990s, India's population increased by 1.9 percent per annum, GDP by 6.1 percent, and per capita GDP by 4.1 percent. Life expectancy in India is now comparable to the world average.
7. According to the Maharashtra Industrial Development Corporation.
8. The average yield is half the potential yield in the case of Thompson Seedless and Aneb-e-Shahi (Shikhamany 2001).
9. Grape Guard slowly releases SO_2, which prevents diseases during storage and transportation.

References

APEDA (Agricultural and Processed Food Products Export Development Authority). 2000. "Grapes." [Retrieved from http://www.apeda.com/html/grapes1.html]

Chaturvedi, Sachin, and Gunjan Nagpal. 2003. "WTO and Product-Related Environmental Standards: Emerging Issues and Policy Options before India." Research and Information Systems for the Non-Aligned and Other Developing Countries Discussion Paper 36/2002. New Delhi.

FAO (Food and Agricultural Organization). 2001. "The Market for 'Organic' and 'Fair-Trade' Bananas." Paper presented at the Intergovernmental Group on Bananas and on Tropical Fruits, Committee on Commodity Problems, December 4–8, San Jose, Costa Rica.

Ministry of Agriculture. 2002a. "Agricultural Statistics at a Glance 2002." New Delhi: Directorate of Economics and Statistics, Government of India.

Ministry of Agriculture. 2002b. "Report and Data Base of Pilot Scheme on Major Fruits & Vegetables—1982–1983 to 2000–2001." New Delhi: Directorate of Economics and Statistics, Government of India.

Ministry of Finance and Company Affairs. 2003. "Economic Survey 2002–03." New Delhi: Government of India.

National Sample Survey Organization. "Level and Pattern of Consumer Expenditure: Fiftieth Survey: 1993–94." Report 402. New Delhi: Department of Statistics, Government of India.

Rakshit, Mihir. 2004. "Some Macroeconomics of India's Reform Experience." In Kaushik Basu, ed., *India's emerging Economy: Performance and Prospects in the 1990s and Beyond*. Cambridge, MA: MIT Press.

Shikhamany S. D. 2001. "Grape Production in India Grape." In Minas K. Papademetriou and Frank J. Dent, eds., *Grape Production in the Asia-Pacific Region*. Bangkok, Thailand: Food and Agriculture Organization of the United Nations, Regional Office for Asia and the Pacific.

Singh, R. B. 2000. "Welcome Address." In Minas K. Papademetriou and Frank J. Dent, eds., *Grape Production in the Asia-Pacific Region*. Bangkok, Thailand: Food and Agriculture Organization of the United Nations Regional Office for Asia and the Pacific.

Swaminathan, Monkombu S. 1985. *Food and Nutrition*, vol. 2. Bangalore: Bappco.

CHAPTER 9

Closing the Yield Gap: Maize in India

Gopal Naik

FOR MORE THAN 20 YEARS, THE government played a key role in creating a vibrant maize industry in India. By deregulating seed imports; generating improved seed varieties; attracting private investment; and supporting poultry, livestock, pharmaceutical, and industrial markets that consume maize, the government helped increase production to 13 million tons.

However, a lag in actual yield per hectare limits the actual benefits received by India from this production increase. Due to relatively low yields, India's performance is poor relative to other maize producers. Average maize yields in India are two tons per hectare, less than one-quarter of those obtained in the United States and Europe. China, which had yield levels similar to India's in the early 1960s, has made fast progress; today India's yield levels are less than half of China's. Though the country has been able to meet growing domestic demand and narrow the yield gap in certain production centers, India's international trade in maize is negligible.

This chapter explores how India could further narrow the maize yield gap, helping to reduce poverty and position the nation favorably in the competitive global maize market.

Introduction

In the last decade, the increasing international demand for maize for livestock feed and for pharmaceutical and industrial uses has drawn attention from policy makers and private agribusiness in India. Although maize is the most widely grown grain in the world, India's efforts since the Green Revolution have focused on rice and wheat. Maize was considered an inferior, coarse grain; the government's focus was on self-sufficiency in grains preferred for human consumption, such as rice and wheat.

The global importance of maize is driven by its broad range of uses— as human food, livestock feed, industrial input, and seed. In 2000 the world produced about 590 million tons. Maize is the main cereal in the United States, China, Latin America, and Africa, and ranks third after rice and wheat in Asia. Global per capita use in 1994–96 was almost 100 kg/year. Asian consumption remains lower, at about 12 kg/year, but demand is expected to grow. Consumption in India, at 9 kg/year, is expected to increase significantly (Gerpacio 2001).

Maize as a food crop has declined as wealthier consumers prefer rice and wheat. However, rising incomes have increased demand for meat and poultry. Maize is used in poultry feed production, which grew from 0.41 million tons in 1985 to 1.94 million tons in 2001. Demand for maize as an industrial material, such as pharmaceutical starch, is also growing, and the current estimate of 1 million tons of starch maize is expected to increase.

Singh (2001) reports that the amount of Indian maize consumed as food is steadily declining, while feed and industrial uses are increasing. But precise data on maize use are lacking. Based on government and industry sources, as well as expert judgment, Landes and others (2004) report that nonfeed consumption remained relatively steady at six to seven million tons throughout the last 30 years, while feed consumption grew from 158,000 tons in 1969–71 to almost 5 million tons in 1999–2001. In 1999–2001, feed accounted for 42 percent of total maize use in India (table 9.1).

Table 9.1 Trends in Maize Use in India, 1969–2000
(Thousands of tons)

Year	Total	Nonfeed	Feed	Feed share (%)
1969–71	6,405	6,246	158	2.5
1979–81	6,521	5,921	600	9.2
1989–91	8,956	6,839	2,117	23.6
1994–96	9,553	6,987	2,567	26.9
1999–2001	11,717	6,783	4,933	42.1
Growth rates (percent)				
1970–80	0.2	−0.5	14.3	—
1980–90	3.2	1.5	13.4	—
1990–2000	2.7	−0.1	8.8	—
1995–2000	4.2	−0.6	14.0	—

Source: Landes and others 2004.
Note: — = not applicable.

Changing Use Patterns

Maize is a major feed grain for livestock in developed countries. In India, too, its use as feed is increasing. Maize constitutes 50 to 65 percent of poultry feed and 10 to 35 percent of overall animal feed, depending on feed type.

Poultry consumption in India is growing much faster than that of other meats (table 9.2). While the Food and Agriculture Organization (FAO) of the United Nations and the U.S. Department of Agriculture (USDA) differ in their estimates, growth rates for the 1980s and 1990s are substantial. The share of poultry in total meat consumption was only 6 percent during 1997–99, well behind fish (51 percent) and beef (29 percent).

The per capita poultry consumption growth rate (6 to 9 percent) in India during 1990–99 is in line with the growth rate in developing countries (table 9.3). Because per capita consumption is already high in the developed countries, annual growth rates there declined from 4.8 percent in the 1970s to 1.9 percent in the 1990s. China's growth rate of 13.6 percent suggests that the high growth rate is likely to prevail in India in the coming years.

A proper assessment of growth in the poultry sector is difficult due to poor and unreliable data. There are discrepancies among FAO, USDA, and industry data (table 9.4). For example, estimated 1997–2002 growth rates range from 2.5 (FAO) to 18.6 percent (USDA). USDA estimates of 1.4 million tons of poultry in 2002 would make India the sixth-largest poultry producer in the world. The USDA

Table 9.2 Growth in Consumption of Animal Products in India, 1980–2000
(Percentage increases)

Product	FAO estimates		USDA estimates	
	1980–90	*1990–98*	*1980–90*	*1990–2000*
All meat	4.0	2.3	8.8	—
Bovine	3.5	1.7	11.6	−2.5
Eggs	7.4	4.4	0.7	4.6
Fish and seafood	4.3	4.2	—	—
Milk	5.4	3.8	—	—
Mutton and goat	3.0	1.5	7.3	1.0
Pig	4.7	3.4	—	—
Poultry	11.5	6.4	6.3	10.3

Source: Landes and others 2004.
Note: — = not available.

Table 9.3 Growth of Per Capita Poultry Consumption,
by Region, 1970–99
*(Compound annual growth rates between three-year averages,
centered on the years indicated)*

	Growth rates (%)		
Region	1970–80	1980–90	1990–99
Developed countries	4.8	3.6	1.9
Developing countries	7.8	6.5	9.1
Asia, developing	7.2	7.5	9.8
South Asia	3.5	9.6	6.3
China	5.4	9.0	13.6
East and Southeast Asia	6.3	6.9	5.9
India			
FAO data	3.2	11.5	6.0
USDA data[a]	5.9	6.3	9.2

Source: Landes and others 2004.
a. 1975–80 growth rates instead of 1970–80.

Table 9.4 Growth in Poultry Production in India, 1980–2002
(Percentage increases)

Period	USDA	FAO	USAPEEC	Industry average[a]
1980–90	6.0	11.7	—	—
1990–2000	11.0	5.3	7.7	—
1990–96	9.8	5.8	8.3	—
1997–2002	18.6	2.5	—	14.8

Source: Landes and others 2004.
Note: USAPEEC is USA Poultry and Egg Export Council. The growth rate is for
1990–99 instead of 1990–2000. — = not available.
a. Industry average is 1997–2001 instead of 1997–2002.

data indicate significant growth in poultry production in the 1990s,
and more rapid growth in poultry consumption than in consumption
of other meats.

Indian government data also indicate high growth of broiler and
poultry meat production (table 9.5). Commercial poultry production—
mainly of eggs—started in the 1960s, and broiler production in the
1970s. Broiler production grew at a rate of 12 percent, and poultry
meat at 9.3 percent during 1995–2003. This high growth is an indication of the growing competitiveness of the Indian poultry industry.

Table 9.5 Production of Eggs, Broilers, and Poultry in India, 1995, 2000, and 2002

Year	Eggs (millions)	Broilers (millions)	Poultry, meat (thousands of tons)
1995	27,275	350	578
2000	32,500	600	775
2002	35,000	800	975
CGR (%)	3.40	12.03	9.34

Source: Ministry of Food Processing Industries 2003.
Note: CGR = compound growth rate.

The competitiveness of the poultry sector with other meat sectors, both within the country and with external poultry markets, depends on feed costs (which make up 55 to 64 percent of the variable cost of production) as well as the extent of vertical integration. Vertical integration helps reduce production costs by increasing technical efficiency and eliminating the margin on feed and one-day-old chicks (Landes and others 2004). The Indian poultry industry is increasingly vertically integrated.

Increasing demand for poultry products signaled higher profitability to private entrepreneurs, who spotted the opportunity and invested in additional production capacity, mainly additional hatcheries and integrators. The organized poultry sector is heavily dependent on manufactured feed, of which maize is a major ingredient. The demand for maize therefore increased along with the growth in the poultry industry. The relationship is reflected in the pattern of price increases in poultry products and maize (table 9.6).

Table 9.6 Growth in Wholesale Prices and Income in India, 1980–99
(Percentage increases)

Period	Wholesale price indices				Per capita GDP
	All food	Poultry	Eggs, fish, and meat	Maize[a]	
1980–90	8.1	4.3[b]	8.3	6.8	2.9
1990–99	9.5	7.0	10.9	8.4	3.3
1990–95	10.8	10.9	14.3	11.6	2.8
1995–99	7.8	2.3	6.8	4.5	3.8

Source: Landes and others 2004.
a. Index of average wholesale prices in Bihar, Karnataka, and Uttar Pradesh.
b. 1983–90 growth rate.

Rising incomes, a growing population, faster urbanization, and declining prices led to poultry's rapid growth (Landes and others 2004). Also favorable is the fact that poultry price increases have been slower than increases in other food prices (table 9.6), and that poultry is acceptable to all religions, unlike beef or pork. The demand for poultry should increase at a high rate. The estimated elasticity for income (1.7) and price (−1.5) is high (Landes and others 2004), and incomes are rising and prices falling. As income, population, and urbanization increase, and as the sector becomes more efficient, high growth will continue for decades. With the current per capita poultry consumption at less than one kilogram per year, as compared to Pakistan (2.3 kg), China (4 kg), and Thailand (9 kg), the potential is considerable. In fact all these countries have far to go to achieve the 44-kilogram consumption level of the United States (Speedy 2001).

Annual growth rates for feed from 1965 to 2001 were 12.3 percent for cattle and 14.7 percent for poultry (table 9.7). The International Model for Policy Analysis for Commodities and Trade (IMPACT) developed by the International Food Policy Research Institute (IFPRI) predicts that per capita feed consumption will increase from 4 kg in the early 1990s to 11 kg in 2020 (IARI 2004). The current estimated consumption of 1 million tons of maize for starch production will also increase in coming years. New segments of human consumption such as sweet and baby corn are emerging. These developments will drive up demand and it will be critical for India to meet it with competitive prices.

Table 9.7 Production of Compounded Livestock Feed in India, by Sector, 1965–2001
(Millions of tons)

Feed type	1965	1975	1985	1995	2000	2001	Growth rate (%), 1965–2001
Cattle	0.025	0.26	0.75	1.50	1.64	1.65	12.3
Poultry	0.014	0.17	0.41	1.26	1.94	1.94	14.7
Others	—	—	—	0.05	0.03	0.04	—
Total	0.039	0.44	1.16	2.81	3.61	3.63	13.4

Source: IARI 2004; growth rates are recomputed.
Note: — = not available.

The Production Environment

The global market for maize is dominated by a few large producing countries. Of the 604 million tons produced worldwide in 2002, the United States accounted for 38 percent, China for 20 percent, and the European Union for 7 percent. India contributed 2 percent, ranking seventh in the world. India ranks fourth in land area under production, however, with almost 7 percent of the world's land area in maize. India's poor production performance relative to the other major producers has been due to relatively low yields.

Maize is a versatile crop and can be produced in a wide range of production environments, including temperate and tropical climates, semi-arid conditions, and heavy rainfall areas. It is cultivated under widely varying conditions in India, where temperatures range from 10°C to 45°C and rainfall from 200 to 2,500 mm per year. Maize is grown in all four of India's regions: the Northwest Plain, the Northeast Plain, the Central region, and the Peninsular region. In some parts it is grown as a *kharif* (monsoon) crop or as a *rabi* (winter) crop, under both rainfed and irrigated conditions.

While maize production and utilization have grown steadily, if not dramatically, the share of gross cropped area for maize has remained stable over the past three decades, while the area for sorghum and millet has fallen dramatically (table 9.8). Maize production grew 6.23 percent during the 1960s due to area (2.8 percent) and yield increases (3.28 percent). The All India Coordinated Maize Improvement Program (AICMIP) develops and disseminates improved technology for better seeds and other practices. Though the land area producing maize shrank in the 1980s, yields of 2.74 percent still produced an annual growth of 2.54 percent. Improved hybrid varieties distributed by AICMIP significantly contributed to this increase. In the 1990s, the

Table 9.8 Trends in the Share of Crops in Gross Cropped Area, 1970–2000
(Percent)

Year	Maize	Pearl millet	Rice	Sorghum	Wheat
1970–71	3.53	8.08	22.56	10.18	11.04
1980–81	3.50	6.76	23.32	9.51	12.88
1990–91	3.17	5.78	23.01	7.62	12.95
1999–2000	3.43	4.68	23.90	5.40	14.60

Source: Indiastat.com.

Figure 9.1 Maize Yields in Selected Countries, 1981–2003

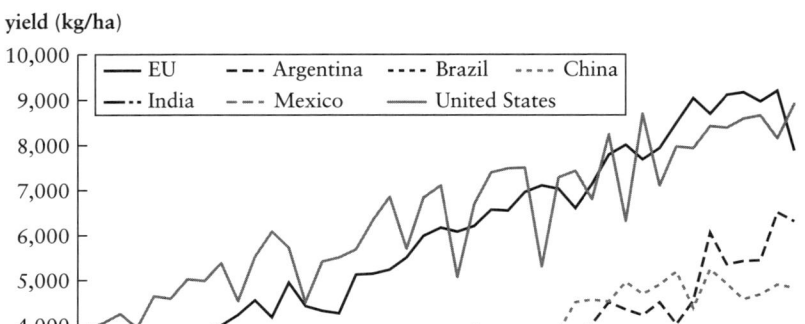

Source: United Nations Food and Agriculture Organization.

land area producing maize actually expanded, and although growth in yields slowed to 1.84 percent annually, overall production grew 2.92 percent, reaching 13 million tons in 2001 (table 9.9).[1] However, average maize yields in India are around 2 tons per hectare, which is still *less than one-quarter of the levels achieved in the United States and Europe* (figure 9.1). China, which had yield levels similar to those of India in the early 1960s, has made substantial progress in increasing yield. In India, though favorable growth trends continue, *yield levels are less than half those of China.*

Improved seed, proper cultivation practices, and optimum and appropriate level of inputs and irrigation can improve crop yields. Private and public institutions can play a significant role in providing improved seed, other inputs, and knowledge about cultivation practices. While government investment and policies can help increase the total available irrigated land area, deliberate allocation to individual crops depends on the attractiveness of each crop.

India's international trade in maize is negligible. Restrictions on maize imports have been removed due to the Uruguay Round Agreement on Agriculture; maize is imported under a tariff rate quota. The

Table 9.9 Growth Rates of Area (A), Production (P), and Yield (Y) of Major Crops in India, 1961–2001 *(Percent)*

Crop	1961–71			1971–81			1981–91			1991–2001		
	A	P	Y	A	P	Y	A	P	Y	A	P	Y
Coarse cereals	0.22	2.59	2.33	−0.95	−0.54	0.44	−1.40	1.21	2.62	−1.79	−0.50	1.31
Cotton	0.00	−1.53	−1.64	0.26	3.85	3.67	−0.53	3.42	4.00	1.40	−0.31	−1.68
Food grains	0.73	2.83	2.08	0.19	1.80	1.61	0.09	3.13	3.04	−0.55	1.24	1.79
Gram	−1.74	−1.90	−0.16	−1.66	−1.88	−0.09	1.29	2.30	0.81	−3.60	−3.20	0.44
Groundnut	1.17	2.43	1.13	−0.71	−1.97	−1.24	2.01	4.14	2.08	−2.27	−1.57	0.78
Maize	2.80	6.23	3.28	0.34	−0.69	−0.98	−0.17	2.54	2.74	1.13	2.92	1.84
Nine oilseeds	1.86	3.21	1.34	0.59	−0.21	−0.84	3.19	7.06	3.78	−0.55	−0.11	0.49
Pearl millet	1.19	9.37	8.08	−1.01	−4.00	−3.01	−1.06	2.58	3.69	−1.93	−2.32	−0.39
Potato	2.36	5.87	3.24	4.28	7.23	2.89	2.56	4.63	2.06	4.02	5.68	1.54
Rapeseed & mustard	1.30	3.63	2.43	2.19	1.41	−0.59	3.53	8.50	4.91	−2.51	−2.11	0.34
Red gram	1.18	−1.00	−1.79	0.36	0.51	−0.29	2.54	1.84	−0.23	0.00	−0.42	−0.85
Rice	0.98	2.01	1.04	0.65	2.42	1.75	0.63	3.32	2.68	0.46	1.35	0.89
Sorghum	−0.56	−1.89	−1.33	−0.96	2.53	3.54	−0.92	1.18	2.12	−3.68	−4.35	−0.63
Tobacco	2.26	2.92	0.56	0.00	2.26	2.77	−2.21	1.84	2.42	−2.84	−6.70	−0.26
Total pulses	−0.43	−0.73	−0.28	−0.04	−1.07	−1.02	0.94	3.04	2.03	−1.94	−2.59	−0.60
Wheat	3.50	8.02	4.38	2.05	4.31	2.23	0.82	4.26	3.42	0.60	2.38	1.73

Source: Ministry of Agriculture 2002.

Table 9.10 Exports of Maize from India, 1997–2001

Year	Total maize exports (tons)	To Bangladesh (tons)
1997	6,572	2,813
1998	190	0
1999	4,158	0
2000	40,105	32,679
2001	15,611	10,595

Source: FAO.

first 400,000 tons of imports receive a 15 percent duty; quantities over the quota receive a 50 percent duty. Up to 250,000 tons are imported annually by starch and poultry feed manufacturers. Domestic and import parity prices between 1996 and 2001 indicate that in four out of five years, domestic prices were lower than import parity prices (Landes and others 2004). For three years, import parity prices were higher than domestic prices by more than 25 percent, suggesting that, *with increased productivity, Indian farmers would become competitive domestically and would even be able to export maize to growing markets in Asia.* Maize has been exported in small quantities since 1997, mostly to Bangladesh to meet the requirement of its growing poultry industry (table 9.10).

The Macroeconomic Environment in Agriculture

India's growing rural population depends on agriculture, so agricultural growth is critical to reducing poverty. India is the second-largest country in the world in population, supporting 17 percent of the world's people on only 2.4 percent of the world's land mass. India's population passed one billion at the end of the millennium. Although the population growth rate has slowed (table 9.11), the current growth rate of just under 2 percent still adds almost 20 million to the population every year. The population remains 72 percent rural and primarily dependent on agriculture for livelihood. In the last two decades, the share of India's population living in rural areas has declined by less than 5 percent, and the absolute number of people living in rural areas continues to grow.

Between 1989 and 1999, the agricultural population increased by 10 percent, while more than 10 million hectares—or 6 percent—of the

CLOSING THE YIELD GAP

Table 9.11 Trends in Population and Composition, 1981, 1991, and 2001

Year	Population (millions)	Annual population growth rate (%)	Rural population share (%)
1981	683.3	2.20	76.7
1991	846.3	2.14	74.3
2001	1,027.0	1.93	72.2

Source: Ministry of Agriculture 2002.

Table 9.12 Distribution of Employment by Sector, 1977–2001 *(Percent)*

	Year				
Sectors	1977/78	1983	1987/88	1993/94	2000/2001
Primary (agriculture and allied sectors)	80.1	78.5	75.2	75.3	65.0
Secondary (industry)	9.9	10.8	13.9	12.0	12.5
Tertiary (service sector)	10.0	10.2	11.0	12.0	22.5

Source: Institute of Applied Manpower Research, http://www.indiastat.com/.

agricultural land base was converted away from agriculture. With increased pressure on agricultural land, the average size of land holdings dropped from 1.69 (1985) to 1.57 hectares (1991), with nearly 60 percent of holdings now smaller than 1 hectare (2.5 acres). *Such noneconomic farm sizes have led to a decline in the proportion of cultivators* in the rural population and an increase in the share of agricultural laborers and other workers.

In the absence of significant growth in job opportunities in the "organized" sector in recent years, the primary sector—which includes agriculture and allied sectors such as animal husbandry, fisheries, and forestry—still provides two-thirds of total employment in India (table 9.12), although it accounts for barely one quarter of the GNP. In the last half of the 1990s, the labor force grew just over 1 percent, while employment in the organized sector grew only 0.53 percent (Ministry of Finance 2003).

India's economy is one of the fastest growing in the world, reflecting a dynamism not seen in the agricultural sector. The impetus for growth came in the mid-1970s and accelerated with the economic

Table 9.13 Performance of Major Sectors, 1970–2001

Sectors	Growth rates of select indicators	1970/71 to 1979/80	1980/81 to 1989/90	1990/91 to 2000/2001
Services	Share of GDP	34.40	38.60	44.30
	Growth rate	4.5	6.6	7.6
	Contribution to GDP growth	52.70	43.60	57.60
Industry	Share of GDP	22.80	25.00	27.10
	Growth rate	3.7	6.8	5.9
	Contribution to GDP growth	28.70	28.90	27.60
Agriculture	Share of GDP	42.80	36.40	31.15
	Growth rate	1.3	4.67	2.87
	Contribution to GDP growth	18.6	27.50	14.8

Sources: Reserve Bank of India (2001–2) and Government of India (2002), cited in Rakshit 2004.

Table 9.14 Per Capita Income, 1980/81 to 1999/2000 *(In Indian rupees)*

Year	Per capita income (current prices)	Per capita income (constant prices; 1993/94 = 100)
1980/81	1,751	5,398
1990/91	5,366	7,345
1999/2000	15,887	10,207

Source: Planning Commission 2001, www.indiastat.com.

liberalization initiated in the 1990s. Annual growth rates of almost 6 percent in this decade were led, primarily, by private sector investment. Private investment in the 1990s grew at a respectable 8.75 percent, compared to 3.14 percent for public investment.

India's growth in the 1990s was driven by the service sector (table 9.13). Between 1997/98 and 2000/2001 the service sector grew by almost 9 percent annually, and the industrial sector by almost 5 percent, while the agricultural sector grew only 1.23 percent. Agriculture's share of GDP has consistently declined for three decades, from 43 percent in the 1970s to around 24 percent in 2003.

Along with economic growth has come a steady increase in per capita income. Income grew 39 percent, from Rs 7,345 in 1990 to 10,207 in 2000 (table 9.14). The poverty ratio declined sharply, from more than a third in 1994 to about a quarter in 2000. However, the number of people below the poverty line was still a staggering 260 million in 1999/2000, with two-thirds in rural areas (table 9.15).

Table 9.15 Incidence of Poverty in India, 1993/94 and 1999/2000

	Head count poverty ratio (%)			Number (millions)		
Year	Total	Rural	Urban	Total	Rural	Urban
1993/94	36.0	37.3	32.4	320	244	76
1999/2000	26.1	27.1	23.6	260	143	67

Source: Ministry of Finance 2002.

The Policy Environment

Self-Sufficiency in Staples Through Improved Seed Technologies

The Bengal Famine of 1943 influenced goals for India's agricultural sector for many decades. After independence, India aimed for food self-sufficiency and meeting raw material requirements. Toward this goal, the government formulated policies, set up new institutions, strengthened existing ones, and created new infrastructure.

In the "Green Revolution," technology for high-yielding seed varieties initiated new agricultural growth. The key role of improved seed was recognized in the Central Seed Act of 1966, which assigned responsibility for commodity research and seed production to public organizations, ensuring that breeding research and seed production activities for most staple food crops remained firmly under public domain (Morris and others 1998).

As Green Revolution technologies led to commercial farming, the private sector entered the seed industry (Singh 2001). Some companies established their own research capacity and developed hybrids, while others specialized in duplicating seed from public open-pollinated varieties (OPVs) and hybrids. The private seed sector flourished in some segments, such as vegetable seeds.

In the 1980s, government policies in agriculture focused on improved technologies such as fertilizer, water, and power, and on investments in irrigation, electrification, and roads. By the 1990s, India's economy opened up, and its agricultural policy liberalized trade and encouraged private sector production and marketing activities.

Early Government Efforts to Improve Maize Production Technology

On the suggestion of the Rockefeller Foundation's experts on breeding local hybrids, a coordinated system to integrate research activities was

devised in 1957 as the AICMIP. By the mid-1960s, hybrids were successfully developed but were not widely adopted. Initial success was
achieved when OPVs were developed with the germplasm provided by
CIMMYT (Centro Internacional de Mejoramiento de Maíz y Trigo)
under a Rockefeller Foundation program. A number of hybrids were
developed by the Indian Council for Agricultural Research (ICAR)
and state agricultural universities, but these did not constitute a break-
through (Evenson and others 1999).

Maize is one of 22 commodities regulated by the Indian government's minimum support prices (MSPs). The MSP is expected to cover
the average cost of production in order to insure farmers against price
risk. Though the MSP on maize has not been as good, over time, as for
wheat (figure 9.2), it has helped to cover the risk for farmers. The MSP
for maize was lower than the farm harvest price between 1963 and
1998, supporting a well-developed and growing market. As a result,
large-scale government procurement is not taking place.

Government Policies to Support Adaptation of Superior Technologies and Promote Private Sector Maize Production

Major policy reforms have facilitated private sector participation in
the maize seed industry and resulted in increased productivity. The

Figure 9.2 Minimum Support Prices for Maize and Wheat in India, 1960–2003

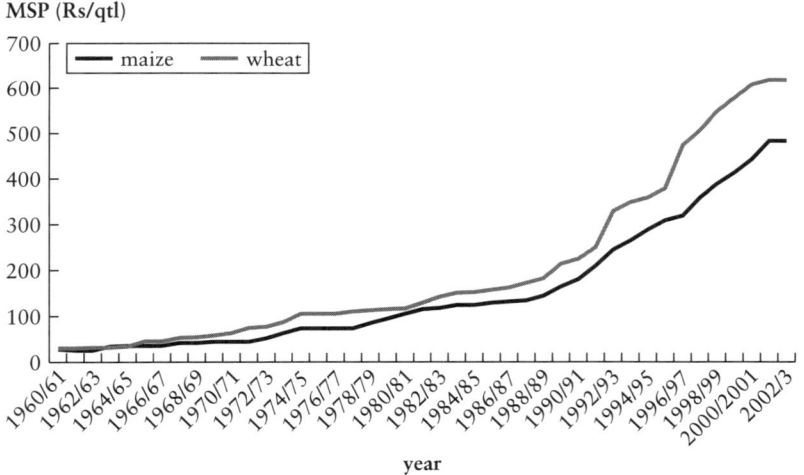

Note: MSP = minimum support price, Rs/qtl = rupees per quintal.

New Policy for Seed Development, enacted in 1988, changed licensing policies to encourage investment from domestic and multinational companies. This permitted the import of improved varieties of coarse grains, pulses, and oilseeds, if the foreign supplier agreed to cede control over parent lines, breeder seeds, and seed production technology to the importing company within two years. Import duties were also lowered from 100 to 15 percent (Morris and others 1998). The subsequent 1991 Industrial Policy, which identified seed production as a priority investment, further facilitated the entry of multinationals into the seed market.

Private Sector Participation in Maize Seed Technology

Companies responded quickly and many initiated research and development activities. The Indian seed market tripled from Rs 6.8 billion in 1990 to Rs 22.5 billion in 1999 (Gadwal 2003). By 1998, only 10 years after the new policy, an estimated 218 private domestic companies and 10 multinationals were supplying maize seeds to India, along with 27 public agencies and six universities (Gerpacio 2001). By 1997, growers offered 33 improved varieties and 88 hybrids to the market, with as many varieties coming from private as public sectors. One source (Singh 2001) estimates that, in 1997, the private sector had a market share of 87 percent of seed sales (multinationals 53 percent; national companies 34 percent), and public agencies only 13 percent (table 9.16). Industry sources estimate the share of private companies in seed sales more conservatively, at 75 percent in 2003. Private sector growth has been dramatic and is likely to continue.

Investments in Research and Development by the Public Sector and by Domestic and Foreign Firms

While most companies are engaged primarily in producing and selling seeds, some invest in research as well. In 1997, out of the 228 companies supplying maize, 28 managed breeding efforts. Private companies had 74 scientists employed in maize breeding, compared to 56 in the public sector. Even multinational companies conduct varietal development research locally rather than importing seed that might be unsuitable. Private research emphasizes germplasm improvement. The private sector spends twice as much on maize research as the public sector ($2 million compared to $1.2 million) (Gerpacio 2001). A few companies are involved in maize biotechnology research as well. India could, potentially, become a hub for regional breeding; hybrids bred in India are already sold in other countries, such as Nepal and Vietnam.

Table 9.16 Maize Seed Sales by the Public and Private Sectors in India, 1981–97
(Tons)

Agencies/type of seed	Years			
	1981–85	*1986–90*	*1991–94*	*1995–97*
Public seed agencies				
Improved OPVs	278	817	479	388
Hybrids	3,659	2,598	3,622	4,382
Private seed companies				
Improved OPVs	0	332	434	653
Hybrids	427	2,063	14,494	27,130
All Sales				
Improved OPVs	278	1,150	914	1,040
Hybrids	4,087	4,661	18,116	31,512
All materials	4,365	5,811	19,030	32,553
Market share of private sector (%)				
Improved OPVs	0	29	48	63
Hybrids	10	44	80	86
All materials	10	41	78	85

Source: Singh 2001.
Note: OPV = open-pollinated variety.

Public sector research and technology dissemination have complemented private breeding efforts. ICAR, which plans and coordinates crop research, has established directorates for focused research on individual crops. The Directorate of Maize Research (DMR), which absorbed the AICMIP, has a national mandate to undertake basic and strategic research for genetic improvement of maize. With more than two dozen research centers in different agro-climatic environments around India, the DMR conducts multilocational testing of germplasm, develops packages of practices, and generates technologies tailored to specific end uses (Morris and others 1998). State agricultural and nonagricultural universities, various government departments, and research organizations such as the International Maize and Wheat Improvement Center, also conduct maize research (Pal and others 1998). The DMRs and state agricultural universities have 102 full-time researchers studying maize, nearly 80 percent of them in germplasm improvement (Singh 2001).

Private sector hybrids are tested under the AICMIP in different centers, and promising hybrids are released by the Central Subcommittee on Crop Standards, Notification, and Release of Varieties. Such varieties are included in the field demonstrations of various government programs. Besides the parental lines and varieties developed by public sectors and made available to private firms on demand, new pesticide molecules uncovered by private researchers are tested for their efficacy.

Special Government Programs to Accelerate Technology Diffusion and Learning Among Farmers—With Nationwide Targets

In response to the increasing demand for maize, the government strengthened its efforts to increase production and productivity with the initiation of the Accelerated Maize Development Program (AMDP) in 1995/96. The program emphasized technology transfer through field demonstrations, officer training at the national and state levels, farmer training, production of certified seed, and interaction and education through conferences and workshops. A major expense was field demonstrations, considered an effective way of communicating new technology to farmers. Other important measures were the popularization of integrated pest management, farmer training, distribution of agricultural implements, and public media campaigns. In some states, production targets were achieved. Area planted in maize had increased by nearly 5 percent by 2001, while production increased 30 percent and productivity 24 percent in Karnataka. Currently, the government and public agencies involve private agencies, including NGOs, farmer societies, and self-help groups, in implementing the program, particularly in seed production, extension support, conduct of frontline and block demonstrations, and integrated pest management (IPM) demonstrations. The pattern of expenditure in Karnataka, a major maize-producing state, is shown in figure 9.3.

Private Sector Participation in Technology Adaptation Among Maize Farmers

Commercial success also depends on a company's ability to develop appropriate and better performing hybrids. Companies use a variety of farmer outreach methods, such as strong production and distribution networks, demonstrations, dealer promotions, and direct farmer contact. Companies also participate in the DMR's All India Coordinated Program trials by providing free seeds and technical support to farmers.

Figure 9.3 Major Expenses of Accelerated Maize Development
Program in Karnataka, 1999–2004

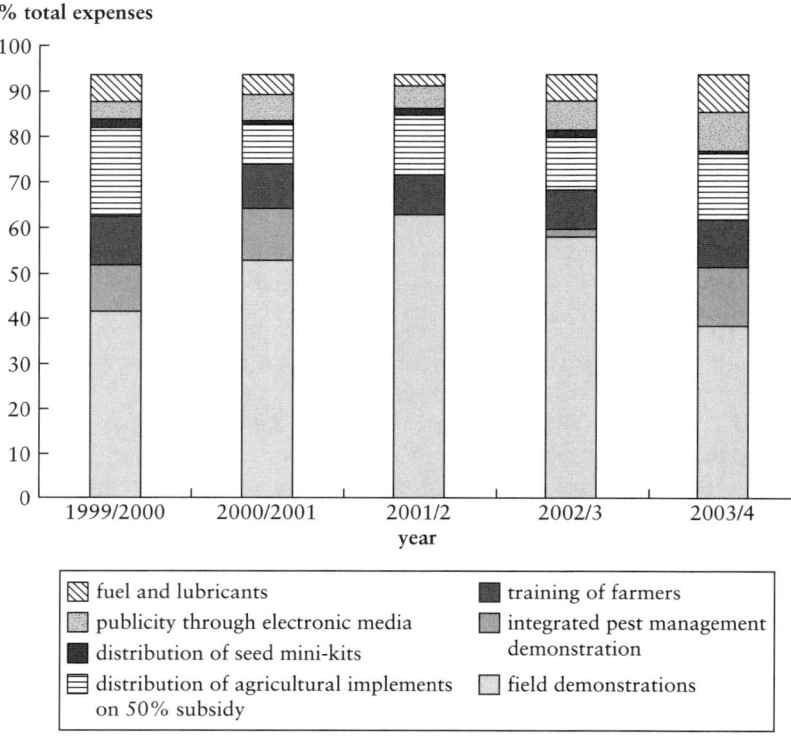

% total expenses

Source: Karnatka Department of Agriculture 2004.

Public and Private Sector Seed Technologies Compete

With privatization, the proportion of hybrids among improved varieties has increased. The private sector typically focuses on the high-
profit hybrid varieties. Maize yields can be increased either from the
seed of OPVs or hybrids. Seed from OPVs can be saved by farmers and
replanted without loss of yield potential. Indian farmers who grow
OPVs often save their own seed for the next year, purchasing new seed
only when the market price is not much higher than the grain price. In
contrast, hybrid seed must be purchased every year to retain yield potential. As a result, seed prices and returns to breeding are low on
OPVs compared to hybrids, and no private research is conducted on
OPVs. Between 1966 and 1998, 80 improved OPVs were released in
India, all the product of publicly funded research. In contrast, 31 hybrids were released out of the public sector between 1966 and 1998

Table 9.17 Characteristics of Maize Varieties Released in India, by Sector

	Public sector	Private sector		Total
		Private national companies	Multinational companies	
Type of maize				
Improved OPVs	80	0	0	0
Hybrids				
Single cross	8	14	2	16
Double cross	5	19	4	23
Three-way cross	10	40	19	59
Other	8	0	4	4
Characteristic (percent)				
Ecological adaptation				
Lowland tropical	20	1	3	4
Subtropical/mid-altitude	55	5	0	5
Temperate	3	0	0	0
Grain color				
White	20	4	8	12
Yellow	90	69	20	89
Grain texture				
Flint	71	19	5	24
Semi-flint	11	26	12	38
Dent	9	9	7	16
Semi-dent	14	19	5	24
Maturity range				
Extra-early (<100 days)	23	28	9	37
Early (100–110) days	34	32	8	40
Intermediate (110–120 days)	8	9	8	17
Late (120–135 days)	0	1	0	1
Extra-late (>135 days)	5	0	4	4

Source: Gerpacio 2001.

and, just since 1988, another 102 from the private sector (table 9.17). Public sector institutions are also focusing research on hybrids, especially those that have special characteristics, such as disease resistance or higher-quality protein, as well as trying to address new segments such as sweet corn.

Ineffective property rights protection in India complicates private breeding efforts and reduces benefits from private research. In the absence of strong intellectual property rights, private firms go to great lengths to protect hybrids from illegal duplication. Maize hybrids are produced by crossing different parental lines in single, double, three-way, or other crosses. Single crosses usually produce the highest yields. But because seed is typically reproduced off-site through contract farming, private companies must use more complicated crosses to conceal the parentage of the hybrid. Out of 102 private hybrids released by 1998, only 16 percent were single crosses. By contrast, of 31 public hybrids, 26 percent were single crosses (table 9.17). Multinationals in particular are not expected to release their highest-yielding single crosses in India until intellectual property rights are strengthened.

Both public and private breeding efforts are needed because the private sector is likely to focus only on higher-profit markets where production conditions are most favorable. Maize in India is grown in a wide range of conditions, from lowland tropical to subtropical/midaltitude to temperate. Varietal research focuses on breeding for ecological adaptation, grain color, grain texture, and crop duration. Although, in both the public and private sector, development of OPVs and hybrids has focused on improved varieties for favorable production conditions, the public breeding programs have, thus far, produced a broader range of materials than the private companies. On the other hand, companies are very active in identifying marketing opportunities and may be more responsive than the public sector to producer preferences.

Measuring Technology Adoption: Some Indicators

The contribution of improved cultivars to higher yields has been substantial. In states such as Karnataka, Andhra Pradesh, and Bihar, *wherever improved cultivar adoption has been higher, average yields have also been higher* (figure 9.4). State-level trials of improved cultivars indicate that maize hybrid yields were higher than improved OPVs, and both outperformed traditional varieties. *Hybrid varieties at their best, such as those in Bihar, produced yields comparable with average yields worldwide.* Hybrid maize varieties yielded as much as 6 tons per hectare in winter in Bihar. With improved varieties, the relative *profitability of maize was also higher.* In Bihar, the returns from maize were higher than alternative crops in both seasons, and twice as high as returns from wheat in winter (IARI 2004). In Karnataka and Andhra Pradesh, net returns from maize were higher than other monsoon crops.

The availability of improved germplasm may have contributed to rapid adoption of technology. Annual sales of maize seed increased dramatically from 4,500 in 1981–85 to 33,000 tons in 1995–97 and about 60,000 tons by the turn of the century (Singh 2001). With the availability of quality hybrids and OPVs in the 1990s, the land area under improved maize varieties grew from 40 percent in 1991 to 60 percent in 1999 (figure 9.5).

Figure 9.4 Maize Yields in Major Indian States, 1990–2001

yield (kg/ha, thousands)

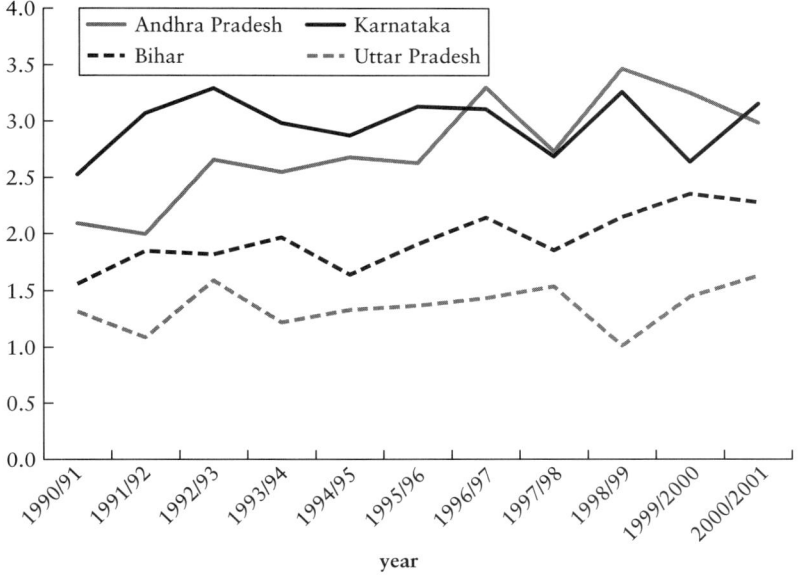

Source: Ministry of Agriculture, government of India.

Twenty-nine percent of India's maize land area was planted with hybrids by 1995, with another 22 percent in improved OPVs and the remaining 49 percent in traditional varieties (table 9.18). Within these numbers, however, there are significant differences across growing regions.

With the increased demand for maize in feed and the availability of improved varieties, farmers in favorable production areas such as Karnataka, Andhra Pradesh, and Bihar are enthusiastically adopting better plants for commercial production. In Andhra Pradesh and Karnataka, more than 90 percent of the maize land area is planted with improved varieties.

In Bihar, where private seed companies have been active, cold-tolerant, long-duration varieties have been introduced that produce very high yields in winter production. Industry sources suggest that winter maize in Bihar can yield as much as 9 to 12 tons per hectare. For the winter crop, which is mainly marketed, the adoption of hybrids has been 100 percent (Singh 2001).

In the leading areas with high adoption of improved maize varieties, the use of irrigation has also increased. An expansion in the irrigated

Figure 9.5 Area Under High-Yielding Varieties of Maize in India, 1966–99

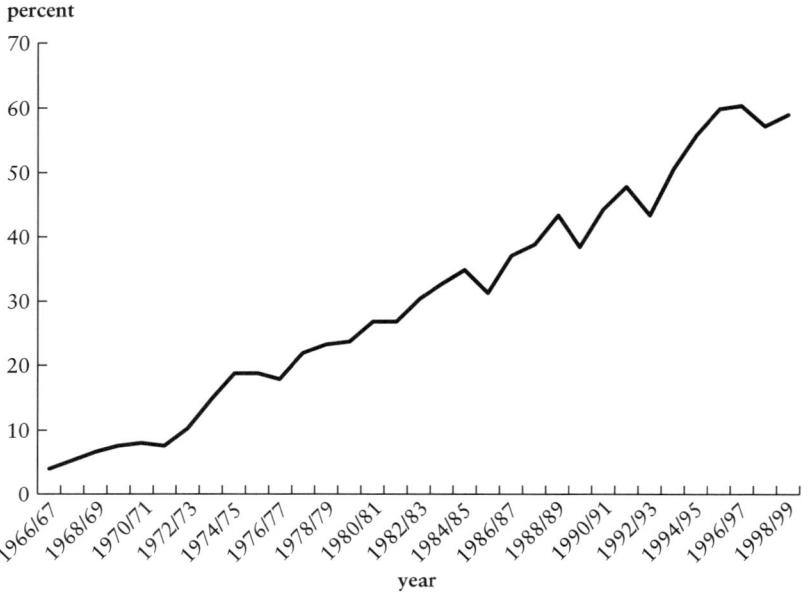

Source: Fertilizer Association of India.

Table 9.18 Proportion of Maize Area Under Improved Germplasm in Selected States and Nationally, 1995 *(Percent)*

| | Improved Varieties | | | |
State	Improved OPVs	Hybrids	All improved	Local varieties
Andhra Pradesh	15	79	94	6
Bihar	26	55	81	19
Karnataka	12	86	98	2
Madhya Pradesh	23	15	38	62
Rajasthan	18	8	26	74
Uttar Pradesh	31	8	39	61
Total	22	29	51	49

Source: Gerpacio 2001.

area for winter maize cultivation in Bihar and Andhra Pradesh has contributed to the increase in production.

In contrast, more than 60 percent of the crop area in the relatively marginal maize-growing states of Madhya Pradesh, Rajasthan, and Uttar Pradesh remains planted with traditional varieties, and the use of irrigation is very limited.

Public-Private Coordination

Public and private organizations engaged in the development of maize technology exchange products, services, and information through a variety of mechanisms, including international germplasm exchanges, public-private germplasm transfers, and collaborative varietal testing networks. CIMMYT coordinates the international maize germplasm distribution and exchange network that provides maize researchers in the public and private sectors with free access to promising experimental results. Public sector breeding programs provide private seed companies with access to their germplasm. The Tropical Asian Maize Network (TAMNET) provides a structure for comparing promising experimental materials and exchanging information. The network manages a multisite varietal evaluation program, conducting field trials and sharing performance data. This is in addition to the Indian trials system, which provides the basis for extension recommendations.

India's public investments in maize research have encouraged a vibrant private seed sector. Access to publicly developed breeding material has enabled domestic companies to enter the seed sector and to compete effectively with multinationals. While some firms have invested in in-house research, the bulk of smaller companies are marketing publicly developed varieties. The availability of inbred lines from public programs has enabled small firms to compete with larger ones and to exploit niche markets that large firms may ignore. The rapid growth in research within the domestic private seed sector has been possible because of a large pool of skilled and experienced scientists and technical staff groomed in the public research system.

Key Success Factors

The success of the maize sector in India is due to four main factors.

A strong research base, developed through a National Agricultural Research System that provided the structure for technology development and dissemination, was crucial. The state agricultural universities set up in the 1960s and 1970s helped prepare skilled manpower to conduct advanced research on crops. The private sector has effectively used this knowledge base and the skilled manpower to commercialize activities quickly. The role of public research and the education system continues to be important even after significant private involvement. There is still a need to develop public knowledge on certain knowledge-intensive issues in order to guide policy making and regulation. For example, on issues such as genetically modified crops, the knowledge base developed by public institutions can help alleviate asymmetric information in distribution. The presence of a public system in markets such as seed enhances competition. The public sector can focus research on areas that the private sector is unlikely to enter, such as newer market segments or geographic regions with production obstacles.

Significant demand for the commodity, whether from the domestic or international market, is essential. In the case of maize, the rapid growth of the feed industry generated high demand. This helped to provide remunerative prices and reduce output price risks. Technology adoption was critically dependent on market incentives.

Timely changes in government policies constitute the third critical factor. The change in the Seed Act was a watershed in the development of private sector participation and, therefore, enabled production to meet increasing demand. This increase has helped the poultry industry acquire feed at competitive prices and operate more competitively. The removal of restrictions on storage and transportation also helped the feed and starch manufacturers to source ingredients from anywhere in the country. Poultry industry growth is, in turn, generating more demand for maize. Recognizing market opportunities and devising appropriate and timely policies can help countries take advantage of their resources.

Coordinated public-private participation helped to enhance the growth of the maize sector and therefore meet the many challenges on the path to growth.

Conclusions

In the decades after independence, the public sector was entrusted to play a major role in seed technology development, reproduction, and distribution. This was an appropriate strategy, as all stakeholders—farmers, the government, and the private sector—were on a learning curve. The capacity to invest in research and to develop and devise appropriate technology was lacking in the private sector. Given the infancy of the seed market, few entrepreneurs would have invested. To the extent that risk was high and the risk-bearing ability of farmers and the private sector low, it was necessary that public agencies should serve in a leadership role. For more than two decades, the public sector played a critical role in generating improved OPVs and hybrids suitable for various ecological adaptations and grain-quality requirements. The major impetus came from two developments.

First, was the *New Seed Policy,* which allowed the private sector to import seeds and sell them in India. This enabled private companies to enter the seed marketplace and offer better performing seed materials to Indian farmers.

Second, was the *development of the livestock sector,* which created demand for maize and was able to sustain remunerative prices in the market. Income increases, urbanization, and falling prices had triggered a significant increase in the demand for beef and chicken, and therefore the demand for feed increased significantly. The private sector played a major role in the development of the poultry sector and helped improve its efficiency. These factors created demand for maize all over the country and the prices offered were competitive. Higher yields and reasonable prices changed the economics in favor of maize and encouraged farmers to adopt improved varieties.

Future policies could capitalize on these lessons and close the yield gap that keeps Indian maize cultivation just short of its full potential. In addition to lifting more small-scale farmers out of poverty by enabling them to use new technologies, these policies must open export markets in Asia; ensure food and nutritional security in India; provide legal protection of intellectual property rights so that companies can distribute high-yield, single-cross hybrids to India's farmers; build a foundation for India to become a maize seed production center; and continue all effective policies and the maintenance of a strong research base in public and private maize efforts.

Endnote

1. FAO data are more optimistic than these figures—estimating production growth at 5.1 percent for 1991 to 2001 and yield growth at 3.9 percent in the same period.

References

Evenson, Robert E., Carl E. Pray, and Mark W. Rosegrant. 1999. *Agricultural Research and Productivity Growth in India.* International Food Policy Research Institute Report 109. Washington, D.C.

Gadwal, V. R. 2003. "The Indian Seed Industry: Its History, Current Status and Future." *Current Science* 84 (3): 399–406.

Gerpacio, Robert. V. 2001. "The Maize Economy of Asia." In Robert V. Gerpacio, ed., *Impact of Public- and Private-Sector Maize Breeding Research in Asia, 1966–1997/98.* Mexico City: International Maize and Wheat Improvement Center.

IARI (Indian Agricultural Research Institute). 2004. "Technological Change and Production Performance in Irrigated Maize-based Agro-ecosystem: The Interplay of Economic, Technological and Institutional Factors." Agricultural Economics Research Report 2004-01, Division of Agricultural Economics, New Delhi.

Landes, Maurice, Seresh Persaud, and John Dyck. 2004. *India's Poultry Sector: Development and Prospects.* Market and Trade Economics Division, Economic Research Service, U.S. Department of Agriculture and Trade Report WRS-04-03. Washington, D.C.

Ministry of Agriculture. 2002. "Agricultural Statistics at a Glance 2002." Department of Economics and Statistics, Government of India, New Delhi.

Ministry of Finance. 2003. *Economic Survey 2002–03.* New Delhi: Government of India.

Ministry of Food Processing Industries. 2003. *2003 Annual Report.* New Delhi: Government of India.

Morris, Michael. L., R. P. Singh, and Suresh Pal. 1998. "India's Maize Seed Industry in Transition: Changing Role for the Public and Private Sectors." *Food Policy* 23 (1): 55–71.

Pal, Suresh, R. P. Singh, and Michael L. Morris. 1998. "Country Case Study: India." In Michael L. Morris, ed., *Maize Seed Industries in Developing Countries.* Boulder, CO: Lynne Rienner.

Rakshit, Mihir. 2004. "Some Macroeconomics of India's Reform Experience." In Kaushik Basu, ed., *India's Emerging Economy: Performance and Prospects in the 1990s and Beyond.* Cambridge, MA: MIT Press.

Singh, R. P. 2001. "An Interface in Public and Private Maize Research in India." In Robert V. Gerpacio, ed., *Impact of Public- and Private-Sector Maize Breeding Research in Asia, 1966–1997/98.* Mexico City: International Maize and Wheat Improvement Center.

Speedy, Andrew W. 2001. "The Global Livestock Revolution: Opportunities and Constraints for the Feed and Livestock Industries." Paper presented at the Compound Livestock Feed Manufacturers Association of India, Forty-Third National Symposium: Growth Prospects under Globalised Scenario *vis-à-vis* Livestock Production and Trade, September 29, Goa, India.

CHAPTER 10

Technological Change in Uganda's Fishery Exports

Rose Kiggundu

THE RECENT SUCCESS OF THE FISH-PROCESSING industry in Uganda demonstrates how technological change might be introduced to other industries and countries lagging behind. From the start of private industrial fish processing in 1988, firms enjoyed favorable macroeconomic conditions and contributed tremendously to Uganda's GDP. However, the improvement in economic conditions did not induce investment in technological change. Unable to meet sanitary and phytosanitary (SPS) measures enforced by the European Union (EU), the fish industry plunged into a crisis; in 1997, firms were locked out of their most lucrative export market. The toll of lost jobs, livelihoods, and the commensurate effect on export revenues exerted pressure on industry players to respond.

In an unprecedented manner, a new form of useful interaction (Freeman 1987; Lundvall 1988, 1992) unfolded across fish-processing firms, buyers, suppliers, policy makers, support agencies, and various private firms. Fish-processing firms learned to apply and improve standard procedures for health and food safety. The nutritional value and organoleptic appearance of fish was enhanced, and the conditions under which it was produced and exported improved tremendously. The knowledge and techniques introduced, new to the industry, enabled quality exports to reach high-value markets. We suggest that these improved circumstances derived from the introduction of process-related technological change. How and why this was possible in an African setting is one of the questions explored in this chapter.

The chapter concentrates on the factors that guided the introduction, adoption, and diffusion of technological change in the fish-processing industry of Uganda. Part one examines Uganda's macroeconomic environment and its contribution toward the revamping of its fish-processing industry. Part two analyzes the key factors behind technological change. Part three explores whether technological change was,

in fact, the main factor underlying the industry's success and concludes
by offering directions for further research.

Uganda's Macroeconomic Environment

Uganda has made remarkable economic progress since 1987 (Bigsten
and Kayizzi-Mugerwa 1999; Reinikka and Svensson 1999; Kasekende
and Ssemogerere 1994; Bigsten 2000; Holmgren and others 2001).
Instead of financing fiscal deficits through bank borrowing and print-
ing money, money supply was strictly controlled through restricted
government borrowing and selling of government paper at market
interest rates. Better revenue collection (through the establishment of
the Uganda Revenue Authority in 1991) coupled with better fiscal discipline resulted in reduced pressure on monetary growth. Authorities
managed to bring down inflation from its three-digit level of 196 percent in 1988 to 6 percent in 1993, which reduced real rates of interest.
Other indicators showed improvement as well: the GDP growth rate
was 8 percent in 1988, fell to 6 percent from 1989 to 1991, but increased again to 8 percent by 1993 (figure 10.1). However, between
1987 and 1993, there was a severe depreciation of the actual exchange
rate—from 558 to 1,333 Ugandan shillings per U.S. dollar—which led
to an increased cost of imports, since tariff levels were still relatively
high (Morrissey and Rudaheranwa 1998). But due to the reconstruc-
tion effort, demand remained high. Because export revenues could not
match a rising import bill, the deficit was financed by growing external debt. Total debt service dramatically increased from 39 percent
in 1987 to 73 percent in 1991, fell to 64 percent in 1993, improved
to about 18 percent in 1997, and fell to a low of 7 percent in 2001
(figure 10.1).

Before the 1987 Economic Recovery Program (ERP), prices for
agricultural commodities were determined by the state but were often
priced below the inflation rate, resulting in low real returns and disin-
centives to farmers. The export base was narrow; state-owned processing
and marketing boards were inefficient and crowded out private sector
activity. Foreign exchange was strictly allocated, and cash-crop exporters were required to relinquish their export proceeds at the official
overvalued exchange rate. The export-led growth strategy under the
ERP reversed all this: inefficient marketing boards were dismantled
and privatized, and trade licenses were replaced by a certification system. Foreign exchange gradually improved. First, authorities carried
out repeated devaluation and rationed foreign exchange through the
Open General License in 1988, then introduced the Special Import

Figure 10.1 Selected Economic Indicators, 1980–2002

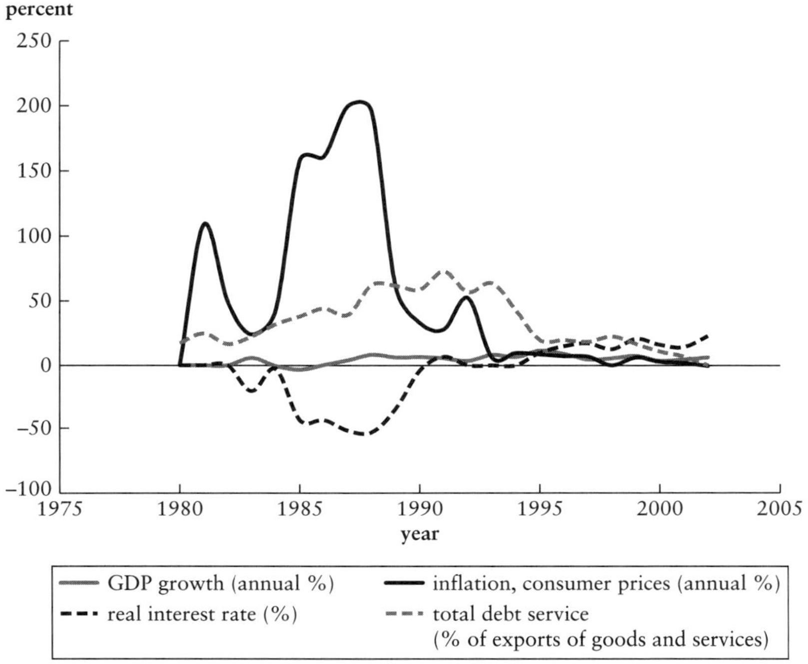

Source: World Development Indicators 2005, World Bank.

Program in 1989. But the existence of these schemes meant, in effect, that the exchange rate was not yet market-determined. In 1990, a decision was made to legalize the parallel foreign exchange market, resulting in the entry of a large number of private foreign exchange bureaus. Tariffs on imports were gradually rationalized. By 1993, a series of measures to attain a market-based exchange-rate management system, accompanied by continued liberalization, had further opened the economy.

In 1987, the Ugandan government introduced initiatives aimed at boosting the volume and diversity of high-value exports. An Export Policy Development and Analysis Unit was established, and research and agricultural extension services were directed to the effort. Emphasis was placed on nontraditional agricultural exports (NTAEs), including spices, cut flowers, fresh and dried vegetables, and fish. As a result of these efforts, between 1991 and 2002, the percentage share of traditional export crops declined, while the exportation of NTAEs increased.

In 1986, in the effort to expand exports and promote economic growth, a number of credit schemes were set up at the Bank of Uganda Development Finance Department. While some of these schemes undoubtedly performed fairly well, long loan-processing periods and high interest rates were problematic (Morrissey and Rudaheranwa 1998).

In 1991, in addition to changes in economic policy, a new investment code became law. It established the Uganda Investment Authority (UIA), which became responsible for the promotion of local and foreign investment. The code provided for tax relief, expatriation of funds, and first-arrival incentives (UIA 1991). Investment in fish processing qualified for incentives; however, in 1994, a survey of licensed investors found that lack of awareness and red tape prevented many from accessing these incentives (UIA 1995: 45). In 1997, a new code was introduced to replace the tax break aspect of the old regime. The new regime included allowances for plant and machinery, start-up costs, scientific research, training, and mineral exploration. However, both the old and new incentive regimes in Uganda played only a passive role in the development of local fisheries technology.

History and Organization of Uganda's Fisheries Sector

The bulk of the fish exported from Uganda—mainly Nile perch, tilapia, and silver cyprinid—comes from Lake Victoria, an inland lake with a surface area of about 68,800 square kilometers and bordered by three countries: Tanzania, Uganda, and Kenya. Fish from the lake provide livelihoods and an important source of protein. In Uganda, about 700,000 people work in the fisheries industry (Nsimbe-Bulega and Akankwasa 2002). Uganda's Fisheries Master Study (DCI 1999) identified five key segments of the industry: the fishermen; the collectors, who transport the fish to sales outlets at major landing sites; the boat builders and suppliers of fishing gear; fisheries-management personnel (including those in research, extension services, and training); and fish distributors. Until 2002, government policy permitted only local firms to engage in fishing, which explains the dominance of local operators.

Traditional processing, which involves sun drying, smoking, and salting fresh fish, has contracted as artisans have been reduced to using lesser quality and juvenile fish often rejected by industrial fish-processing plants, forcing a majority of the women who previously dominated this segment of the industry to switch to nonfishing activities (Olinga 2000). The change has attracted only minimal attention from government and support agencies.

Industrial fish processing began in 1948 with the establishment of the state-owned Uganda Fish Marketing Corporation, which focused on providing salted and frozen products for Europeans living in East Africa, but the domestic and regional markets were undemanding. Health officials regularly visited fish-filleting plants to monitor hygiene conditions. There was also some effort to "formulate regulations governing inspection of fish and fishery products . . . to make them conform to international standards" (Semakula 1967: 23). However, this effort remained totally voluntary. In 1975, the state enterprise collapsed, and, facing a deteriorating political and economic climate, the private filleting plants closed.

In the 1950s, the Food and Agricultural Organization (FAO) of the United Nations helped the government of Uganda assess the stock of Lake Victoria. Results showed that close to 80 percent of the lake's resources consisted of a species known as Haplochromines, considered by local officials to be "trash fish," as they had "failed to make it to the kitchens and dining places of domestic buyers in Uganda" (field interviews, 2002). The FAO study sparked a debate similar to the earlier one over Haplochromines in Lake Kyoga. To solve that problem, the government introduced Nile perch to feed on the Haplochromines. Indigenous to Lake Albert, Nile perch was already considered a table fish in northwestern Uganda. Its introduction into Lake Kyoga helped to reduce the Haplochromines, while converting the Nile perch into table fish in the communities around Lake Kyoga.

The question was whether or not the Lake Kyoga experiment could be emulated on Lake Victoria. Geheb (1997) discussed some of the arguments for and against the introduction of exotic species. Those in support of its introduction cited the success of the Lake Kyoga experiment, while those against it called for scientific investigation first, a debate that dated back to the 1920s (Graham 1929; Worthington 1932), but this debate was never resolved. In the late 1950s, amid great controversy, Nile perch was surreptitiously introduced to Lake Victoria from the Ugandan side of the lake,[1] with further introductions in the early 1960s.

Between 1983 and 1989, Nile perch landings rose from 1,400 tons to 100,000 tons (Namisi 2000; Nsimbe-Bulega and Akankwasa 2002), and, as the population grew, opportunities for private fish processing grew as well. On the Kenyan side, Nile perch were exported mainly to the European Union. However, the political and economic climate during the Idi Amin regime did not allow Uganda to follow her neighbor. Private industrial fish processing was introduced in Uganda much later, in the 1980s and early 1990s, after the overall political, investment, and economic outlooks had improved.

The first private industrial fish-processing plant in Uganda was established in 1988. In 1991, after Uganda's Ministry of Trade imposed a ban on the export of unprocessed whole Nile perch, an influx of investors from neighboring Kenya crossed into Uganda to process perch that previously had been transported whole for processing in Kenya (Nsimbe-Bulega and Wadanya 1999). The years from 1988 to 1993 were marked by the rapid rise of fish-processing plants and a steady improvement in macroeconomic indicators (figure 10.1). It is likely that a positive macroeconomic outlook attracted investors and accelerated a private sector response to the new business and export opportunity that arose from the surge in Nile perch landings. But it is also possible that without a deliberate policy to stop raw material transportation out of Uganda for processing in neighboring Kenya, some processing firms might not have located in Uganda.

By 1990, three firms in Uganda exported frozen (–18°C or below) fillets. Initially, filleting plants sent frozen exports to the European market, but in March 1989 experiments with fresh, or chilled (−0.5°C to −1°C), exports began, fetching a better price. Initially, 20 fish-processing plants began to set up in Uganda, but only 14 became operational (Nsimbe-Bulega and Akankwasa 2002). Of these, only 11 were still in business in 1996/97. Nine survived the EU bans imposed after 1997, but the industry has expanded since, with the arrival of six new entrants as of April 2004 (Kiggundu 2005).

Failure to Support Technological Change

The rapid diffusion and adoption of Nile perch processing technology was a spontaneous development driven by private firms. However, the technology was relatively simple and did not meet international standards.

Up until the early 1990s, the Ugandan government managed a number of projects in the fisheries sector, including the production of fish nets, fish trawling, fish processing, distribution centers, a truck fleet, and the supply of fishing inputs. Many of these ran into management and financial trouble prior to their eventual privatization or closure.

In the early 1990s, new donor-led strategies aimed to scale down government involvement to sectoral planning and monitoring, resource evaluation and statistics, management measures and enforcement, adaptive research and extension, export promotion and quality control, education and training, and rural credit (Frielink 1990). The Artisanal Fisheries Rehabilitation Project suggested reorganizing the fisheries department into four independent divisions: statistics and planning, management and law enforcement, research and development, and extension and training.

Although these efforts were a good attempt at improving policy coordination and strengthening support systems, they were not attentive to technological change. The promotion, diffusion, adoption, and adaptation of technology soon fell into disarray. Research on lake resources was scattered and its dissemination problematic. Data and information on sustainable yields were sparse. Often, fish-processing firms and regulators were unable to access results generated through donor-funded studies. Even when the Fisheries Research Institute increased its capacity, access to knowledge remained limited (Kiggundu 2005). Management measures and regulations were often inconsistent or inadequately enforced. For example, mesh size requirements differed from lake to lake, and the Fish and Crocodile Act of 1964 had not yet been upgraded to meet changed conditions and requirements of the fish sector (Frielink 1990: 3).

For years, the sole source of undergraduate training for fishery officers was offered by the zoology department of Makerere University, but it provided only generalized training in animal science, not specific skills for fisheries. Training in aquaculture was minimal, and collaboration with other departments, such as food science or veterinary science, simply did not exist.

The Fisheries Training Institute (FTI) offered two-year diploma and certificate programs in boatbuilding and yacht making, but the courses were primarily for future government officials. Given what it had set out to do, Frielink (1990) argued that the institute would have done well to provide shorter courses to boat builders, fishermen, beach mechanics, and traders (Kiggundu 2005). The much-needed diffusion of skills to the fishing communities remained extremely low.

By international standards, sanitation was poor. Fish-landing sites lacked elementary infrastructure; ice, potable water, adequate shelter, and electricity and lavatories were often unavailable. At the factories, sanitary, health, and environmental conditions were inadequate, and the layout and structural design were unsatisfactory. While the Uganda National Bureau of Standards (UNBS) was aware of international guidelines to rectify these matters, translation of those guidelines into good fish-handling and -processing practices was difficult. Attempts by the government's department for fisheries resources (DFR) to improve and maintain higher standards fell short of their aims.

Even the emergence of the Uganda Fish Processors and Exporters Association (UFPEA) was not enough to increase technical support to the sector. Only strict enforcement of EU sanitary and phytosanitary regulations eventually galvanized the industry into making radical changes.

The Introduction of Technological Improvements

A number of factors and relationships were critical to the transformation of Uganda's fisheries industry—notably joint efforts across firms; action by the public sector; the involvement of quasi-public, private, and international standard-setting bodies; and the requirements of global buyers (table 10.1).

More Demanding Markets

The need to upgrade technology was dictated by developments in the international trading environment, and the technological standards were set by the European Commission (EC) through fisheries legislation, particularly Council Directive 91/493/EEC of July 1991. In 1997, the EC required (in Council Directive 97/296/EC) all developing countries to obtain authorization from the commission before exporting fishery products to the European Union. An authorization system was established whereby one list (List II) authorized countries on a "definitive basis to export to the European Union after a compulsory on-the-spot evaluation of the country's system," while the other (List I) authorized countries to export to the European Union on a temporary basis (Nathan and Associates 2000).

Kiggundu (2005) provides a detailed discussion of what it took to comply with the new fisheries legislation. Briefly, a local authority was appointed by the Ugandan government and approved by the European Union to oversee and manage the inspection process. Second, laboratories were designated and approved. Third, infrastructure and sanitary conditions at landing sites had to be improved. Fourth, the government had to ensure better hygiene and handling of fish throughout every link in the supply chain. Fish-processing firms would be certified only if they fulfilled requirements in three areas: plant layout, operations, and Hazard Analysis and Critical Control Point (HACCP) inspection. The list of compulsory requirements was long and complicated, and it required the restructuring of plants and huge financial resources. Henson and Mitullah (2004) provide an account of the considerable differences in the requirements applied to fish and fishery imports by the European Union, the United States, Japan, and Australia.

Table 10.1 Explanatory Scheme for Technological Adaptation in Uganda's Fisheries

Key factors	Before 1997–99	After 1997–99
More demanding markets via stringent standards and an effective framework for enforcement	No compelling reasons or incentives to introduce or implement technological change	Access to high-value markets only for exporters in full compliance with international and local standards
Enhancement of SPS management capacities for food safety and the associated knowledge infrastructure	Weak SPS management capacity and a lack of technical support	Strong coordination and accessibility of required knowledge as well as technical assistance through public sector agencies; increased SPS capacities in terms of awareness, setting and maintaining mandatory standards, legislation, surveillance, and inspection and diagnostic abilities
Sector-based policies and practices of technological change	Scattered efforts toward improvement; weak political interest and leadership	Greater effort by public agencies and the state to strengthen the sector; enhanced political will and decisive and open leadership
Firm-level competencies to respond to conditions imposed by more demanding markets	Relatively large firm size, making it possible to attract skilled personnel and benefit from technical assistance; favorable ownership conditions with financial backing and access to support from parent firms	

Note: SPS = sanitary and phytosanitary.

Almost six years after Council Directive 91/493/EEC, the European Union had not yet visited Uganda to monitor compliance, nor had Uganda made efforts to comply. Then, in February 1997, Spain detected salmonellae bacteria in Uganda's fish exports[2] and, with Italy, imposed a bilateral ban on the fish. Shortly thereafter, the European Union made a decision (in April 1997) requiring fresh and frozen Nile perch exports from all three countries sharing Lake Victoria to be systematically checked for salmonellae as they entered the EU market. As Uganda began to deal with new exporting standards, a cholera outbreak hit the country in December 1997, and on December 23, the European Union placed a ban on imports of fresh fish from Uganda, Kenya, Tanzania, and Mozambique (Nathan and Associates 2000). After an EU inspection mission in November 1998, the industry suffered another setback when anecdotal evidence suggested that pesticides may have been used to poison and capture fish. In March 1999, Ugandan authorities responded with a self-imposed export ban until the safety of the fish for human consumption could be guaranteed. One month later, the European Union suspended imports of fish products from Uganda, Kenya, and Tanzania. Only when all compliance requirements were fulfilled were exporting countries and firms permitted to reenter the EU market.

The importance of the fish sector to Uganda's economy created a sense of urgency to support fisheries' compliance with the EU-imposed SPS conditions. The response to the export crisis was a spontaneous, fire-fighting process. The government, the private sector, and international development agencies worked relentlessly to facilitate compliance with the European Union, but not necessarily to upgrade the technology support system, as they did not set out either to achieve or deny technological change. But the very process of complying with EU standards facilitated some technological change.

Enhancement of Sanitary Capacities for Food Safety and the Associated Knowledge Infrastructure

The public sector played two complementary roles in supporting the technological change: it enabled the compliance process, while also investing in the enhancement of food-safety management capacities and the associated knowledge infrastructure.

The government created a legal framework that effectively granted powers of surveillance and enforcement to the fish-processing plant inspectors in DFR. Previously, UNBS had been unable to fulfill its role as the competent authority. UNBS and DFR were both government agencies with responsibilities in the fish sector, but they fell under different ministries, making it difficult to have a clear line of command. UNBS fell under the Ministry of Tourism, Trade, and Industry, while DFR fell under the Ministry of Agriculture, Animal Industry, and Fisheries (MAAIF). Soon after DFR took over from UNBS, it also took on overall responsibility for responding to the export crisis. It set up task committees and developed national standards based on those stipulated by Council Directive 91/493/EEC. With the change in the legal framework, DFR now had powers to enforce compliance. Most important, the new local and national standards applied to all participants in the fisheries supply chain. "These efforts included formulation of national standards in line with international norms" (Ecaat and Odong 1999). However, DFR's financial and human resources fell short of its new responsibilities.

As mentioned above, the public sector also played a leading role in the direct enhancement of SPS capacities and the associated knowledge infrastructure. The government-based Uganda Integrated Program (UIP), supported by the United Nations Industrial Development Organization (UNIDO), which we will refer to as UNIDO UIP, provided technical assistance to the government, with quick and effective communication with the EU commission. It identified and hired consulting firms based in Europe to strengthen HACCP audit systems at DFR and to train fisheries inspectors and quality-assurance managers in fish-processing and -exporting firms, steps that were key to the reentry of the fish industry into the European market. The UIP also assisted DFR in obtaining office equipment and helped publish a manual on fish inspection that was developed by local fisheries scientists within DFR. Enhanced inspection made it possible to undertake regular and close surveillance of compliance. It also became possible to arrange joint inspection missions in collaboration with other specialist organizations. The strengthened fish-inspection service also facilitated entry to the U.S. market, where public authorities demanded approved HACCP systems.

The government faced a critical problem in the lack of suitable in-country laboratories for pesticide residues in fish products. A UNBS lab had been put in charge of this, but its capabilities could not meet EU standards. The European Union required Uganda's fish exports to be tested in Europe until laboratories in Uganda could be upgraded. Meanwhile, UNIDO UIP financed the process of upgrading Uganda's laboratories and assisted in introducing an internationally acceptable quality-management system. UNIDO UIP also provided support to a private lab owned by a locally incorporated Belgian firm, which was later approved by the European Union to conduct in-country analyses. The government's analytical laboratory was simultaneously upgraded (interviews at UNIDO UIP, July 2001).

The Lake Victoria Environmental Management Project (LVEMP), a government project funded by the World Bank, provided duty allowances, transportation, and logistical support that were critical for effective implementation of inspection and law enforcement services.

With financial support from UNIDO UIP and LVEMP, traditional boat builders were trained at the FTI. Out of this training process, two pilot boats designed to meet food safety and SPS conditions were handed over to UFPEA for trials and further assessment (author's interviews with FTI and UNIDO UIP, 2001/2).

The method of delivering services also changed. For example, the DFR, the UNBS, and the National Agriculture Research Organization's Food Science Research Institute conducted joint evaluation visits. These changes enhanced cross-learning and the flow of knowledge within the industry, even though further technological upgrades were still needed.

In 2000, with support from agencies such as LVEMP, a new formal course in fisheries and aquaculture was introduced at Makerere University to train specialized fisheries officers and strengthen collaboration between the zoology department and the faculty of veterinary science, which together developed and delivered the course.

The local private sector also participated. A new firm emerged to provide ISO quality-certification training. Its services were paid for by fish processors and exporters. Local firms were subcontracted to provide pest control and fumigation. Banks and leasing companies provided vital investment and working capital loans for restructuring and day-to-day operations. Fish-processing firms jointly searched and applied for technical assistance from the Center for the Development of Industry (CDI), a Brussels-based agency working within the European Union's Africa, Caribbean, and Pacific framework. CDI assisted with the implementation of EU requirements. Its experts visited plants to conduct practical demonstrations in hygiene control, chemical and microbiological testing, the optimal arrangement of facilities, waste treatment, and product flow.

The Role of Overseas Buyers in Developing Local Phytosanitary Capacity and Knowledge Infrastructure

Overseas buyers often play an important role in upgrading developing-country exports (Gereffi 1994; Sandee 1995; Schmitz and Knorringa 1999; Dolan and others 1999; Humphrey and Schmitz 2000; Schneider

1999). In 2004, interviews conducted with European buyers of Uganda's Nile perch as well as with fish-processing firms in Uganda highlighted the important advocacy role played by European importers. They formed an association (ENPIA) that served as an information broker between fish processors in Uganda and the EC in Brussels. ENPIA updated EU officials in Brussels on the improvements made by fish-processing firms in Uganda and put pressure on officials to schedule assessment missions to Uganda. It disbanded after the European Union lifted the ban on fish exports from Lake Victoria. Although buyers supported the restructuring and upgrading effort of a few Ugandan firms, they generally played a minimal role in helping firms comply with EU rules. The exchange of information between ENPIA and the UFPEA, however, proved useful (author's interviews, 2004).

The firms that received support from their overseas buyers obtained preshipment financial backing as well as soft loans to upgrade vehicles and equipment required to ship exports. Overseas buyers made limited investments in the capture and delivery of fish to landing sites, mainly because prevailing policy permitted only locals to exploit the fisheries. Through contractual arrangements, many buyers (processing firms) invested in normally unaffordable fishing equipment and materials. Some also supported their Ugandan suppliers with inland fish transportation.

Interviews revealed that buyers also were a source of product development ideas, designs, and "marketing know-how."[3] If one looks at fish processing as having three distinct stages—capture and delivery, transport to processing plants, and processing and transport to markets—it is not difficult to see why the circumstances for buyers' investments would be more favorable at the end stage, largely because this is the most opportune time to appropriate benefits associated with superior technical knowledge.

Sector-Based Policies and Practices of Technological Change

Perhaps the most effective incentives to Uganda's fish-processing industry were the efforts made on behalf of the government to promptly comply with the new EU regulations. When enforcement of Council Directive 91/493/EEC started in 1997, it was imperative that the state demonstrate clear vision, leadership, and a readiness to institute change. Changes were made in the legal framework, statutory powers were transferred to a new competent authority, and national standards were developed. It was also essential that the government create conditions that would allow private firms to invest in technological change. Working with two international development agencies and the

UFPEA, standard operating procedures for inspectors were established, eventually rescuing the fisheries industry in Uganda.

In Uganda, firms need an annual license to export fish. For each export consignment made, they are required to obtain a health certificate from DFR explaining the conditions under which the export product had been processed. Following the fish-export crisis, obtaining a license was possible only after DFR had inspected the plant and was satisfied that all required structural and operational standards had been met.

Investment incentives in Uganda are uniformly applied and not targeted toward particular industries or sectors. Fish-processing firms had always produced exclusively for the export market, making it unnecessary to have special incentives for the promotion of exports within this particular industry. Fish-processing and -exporting firms enjoyed the same capital allowances and tax breaks as any other industry, and although it is possible to have incentives and export performance targets related to raw-material diversification and value-added products, we did not find any evidence that these existed. It is important to realize that Uganda's investment incentive regime is still very much "hands-off" and thus inadequate for supporting technological change.

Firm-Level Competencies in the Fish-Processing and Fish-Exporting Industry

Processing fish is labor-intensive. When the fish arrive at the plant, they are sorted, weighed, descaled, and washed. Inevitably, this process loads wastewater with scales, off-cuts, fat, and a high biological oxygen demand (Nsimbe-Bulega and Akankwasa 2002). The fish are then graded and put into cold storage to await processing. Experienced cutters fillet and separate the fish flesh from the skeletons; the skeletons are placed in a byproducts collection container. Next, most fillets are skinned, though depending on the buyers' specifications, some are prepared with the skin on. Finally, the fat is trimmed off, they are weighed, graded again, and then either frozen or chilled in cold storage.

A key change introduced after the EU restrictions was the arrangement of processing operations according to the forward-motion principle. This permits a consistent product flow from the dirtier physical areas, such as descaling, gutting, and washing, toward cleaner areas, such as packing. The arrangement protects cleaned, prepared, and checked items from contamination. By 2002, the nine operating plants had ice-making facilities, refrigeration equipment, and fish-transportation trucks. Fresh/chilled products were transported to cold storage and handling facilities at Entebbe Airport to await shipment to foreign markets. Frozen fish were transported by trucks through Mombasa Port in Kenya and then by sea.

Using a set of qualitative scores assigned by fisheries scientists, we observed that firms differed in the proficiency with which they had implemented change. The nine firms that survived the European Union–imposed restrictions between 1997 and 2000 were grouped based on export-performance data, intensity of innovation activities introduced, and data on proficiency. From this, we roughly derived three performance categories: the high achievers, underachievers, and low achievers.

Finance facilitates innovative change better when certain factors are present. Firm size, the proportion of educated personnel, technical assistance, and market orientation all influenced the intensity of innovation activities. Among the high achievers, export performance was lower for firms that made heavy use of high-cost loans (such as those typically available in underdeveloped financial systems as in Uganda). Lower export sales make it harder to increase profits and so finance learning and innovation activities. On its own, the competition for market share in exports failed to bring about learning and innovative change across these relatively large firms. Instead, the pressure to introduce technological improvements came from the imposition of enforceable standards by demanding markets. Similarly, nonmarket coordinating mechanisms played a vital role.

Among the underachievers, a number of factors important for technological change were either absent or limited. These firms were early champions of upgrading; however, they later fell behind either in export sales or in their level of interactive learning (Lundvall 1988, 1992). Their financial position was not always appropriate for heavy investments.

Among the low achievers, many of these important factors were absent. Firms in this category were small, locally owned, and had only a few links to information and knowledge outside Uganda. There were few workers with higher education. For financing, one of the firms relied entirely on preshipment payments from its buyers in Europe, while another had a high debt-equity ratio. Operating margins were limited, as was the ability to undertake some of the learning activities. All Ugandan-owned survivors of the EU bans fell into the low-achiever category.

By 2004, the industry was dominated by a certain kind of internationalized firm, headquartered in Kenya and not the type of global foreign firms associated with superior technology. The technological knowledge they shared was neither state of the art, nor necessarily better than that adopted by local firms, but it was these firms that initially demonstrated the potential for industrial fish preparation and export.

A distinguishing feature of the regional foreign firm found in Uganda's fish-processing sector is that it often lacks the competencies that typically separate the more competitive global firms from local firms in areas such as research and development, product design, and global marketing. However, given their wider networks, capital, and operational bases, the regional firms were much more likely candidates for partnering with foreign firms. But even where such partnering opened up new possibilities to access new knowledge, the flow of knowledge from the global to the regional firm, especially in the guarded areas of design and marketing, was difficult in the absence of strong local systems and the capacity to absorb, produce, and use knowledge.

Technological Change and Export Competitiveness

At the landing sites, traditional methods of capture and delivery are still widely used. According to Nsimbe-Bulega and Akankwasa (2002), there are approximately 597 fish-landing sites along the Ugandan shores of the lake, about 30,000 fishermen, and 15,000 fishing boats, of which only 17 percent are motorized. Following the EU restrictions, landing sites were upgraded, some by private fish-processing plants, others by government. Meanwhile, public efforts have been made to encourage the adoption of good fishing and handling practices. However, many observers are concerned that the "increasing use of small-meshed nets and nonselective and environmentally damaging fishing practices (beach seine and mosquito nets) is undermining the potential of the resource to renew itself" (Kaelin and Cowx 2002: 4). The reversal of such practices will likely require a better system of incentives and support in addition to the strict enforcement of gear surveillance. For instance, reduction of postharvest losses at the point of capture will require the acquisition of motorized fishing boats that are easy to clean, durable, and able to carry ice. An effective support system would have to include, among other services, an awareness and training program to inform fishing communities and boat builders of the benefits they would derive from the adoption of specialized boats.

In response to the EU directive, some fish-processing firms introduced computer-aided devices for critical procedures, such as tracking yield and temperature. Others introduced automated methods and upgraded equipment, which opened up a further process of technological learning through product diversification. All firms substantially improved their in-house laboratory capabilities. The industry acquired strong knowledge related to HACCP and plant layout, as well as industrial fish preparation, handling, and exporting to complex and demanding markets. Firms diversified their buyers within the EU market and entered new markets in the United States and Asia. Some began to process byproducts that were previously disposed of. Private landing sites were developed to EU standards, while artisans acquired ice, boats, outboard engines, and fishing nets. Some firms invested in their own fish-collection boats, while others made preparations to engage in aquaculture. Most of the technological changes were in process and organization, rather than product-related. By the end of 2002, two years after Uganda's fish exports were permitted to re-enter the European Union, fish-processing firms had not yet learned to produce value-added products, but a publicly financed research and development effort was under way to support trial production of value-added products (Kiggundu 2005).

While these improvements are significant, they are insufficient to sustain competitiveness. Generally, Nile perch imports enter the European Union as a minimally prepared, semiprocessed commodity. European firms add value to these imports by reprocessing them or by simply branding and repackaging them for consumer convenience. Reprocessing opens up upgrading possibilities, such as retail packaging, for Lake Victoria perch producers (author's interviews, 2004). The fish industry in Uganda needs to make continuous technological improvements to compete in ever-changing global markets; we speculate that the fish industries in Kenya and Tanzania need to take similar actions. Given overexploitation of the natural fishery resources and subsequent efforts toward regional conservation measures, improving the export value of the existing harvest in all three countries is vital.

Two important factors should be emphasized. First, it is useful to take a holistic approach in determining the nature and range of technological changes that must be made by the fish industry in the Lake Victoria region. To be fully effective, technological improvements must extend over all stages of fish production. The experiences of the fisheries sectors in Kenya (Henson and Mitullah 2004) and in Uganda (Kiggundu 2005) suggest that public agencies play a vital and multifaceted role in enhancing technological improvement for greater competitiveness. This role includes, but is not limited to, the creation of an enabling framework. In addition, the public sector needs to intervene by directing and supporting capacity-building initiatives. In fact, the Ugandan case underlines the vital role of public sector efforts (UNIDO UIP, for example) in providing overall leadership and coordination of systemic learning, institutional change, and continued interaction among the various players.

A shift from the mere preparation and export of whole and semi-processed fish products to the export of products such as crumbs, juice, marinated fish, fish pellets, flours, fish meal, and tray packs would require the introduction of new equipment and technology, and possibly a modified plant layout. But with the support of overseas buyers and public agencies, producers could take over the production of these consumer-ready products from the reprocessors in Europe. Public agencies could assist by investing in the development of vocational training and in identifying experts to provide technical assistance. In addition, publicly financed research will have to provide critical information on the reproductive biology, spawning habits, and growth potential of the Nile perch (and other species with export potential), given that "current research on the Nile perch has [so far] concentrated only on their stock abundance and distribution, exploitation patterns and general biology (growth and feeding habits)" (Kaelin and Cowx 2002). Public efforts in all three countries will have to continue building awareness so the industry can take advantage of fish farming, which has been successful in Southeast Asia (Kaelin and Cowx 2002: 5). A set of enforceable product- and process-related standards would also be needed.

The government would have to create better incentive structures and have the ability to absorb risks and costs associated with translating the fish-farming research undertaken at the government's Kajjansi fish-farming research center (KARDC) into private sector innovations for export-oriented aquaculture.[4] For this to occur, more and better linkages will have to be built between private sector agents (local and foreign) and relevant overseas technology institutes, universities, and productivity centers. Public agencies can assist by training and demonstrating more efficient uses of energy and water as part of the broader conservation effort. Through publicly financed research efforts, firms will also have to implement means of obtaining better returns from byproducts.

Indicators such as yield need to be interpreted cautiously.[5] For instance, at the European Seafood Exposition of May 3–5, 2004 (Brussels), most buyers looked favorably upon process-related improvements in the countries that produce Nile perch, but also complained that as the effort to achieve higher yield increased, so did the discoloration of the fillets. In the end, an increase in yield might be a good measure of producers' efficiency, but not necessarily a good competitive strategy across quality-driven markets.

Overexploitation will need to be reversed through regional conservation measures, such as enforcement of the Lake Victoria Fisheries Management Plan, which requires effective surveillance systems and associated gear. While improvements in catch efficiency are critical to the viability of the industry, Kaelin and Cowx (2002) draw attention to the potential conflict between the need for diesel-powered boats for firms and sail-powered boats for artisan fishers, and the urgent need to reduce fishing pressure on an already overexploited population. Meanwhile, a more effective strategy for the management of public landing sites will have to be developed, and, as many commentators observe, most of these sites will require further improvement in infrastructure in the form of accessible roads and basic hygiene.[6]

International marketing is another area in which the industry is still weak. Firms in Uganda do not participate in the distribution logistics and marketing stages of Uganda's fish exports.[7]

A complex set of logistics and market know-how is required in order to manage the fish supply chain successfully. This places a heavy burden on the African producer, whose resources and marketing capabilities are limited. A majority of importers, particularly those in Europe, play an active role in organizing air transportation and the timely delivery of Uganda's fresh fish exports. Air freight services are generally arranged by importers/buyers either through a chartered service or through contracts with airlines. Importers manage and organize an efficient logistics system that ensures prompt deliveries to the next player in the supply chain until the product reaches its final buyer. As previously mentioned, frozen products are often transported by truck by independent large firms specializing in transportation.

To obtain a crude analysis of the prospects of Nile perch in the EU market, we asked importers and other buyers if in five years (from 2004) the percentage of Nile perch they bought from Uganda, Kenya, and Tanzania would increase or decrease and why. We also asked for their assessment of what specific aspects each of these three countries had to improve.[8] Apart from its poorer ranking on export handling and airport facilities, Tanzania was ranked higher than Uganda and Kenya in areas such as consistent product quality, filleting and skinning skills, and availability of fish. All three countries ranked low on product development. On market trends, the picture is that of a positive outlook, especially if Nile perch are presented in a form that meets customers' needs. As a commodity, prospects for fresh Lake Victoria perch are good largely because it is considered to be a close substitute for cod. Prospects are also good for value-added products, such as breaded or frozen fish fillets. However, some importers considered the ongoing overexploitation of the Lake Victoria perch as a good reason to look elsewhere for cod substitutes. In other words, various fish species, some of them close substitutes, are marketed such that market share can quickly erode if customers' requirements are not met satisfactorily. It is therefore essential for producers of the Nile perch from Lake Victoria to continuously innovate in order to meet the changing needs of their customers overseas.

Did Fish Exports Grow Because of Technological Change?

There has been an increase in the value of Uganda's fish exports. Export values dropped from $39 million in 1996 to a low of $28 million in 1997, rose to $29 million in 1998, and were about $34 million by 2000. There was a sharp increase to $79 million in 2001, and the exports rose even higher, to about $88 million, in 2002 (table 10.2).

The increase in value between the years 2000 and 2001 is associated with higher export quantities, but this does not appear to be the only reason, because data also indicate that the export quantity fell from 28,000 to 25,000 tons, but value still increased from $79 million to $88 million. The ratio of value to quantity increased from 2.8 in 2001 to about 3.4 in 2002. It was maintained at 3.4 in 2003, in spite

Table 10.2 Changes in Uganda's Fish Exports, 1991–2003

Year	Quantity (tons)	Value ($ millions)	Value/quantity of fish exports	Share of fish in nontraditional exports (%)
1991	4,751	5,309	1.1	12
1992	4831	6,451	1.3	20
1993	6,037	8,807	1.5	13
1994	6,563	14,769	2.3	11
1995	12,971	25,903	2.0	12
1996	16,396	39,781	2.4	17
1997	9,839	28,800	2.9	13
1998	11,604	29,733	2.6	22
1999	13,342	36,608	2.7	18
2000	15,876	34,363	2.2	16
2001	28,153	79,039	2.8	28
2002	25,525	87,945	3.4	31
2003	25,111	86,343	3.4	Not available

Sources: Department for Fisheries Resources, Uganda; Statistical Abstracts, 1996, 2001, 2002, Uganda Bureau of Statistics.

Figure 10.2 Unit Value of Chilled Fish Exports, East Africa and World, 1990–2001

thousands of US$/ton

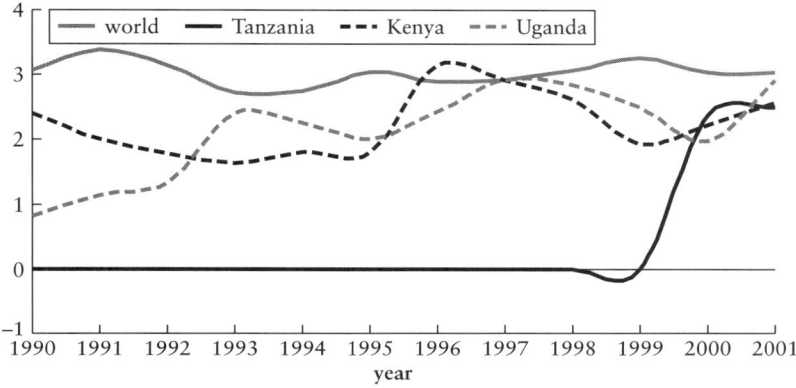

Source: FAOSTAT.

of declining export quantities. Figures 10.2 and 10.3 show a similar trend in Kenya and Tanzania. The average unit value of chilled fillets has been more or less the same in the rest of the world, staying close to $3,000 per ton (figure 10.3).

In the early 1990s, the unit value of chilled fillet exports from Uganda and Tanzania was significantly lower than the worldwide unit value. Between 1995 and 1997, the worldwide unit value and that of Lake Victoria exports almost converged. Thereafter, the unit value of chilled Nile perch exports from the three countries declined due to the restrictions placed on these exports, especially from 1997 to 2000. After 2000, and despite a more or less stable unit value worldwide, the unit value of Nile perch fillets recovered to the level they held before the EU restrictions.

Similarly, since the 1990s, the unit value of frozen Nile perch fillets has fluctuated widely compared with the generally stable trend of fillets in the rest of the world (figure 10.3) for two main reasons: the periodic decline in the supply of Nile perch and the imposition of EU restrictions on Nile perch imports into the European Union at that time. Prior to those restrictions, the major factor affecting the volume of fillet exports was inadequate raw materials. EU restrictions imposed an additional problem, but from figure 10.3, we can see that despite ongoing problems with supplies, the unit value of exports from Tanzania and Uganda (the two countries for which comparable data are available)

Figure 10.3 Unit Value of Frozen Fish Exports, East Africa
and World, 1990–2001

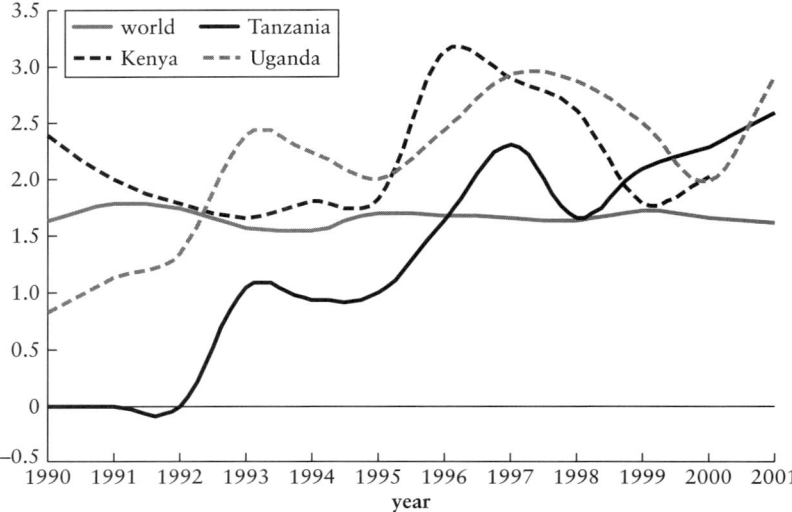

thousands of US$/ton

increased sharply after the year 2000. The unit value of Tanzania's
frozen fillet exports began rising much earlier, around 1998, reflecting
a faster recovery from EU restrictions and a greater ability to export to
high-value markets.

We attribute the ability to export to high-value markets and the
subsequent increase in the value of fish exports to technological
change. It is our contention that improvements in handling and
processing by fish-processing firms enabled producers to place more
valued fish products in markets that appreciate and attach a higher
premium to freshness, safety, and overall quality.

We did not compute an indicator on value added. More than 95 percent of the Nile perch produced by processing firms is exported. Data
on export sales is available for 1991–2003, but data on inputs, particularly fish procurements, are only available for the period during and
after the EU ban. Besides problems of choosing a suitable export defla-
tor, the difference between export sales and inputs might simply capture the favorable effect of a depreciated Uganda shilling on operations
of fish-processing and -exporting firms whose routine inputs for fish
export preparation are largely purchased in Ugandan shillings.

Why Wasn't Fish Farming Successful?

Fish farming in Uganda made some progress in the 1990s, but it has not succeeded in putting farmed fish on global markets. It has also been unsuccessful in relaxing the persistent problem of inadequate fish supplies for export processing. Using an innovation systems approach[9] can enable us to better understand why and how an industry or country was able to successfully introduce technological change while another failed to do so. From an agricultural[10] perspective, there are at least six factors that might explain why the success in fish processing and exporting has not yet spilled over to fish farming: the trigger or incentive to introduce change, structure of the industry, critical knowledge inputs and players for technological change, habits and practices related to technological learning, linkages and investment, and policy environment.[11]

The desire to maintain Uganda's fish export success, concerns over sustained supplies, and aquaculture development projects attracted political attention to fish farming. They also triggered a level of decisiveness that facilitated a policy shift toward fast-tracking of aquaculture. However, this shift did not impose and enforce conditions on the fish-farming subsector that would serve to trigger technological change, as was the case in the fish-processing and -exporting industry. As a result, incentives to introduce technological change in fish farming are weak.

In addition, the fish-farming industry in Uganda is still largely a subsistence affair oriented toward the domestic market. Sales of farmed fish in the domestic and regional markets are a very recent development (since 2000). Even so, the regional market has not yet shown signs of sophistication in quality and other export requirements. As previously discussed, the more demanding markets tend to place substantial requirements on producers, demands that can induce technological change. In terms of size and ownership, the fish-farming industry is composed of subsistence producers, small-scale producers, and a rapidly expanding group of medium-sized commercial producers. The industry had no foreign investors as of March 2004, though some firms from Zimbabwe, South Africa, Iran, and Norway had expressed interest (author's interviews with DFR aquaculture unit, 2004). Due to differences in industry structure, the fish-farming industry might therefore need a broader strategy and supportive measures to realize technological change, increased productivity, and competitiveness. The same policy pack that produced good results in the fish-processing industry might not necessarily produce similar results in the fish-farming industry.

The fish-processing and -exporting sector benefited from the presence of a number of knowledge inputs and players, all of which facilitated technological change. The fish-farming industry also has benefited from a similar knowledge base and critical players. For example, the industry has had national standards as well as aquaculture rules since 2003. Similar to the fish-processing industry, the fish-farming industry had quasi-governmental intermediaries that paid for investment into activities of a public goods character.[12] In addition to the U.K. Department for International Development (DFID), the aquaculture subcomponent of LVEMP has supported data collection and monitoring of aquaculture activities nationwide, as well as the research on the breeding and culture of endangered indigenous species.

More recently, the African Development Bank's Fisheries Development Project has contributed to human resource development at the FTI and to the establishment of a credit fund for fish farmers. FAO supported the development of aquaculture in Eastern Uganda. Similarly, the fish-farming industry has benefited from the establishment of the small-scale fish-farmers' association and commercial-scale fish-farmers' association, which have arranged successful study visits abroad. As in the fish-processing industry, champions among private fish farmers have facilitated technology diffusion by offering their sites for on-farm trials for fish feed and African catfish breeding. Further, the fish-farming industry has benefited from public investment in research and development at KARDC (Cajuns Aquaculture Research and Development Center, Uganda), which has engaged in a number of experiments. Furthermore, the fish-farming industry has benefited from strengthened policy coordination and services, for example, through the fish restocking program from MAAIF, the Uganda National Agricultural Advisory Services program, and the work in progress to establish a fisheries development authority. On human resource development, the situation is not different from that in the fish-processing industry: there are shortages of skilled persons, and the only trained personnel are the few graduates from FTI and Makerere University.

Despite a relatively wide support base, both industries are affected by skill shortages in the country. However, while foreign providers of knowledge played a vital role in the fish-processing industry, this link is still underdeveloped in the fish-farming industry. For example, fish farmers in Uganda are unable to produce fish seed of nonindigenous species, except the mirror carp. DFID supported the rearing of a tilapia strain from Thailand at KARDC, but did not venture into seed production (author's interviews with DFR aquaculture unit, 2004). Technological knowledge on breeding, health, and feeding systems for fast-maturing, nonindigenous, and exportable farmed fish seems to be absent. Unless more productive relationships with other actors are created, KARDC will be unable on its own to provide the knowledge needed for technological upgrading.

Both industries are affected by a national weakness in the science and technology system, insufficient links between local and foreign players, and incoherent measures to sustain investment in technological change. In addition, the focus on aquaculture for export is still conspicuously weak, and the extension and inspection service in the fish-processing industry is much more developed and effective than in the fish-farming industry.

Uganda's fish-farming industry is faced with production lock-in problems in two areas: there is reluctance to switch away from pond culture—a practice that farmers have known and perfected over the years—toward fish pens, cages, or fish tanks; and both artisan farmers and the new medium-sized commercial fish farmers have been reluctant to change from the three species they have traditionally reared— African catfish, tilapia, and mirror carp—to better species (author's interviews with DFR, 2004). There is little evidence of success, so far, in the government's reward systems to encourage a switch to more productive and efficient culturing systems. From an innovation systems perspective, the successful adoption of technologies, their diffusion and adaptation, and increased productivity of the fish-farming industry will require the reversal of traditional problems, habits, and practices through the creation of new reward and support systems.

Summary and Conclusions

In Uganda, public policy on technological change is still ambiguous and incoherent. The fisheries case study provides a good example of the conditions under which technological change might emerge and how it can be supported to flourish in technologically lagging African countries and industries. The case generates a few observations.

Uganda's fish-processing and -export industry was transformed in the late 1990s. Interviews conducted with 9 out of 11 fish-processing firms that survived the EU ban showed tremendous changes after the ban (Kiggundu 2005). Data on fish exports from Uganda show a sharp rise in value beginning in 2000/2001. We associate this increase in export values with the reentry into demanding, but high-value, markets, which was possible only after adhering to the standards specified by the EC. Some of those improvements were compulsory; others were not. But they have transformed Uganda's fish industry and helped it regain its position in high-value export markets. When the ban was lifted, the production of fish fillets and pieces grew by 13 percent per year, compared to a rate of growth in world trade of 4 percent per year, according to international trade calculations based on statistics from the United Nations' COMTRADE database.

A good investment climate and macroeconomic outlook was necessary for the fish-processing industry in Uganda to emerge and grow. However, this coexisted with the failure to invest in technological change—a situation that increased the vulnerability of Uganda's fish commodity exports to changing demands and conditions in global and regional markets.

In the fish sector, technological change did not take care of itself. The pressure and sense of urgency to overcome the long tradition of withholding investment in local technological change came from enforcement of the EU's Council Directive 91/493/EEC. Thus the fisheries case supports the observation that investment in technology change may not occur through invisible market forces. In the case of Uganda's fisheries, these forces were simply too weak to induce technological upgrading. Standards imposed on producers exerted strong pressure to upgrade and stimulated the emergence of an improved support system. The support structure consisted of a combined effort among the government, international development agencies, the industry association, some buyers, and local and foreign private firms. The government played a central role in sustaining the pressure on firms to maintain standards.

By placing the conditional ban on fish imports, the EC did not set out to achieve technological change as an underlying objective of Council Directive 91/493/EEC. However, the standards and requirements for improved hygiene and food safety provided a set of benchmarks that guided players to respond swiftly and successfully to resolve the immediate and structural problems that the fish industry faced. However, the response was *reactive* and not part of a well coordinated *proactive* public policy to catch up or move ahead of the technological standards of developed countries. Hence, a number of improvements occurred in the industry, but critical linkages and further structural improvements were not introduced. It is possible that without well-targeted incentives, continued pressure, and public support, the momentum for technological change in the industry might simply erode with time.

The technological change that did occur in Uganda's fish-processing industry was significantly influenced by orientation to demanding markets, firm size, the high intensity of technical assistance provided, and access to a growing range of financing possibilities (Kiggundu

2005). Orientation to the demanding EU market provided the much-needed incentives[13] for technological change as well as a framework for standard setting and enforcement. By sustaining a good macroeconomic environment and enforcing rules, the government provided excellent leadership in enacting new laws and guidelines to support compliance with EU legislation. Larger fish-processing and -exporting firms gained access to a wider menu of financial sources and thus greater opportunities to invest in technological change. They had well-educated and trainable managers, making it easy to absorb new techniques and to train other workers in the correct implementation of HACCP procedures. In addition, dealing with specialized firms, all engaged in the same activity, must have made it easier for the DFR, UNIDO UIP, CDI, and UFPEA to intensify support services and keep service-delivery costs low while maintaining outreach. These efforts all facilitated the diffusion of knowledge, skills, and resources to foster technological change.

Uganda must invest in SPS capacity and the associated knowledge infrastructure. The nation generally needs to focus on technical change as an important driver of economic growth and development, which requires effective delegation of powers to a high-profile competent authority, perhaps a renewed and upgraded Uganda National Council of Science and Technology (UNCST). Through unambiguous policy and with the involvement of all relevant actors, the national competent authority on technological upgrading should provide leadership in organizing background research across carefully selected sectors; evolving standards and performance targets for technological change; and developing new reward systems, and support, and enforcement mechanisms—all with an appropriate budget. Technological upgrading solutions will have to take account of sector-specific requirements and conditions. In this context, the competent authority should encourage productive relationships across all relevant public and private agencies, such as the investment authority, banks, government ministries, universities, business associations, suppliers, buyers, and research and training institutes.

Directions for Further Research

Highlighted below are several research issues that could not be addressed in this study given the information available. The first issue concerns how Uganda is dealing with overfishing in Lake Victoria. There is an ongoing effort by the East African countries to better harmonize fisheries laws and regulations. For example, control of "open

access to fishing" is an issue that has been widely discussed by stakeholders, and many proposals have been made,[14] including licensing
vessels and fishermen, allocating quotas, enforcing regulations on fishing gear, restricting fishing days (especially during the spawning cycle),
and making greater use of value-added products. A thorough assessment of progress achieved would require field interviews with relevant
officers in all three countries, discussions with researchers and several
stakeholders (including fishing communities), and a comprehensive review of available secondary material.

The second issue is how much of the growth in fish exports was due
to growth in other inputs such as capital investment, domestic FDI,
and labor, and how much to technological change. As mentioned previously, the nine relevant firms in Uganda constitute too small a sample
for such an analysis. A larger data set could be assembled by including
fish-processing firms in Kenya and Tanzania.

Third, we were unsuccessful in our efforts to obtain information
about Uganda's strategy on financing student training in U.S. universities, repatriating skilled Ugandans from abroad, preventing brain
drain, and linking up with foreign universities. Reforms and initiatives
are under way in Uganda based on a white paper on education and
training. Examination of progress and identification of gaps in these
reforms and initiatives require extensive interviewing and careful
scrutiny of study results and progress reports.

Ideally, discussion of the evidence that technology transfer did
occur between foreign and domestic firms would require a conceptu-
alization of knowledge spillovers and a complete mapping of labor
mobility and other mechanisms of knowledge flow. We consider this to
be a full study in its own right, one that would be improved by including fish-processing firms in Kenya and Tanzania. Their inclusion
would make it possible to assess the systemic response of the fisheries
sector to the EU ban carried out in Kenya and Tanzania. Such a comparative study would also cover foreign direct investment, spillovers,
and regional cooperation on overfishing.

It is interesting that in supporting the development of other food industries, UNIDO UIP decided to follow the approach adopted in the
fish industry. It focused on four subsectors with export potential—
dairy, meat and poultry, fruits and vegetables, and baked goods—and,
within them, on 13 food-processing enterprises (other than fish) willing to introduce HACCP and good manufacturing principles. Coordinating the diffusion of lessons from the fish industry to other food
industries is a good example of the type of public-led effort required to
stimulate learning and knowledge flow. However, it is essential that research be conducted to examine factors that might explain the success

or failure of systemic improvements in the four subsectors where the experience of the fish industry was emulated. Such studies are critical for the development of other industries as well. The use of an innovation-systems framework for this type of industry diagnosis is likely to be very useful for policy.

Endnotes

1. See Jackson (1971) in Geheb (1997), and Ogutu-Ohwayo (1999).
2. The chronology of events during the EU export ban was first documented in Nathan and Associates (2000).
3. This is dealt with more extensively in the author's chapter in Oyeyinka and McCormick (eds.) forthcoming.
4. This is discussed further in the last part of this chapter.
5. This ratio is a crude measure of efficiency, since it compares quantities of final production available for export with quantities procured and can be calculated as export quantity/procurement quantity over time, as in Kiggundu (2005).
6. UNIDO UIP supported the introduction of hygienic fish-handling practices on the lake and at landing sites in conformity with EU requirements. The government is said to have invested in the upgrading of these public landing sites as well. Consequently, there was some improvement in basic hygiene at some of the official public landing sites in the form of fish slabs, fish shades, and lavatories. Detailed information on the scope and variety of improvements is not available. Despite the improvements, commentators seem to agree that, unlike private landing sites, many public landings are still in need of further improvement, especially with regard to potable water and other amenities. A recent paper (forthcoming) prepared as part of World Bank–led Diagnostic Trade Integration Study (2005) proposes a shift from the current approach to landing sites as a public sector responsibility to private management and development.
7. As of early 2004, some processing and exporting firms in Uganda were planning to set up marketing units in the European market.
8. Interviews were conducted across 12 buyers based in various European countries. Thus our results should be considered exploratory.
9. See Lundvall (1988, 1992). For recent contributions on application of the innovation systems approach and the learning concept to Africa, see the contributions by Mytelka and others and Wangwe and others in Muchie and others (2003).
10. We draw here on Mytelka (2004), Hall (2004), Mytelka (2000), Hall and others (2001).

11. Before we discuss these factors, it should be emphasized that the preliminary diagnosis presented here is meant to provide directions for further
research and is by no means a complete assessment of Uganda's rapidly
growing fish-farming sector. A more extensive inquiry is required for a better
characterization and diagnosis of technological upgrading and obstacles.

12. A U.K. Department for International Development project supported
awareness creation and communications related to fish farming, renova-
tions of aquaculture research infrastructure and capacity building of
government agencies, poverty-focused advisory services in aquaculture,
research funding to adapt technologies for farm feeds, the farmer-to-
farmer seed-supply program, experimental research in breeding catfish
without hormones, growing Nile perch in captivity, and the introduction
of red tilapia from Thailand.

13. In Uganda, the domestic market is much smaller and weaker (in terms of
purchasing power) than export markets in Europe and North America. In
these high-income markets, better products are rewarded with premium
prices. This was an important incentive for quality improvements that
facilitated technological change.

14. See, for example, Nsimbe-Bulega and Akankwasa (2002); Nanyaro and
Mbilinyi (2002); and Mungai and Obayo (2002).

References

Bigsten, Arne. 2000. "Globalization and Income Inequality in Uganda." Paper
presented at the conference on Poverty and Inequality in Developing
Countries: A Policy Dialogue on the Effects of Globalization, November
30–December 1, 2000, Organisation for Economic and Co-operation and
Development, Development Centre, Paris.

Bigsten, Arne, and Steve Kayizzi-Mugerwa. 1999. *Crisis, Adjustment and
Growth in Uganda: A Study of Adaptation in an African Economy.* New
York: St. Martin's Press.

CDI (Center for the Development of Industry). 1996. "Lake Fishing in
Uganda: The Nile Perch Dossier No. 1." Brussels.

DCI (Development Consultants International) in association with IDC and In-
cafex Consultants, Ltd. 1999. "Uganda Fisheries Master Plan Study." Ministry of Agriculture, Animal Industry and Fisheries, Republic of Uganda.

DFR (Department of Fisheries Resources). 2000. "Manual of Standard Operating Procedures for Fish Inspection and Quality Assurance." Ministry of
Agriculture, Animal Industry and Fisheries, Republic of Uganda.

Dolan, Catherine, John Humphrey, and Carla Harris-Pascal. 1999. "Horticulture Commodity Chains: The Impact of the U.K. Market on the

African Fresh Vegetable Industry." IDS Working Paper 96. Institute of Development Studies, University of Sussex, Brighton, United Kingdom.

Ecaat, Justin, and Ignatius Odong. 1999. "Investigating Quality Control Issues in the Fisheries Sector." In Godfrey Bahiigwa and Eugene Muramira, eds., *Capacity Building for Integrating Environmental Considerations in Development Planning and Decision-Making with Particular Reference to the Fishing Industry in Uganda.* A case study paper for the Economic Policy Research Center, Kampala, Uganda.

European Union, Council Directive 91/493/EEC. 1991. "Laying Down the Health Conditions for the Production and the Placing on the Market of Fishery products." July 22, European Union, Brussels.

Freeman, Christopher. 1987. *Technology and Economic Performance: Lessons from Japan.* London: Pinter.

Frielink, A. B. 1990. "Fisheries Development in Uganda." Working paper for the Artisanal Fisheries Rehabilitation Project, EuroConsult, Uganda.

Geheb, K. 1997. "The Regulators and the Regulated: Fisheries Management, Options and Dynamics in Kenya's Lake Victoria Fishery." Unpublished Ph.D. thesis, University of Sussex, Brighton, United Kingdom. Reprinted as LVFRP Technical Document No. 10. LVFRP/TECH/00/10. The Socio-Economic Data Working Group of the Lake Victoria Fisheries Research Project, Jinja, Uganda.

Gereffi, Gary. 1994. "The Organization of Buyer-Driven Global Commodity Chains: How U.S. Retailers Shape Overseas Production Networks." In Gary Gereffi and Miguel Korzeniewicz, eds., *Commodity Chains and Global Capitalism.* London: Praeger.

Graham, M. 1929. *A Report on the Fishing Survey of Lake Victoria 1927–1928, and Appendices.* London: Crown Agents for the Colonies.

Hall, Andrew. 2004. "Elements of an Agricultural Innovation System." Paper presented at the United Nations University Institute for New Technologies, Technical Centre for Agricultural and Rural Co-operation Workshop, February 8–14, 2004, Maastricht, The Netherlands.

Hall, Andrew, M. V. K. Sivamohan, Norman G. Clark, Sarah Taylor, and Geoffrey Bockett. 2001. "Why Research Partnerships Really Matter: Innovation Theory, Institutional Arrangements and Implications for Developing New Technology for the Poor." *World Development* 29 (5): 783–797.

Henson, Spencer, and Winnie Mitullah. 2004. "Kenyan Exports of Nile Perch: The Impact of Food Safety Standards on an Export-Oriented Supply Chain." World Bank Policy Research Working Paper 3349, Washington D.C.

Holmgren, Torgny, Louis Kasekende, Michael Atingi-Ego, and Daniel Ddamulira. 2001. "Aid and Reform in Uganda." In Shantayanan Devarajan, David R. Dollar, and Torgny Holmgren, eds., *Aid in Africa.* Washington, D.C: World Bank.

Humphrey, John, and Hubert Schmitz. 2000. "Governance and Upgrading: Linking Industrial Cluster and Global Value Chain Research." IDS Working Paper 120, Institute of Development Studies, University of Sussex, Brighton, United Kingdom.

Kaelin, Andrew, and Ian Cowx. 2002. "Outline of the Path Forward in Uganda's Fisheries Sector." Paper presented at the Presidential Conference on Export Competitiveness, The COMPETE Project (USAID), and the European Commission, February 12, Uganda.

Kasekende, L., and G. Ssemogerere. 1994. "Exchange Rate Unification and Economic Development: The case of Uganda, 1987–92." *World Development* 22 (8): 1183–1198.

Kiggundu, R. 2004. "Learning to Change: Why the Fish-Processing Clusters in Uganda Learned to Upgrade." In B. Oyelaran-Oyeyinka and D. McCormick, eds., *The African Cluster: Pattern, Practice, and Policies for Upgrading.* Nairobi: University of Nairobi Press.

———. 2005. "Innovation and Development: The Journey of a Beleaguered Nile Perch Fishery in Uganda." Ph.D. thesis, United Nations University, Institute for New Technologies/MERIT-University of Maastricht, The Netherlands.

Lundvall, Bengt-Ake. 1988. "Innovation As an Interactive Process: From User-Producer Interaction to the National System of Innovation." In Giovanni Dosi, Christopher Freeman, R. Nelson, Gerald Silverberg, and L. Soete, eds., *Technical Change and Economic Theory.* London: Pinter.

Lundvall, Bengt-Ake, ed. 1992. *National Systems of Innovation: Towards a Theory of Innovation and Interactive Learning.* London: Pinter.

Ministry of Finance, Planning and Economic Development. 1996, 1999, 2000, 2001, 2002. Statistical Abstracts. Uganda Bureau of Statistics, Republic of Uganda.

Ministry of Finance, Planning and Economic Development and Ministry of Agriculture, Animal Industry and Fisheries. 2001. "Government Interventions to Promote Production, Processing and Marketing of Selected Strategic Exports." Republic of Uganda.

Morrissey, Oliver, and Nicodemus Rudaheranwa. 1998. "Ugandan Trade Policy and Export Performance in the 1990s." Research Paper No. 98/12. Centre for Research in Economic Development and International Trade, University of Nottingham, United Kingdom.

Muchie, Mammo, Peter Gammeltoft, and Bengt-Ake Lundvall, eds. 2003. *Putting Africa First, The Making of African Innovation Systems.* Aalborg, Denmark: Aalborg University Press.

Mungai, D., and A. Obayo. 2002. "The Status of the Nile Perch Fishery and the Potential for Production and Marketing of Nile Perch Value Added Products in Kenya." CFC/FAO/COMESA Project on the Production and Marketing of Value-Added Fishery Products in East and Southern Africa

Phase I: Kenya Country Report, Lake Victoria Fisheries Organization, Jinja, Uganda.

Mytelka, L. 2000. "Local Systems of Innovation in a Globalized World Economy." *Industry and Innovation* 7(1): 15–32.

———. 2004. "Methodology Notes, Agricultural Innovation Systems in ACP countries." United Nations University-Institute for New Technologies (UNU-INTECH)/Technical Centre for Agricultural and Rural Cooperation ACP-EU (CTA) workshop February 8–14, 2004, UNU-INTECH, Maastricht, The Netherlands.

Namisi, P. W. 2000. "Socio-Economic Implications of the Fish Export Trade on the Fishers and Fisheries of Lake Victoria in Uganda." Unpublished master's dissertation submitted to the National University of Ireland, Cork. LVFRP Technical Document No.14. LVFRP/TECH/01/14. Socioeconomic Data Working Group of the Lake Victoria Fisheries Research Project, Jinja, Uganda.

Nanyaro, G., and H. Mbilinyi. 2002. "The Status of the Nile Perch Fishery and the Potential for Production of Nile Perch Value Added Products in Tanzania." CFC/FAO/COMESA Project on the Production and Marketing of Value-Added Fishery Products in East and Southern Africa Phase I: Tanzania Country Report, Lake Victoria Fisheries Organization, Jinja, Uganda.

Nathan and Associates, Inc. 2000. "Restrictions on Uganda's Fish Exports to the European Union." Unpublished working paper for the USAID-PSF Trade Policy Capacity Building Project, United States Agency for International Development/Private Sector Foundation, Kampala, Uganda.

Nsimbe-Bulega and A. Akankwasa. 2002. "The Status of the Nile Perch Fishery and the Potential for Production and Marketing of Nile Perch Value Added Products in Uganda." CFC/FAO/COMESA Project on the Production and Marketing of Value-Added Fishery Products in East and Southern Africa, Phase I: Uganda Country Report, Lake Victoria Fisheries Organization, Jinja, Uganda

Nsimbe-Bulega and J. Wadanya. 1999. "Fish Processing for Export and Utilization of By-Products." In G. Bahiigwa and E. Muramira, eds., *Capacity Building for Integrating Environmental Considerations in Development Planning and Decision-Making with Particular Reference to the Fishing Industry in Uganda* (pp. 48–57). Kampala, Uganda: Economic Policy Research Center.

Ogutu-Ohwayo, R. 1999. "The Impact of Nile Perch Harvesting on Fish and Fisheries in Uganda." In Godfrey Bahiigwa and Eugene Muramira, eds., *Capacity Building for Integrating Environmental Considerations in Development Planning and Decision-Making with Particular Reference to*

the Fishing Industry in Uganda. A case study paper for the Economic
Policy Research Center, Kampala, Uganda.

Olinga, Forough. 2000. "Gender, Resource Allocation and Fisheries Development: The Case of Buvu and Lulamba Islands, Lake Victoria." Occasional
Paper No. 8, Department of Women and Gender Studies, Makerere University, Kampala, Uganda.

Reinikka, Ritva, and Jakob Svensson. 1999. "Confronting Competition: Investment Response and Constraints in Uganda." Policy Research Working Paper No. 2245, World Bank, Washington D.C.

Sandee, H. 1995. "Innovation Adoption in Rural Industry: Technological
Change in Roof Tile Clusters in Central Java." Ph.D. thesis, Vrije Universiteit, Amsterdam.

Schmitz, Hubert, and Peter Knorringa. 1999. "Learning from Global Buyers."
IDS Working Paper 100, Institute of Development Studies, University of
Sussex, Brighton, United Kingdom.

Schneider, Denise. 1999. "The Role of Buyers in the Development of the Hotel
Furniture Industry in Kenya." IDS Working Paper 93, Institute of Development Studies, University of Sussex, Brighton, United Kingdom.

Semakula, S. 1967. "Survey of the Present Status of Fish Handling, Preserving
and Marketing in Uganda." Occasional Paper No. 1 of the Ministry of Animal Industry, Game and Fisheries, Fisheries Department, Government of
Uganda.

UIA (Uganda Investment Authority). 1991. "The Investment Code, 1991:
Statute No. 1" and 1994, 1997 statutory amendments. Kampala, Uganda.

———. 1995. "Survey of Licensed Investors, 1994." Policy and Planning
Division, Kampala, Uganda.

Worthington, M. 1932. "A Report on the Fisheries of Uganda: Investigated by
the Cambridge Expedition to the East African Lakes, 1930–31." London:
Crown Agents for the Colonies.

CHAPTER 11

Floriculture in Kenya

Meri Whitaker and Shashi Kolavalli

WORLD TRADE IN CUT FLOWERS AND foliage is nearly $4.5 billion per year. In recent years, new flower producers in regions with comparative advantages in growing climate and labor costs, particularly Israel, Colombia, and Kenya, have challenged the historical dominance of North America, Europe, and Japan as producers of floricultural products. Today, Kenya produces approximately $200 million worth of cut flowers and foliage annually, nearly all of it exported, with 94 percent of the exports going to the European Union. Kenya is now the largest exporter to Europe, with about 25 percent of the market share.[1]

Along with tea and tourism, Kenya's cut flower sector is one of the country's three most important sources of foreign exchange, producing about 14 percent of Kenya's total export earnings. Kenya is a major exporter of horticultural products that include flowers and fresh vegetables and fruits, with flowers accounting for more than half of these earnings. Both flower production and total horticultural production have grown remarkably in the past 30 years, a trend that shows no sign of abating. Between 1995 and 2002, Kenyan flower and horticultural export earnings both grew by more than 300 percent, during a period when overall export growth was only 40 percent (table 11.1). In the first three months of 2004 alone, the value of flower exports was about 44 percent greater than the previous year. Growth in the horticultural sector has been responsible for more than half of all export growth in the last decade.

Kenya has been the leader in cut flower production and exports from Sub-Saharan Africa. Today, about 60 percent of the African flower trade originates in Kenya. Kenya's success has had a significant spillover effect, serving as a model for the development of flower industries in other Sub-Saharan African countries, including Zimbabwe, Uganda, and Tanzania. In addition, there have been direct flows of physical and human capital from the flower sector in Kenya to other Sub-Saharan African countries. Farm managers and technical experts trained in the

Table 11.1 Cut Flower Exports, Total Horticultural Exports, and Total Domestic Exports from Kenya, 1995–2003 *(Billions of Kenyan shillings)*

Year	Flower exports	Total horticultural exports	Total domestic exports
1995	3.6	6.5	93.1
1996	4.4	7.7	113.9
1997	4.9	8.7	114.5
1998	4.9	9.7	114.4
1999	7.2	14.2	115.4
2000	7.3	13.9	119.8
2001	10.6	20.2	127.4
2002	14.8	26.7	133.9
2003	16.5	28.8	136.7

Sources: Columns 1 and 2: HCDA (http://www.hcda.or.ke/stats.asp). Column 3: Central Bank of Kenya, "International trade and finance statistics" (http://www.centralbank.go.ke/statistics/statisticsindex.html).

Kenyan flower industry have moved to other Sub-Saharan African countries to start businesses or take jobs in the flower industry.

Most of the estimated 5,000 producers in the Kenyan floriculture sector are smallholders farming a hectare or less of land, but 75 percent of flower exports are produced by several dozen medium- to large-scale enterprises on farms of 20 to 100 hectares. Employment of large numbers of semiskilled laborers on the larger operations and in ancillary industries has had a major impact on poverty. An estimated 40,000 to 50,000 farm workers are employed in the flower industry, with another 60,000 to 70,000 in ancillary industries.

This chapter characterizes the Kenyan cut flower sector and examines the contribution of different factors to technical change and growth in the sector. It is divided into three main sections. The first describes the state of the industry and its evolution, beginning with an overview of critical aspects of the floriculture industry. The second looks at the context in which the sector has grown, the growth of the overall economy and the agricultural sector, and the policy environment. The third discusses the state of technology in the sector and the institutions that facilitate "catching up" with global technological leaders.

Organization of the Industry

The exacting requirements of producing and moving flowers to distant markets, the need to respond to changing preferences of sophisticated consumers in Western markets, and the need to maintain an edge over competing production centers that attract global capital flows require producing countries to have efficient institutions, appropriate policies, and adequate infrastructures, in addition to suitable agroclimatic conditions. Before examining the present state and organization of the Kenyan floriculture industry, we review some critical aspects of the floriculture business.

The Floriculture Industry

Flowers are delicate and, once harvested, highly perishable, requiring careful management of growing conditions and postharvest handling to ensure product quality for consumers. Final product quality—including color, size, shape, and absence of disease and visual defects—determines marketability and price. Quality also depends on how the flower performs after it is purchased, that is, whether the buds open properly and the length of vase life.[2]

Because flowers have exacting growing conditions, moving production to locations with natural conditions that are close to optimal—that is, minimizing environmental modification costs—is a key factor driving the globalization of floriculture. The environment for flower production is primarily determined by agroclimatic conditions, such as temperature, light intensity, length of the day, and water availability. Appropriate growing conditions include mild temperatures with variation between day and night, as well as abundant sunlight and water. Flowers may be damaged by stresses such as high winds, rain, and pests and disease. Some flowers are more tolerant of suboptimal conditions and may be grown in open-field conditions. At the other extreme, some flowers must be grown in highly controlled greenhouse environments to produce top-quality blooms.

The greater the need to cultivate under greenhouse conditions, the higher the costs. The cost of a greenhouse structure depends on the degree to which it modifies the growing environment. Greenhouses range from simple frames with protective covering that simply shield the crops from natural elements, such as rain and wind, to those with the ability to heat, cool, and automatically monitor environmental conditions. The greater the capability to accurately vary conditions within the greenhouse, the greater the complexity of hardware that is needed, thus the greater the capital investment, operating costs, and technical skills required.

Cost of labor is another factor that drives the globalization of production. Labor utilization in floriculture varies from about 6 laborers per hectare in the Netherlands to 15 to 20 laborers in South Africa, and about 25 to 30 laborers in Kenya, reflecting relative labor costs. In addition to technical skills, floriculture requires effective management of semi-skilled labor. Harvesting cut flowers and foliage is highly labor-intensive because it must be done frequently and by hand in order to ensure quality. To gather each stem at optimal maturity, some flower types are harvested as often as three times a day. To minimize wilting, most harvesting is done when temperatures are cooler—in the morning or evening.

Effective and timely management of the supply chain from field to buyer determines the quality of the final product, and therefore the price that can be obtained. After harvest, the cut flowers must be moved to market quickly and with careful handling. Temperature- and humidity-controlled environments are essential to maintaining quality at all stages, from packing and storage to transportation.[3] Cargo transport must be fast and reliable. And the speed with which the flowers reach the final consumer determines the vase life of the product. Because some aspects of quality, particularly vase life, cannot be visually evaluated in the market, buyers put great stock in reliable sources. For that reason, vertical integration has been an important strategy for strengthening competitiveness and market position.

The cut flower market is characterized by a sophisticated consumer base. In recent years, consumers in the United States, Europe, and Japan have been increasingly willing to pay premium prices for high product quality, for certain types of flowers (such as roses), and for new and unusual varieties. Today, only the highest quality flowers are traded on the international market. To maintain competitiveness, growers must produce consistently at the highest level of quality, respond rapidly to changing market demand, and be constantly innovating in production.

Timely and accurate access to market information is essential to reduce risk and remain competitive in flower production and marketing. Increasingly, vertical integration of the supply chain that connects producers to consumers is used to improve information flow and reduce risks. In recent years, European consumers have demanded that flower producers meet global standards for environmental and social practices. A number of organizations connected to the worldwide flower industry have developed standards and enforcement mechanisms to ensure that flower production practices promote good labor conditions and do not harm the environment.

The global floriculture industry is based largely on technologies and systems, from flower breeding and varietal improvement to the systems used to plant, protect, harvest, store, and market the flowers. Rapidly changing consumer preferences and industry marketing strategies have spurred continual investments in varietal research and development, as well as new production, storage, and marketing techniques.

Most of the technological improvements are developed in Europe, the United States, and Israel, where most flower breeding takes place. Varietal development is capital intensive and high risk, requiring relatively long-term investments and, at the same time, the flexibility to respond rapidly to changing preferences. Much of the plant material used worldwide comes from European breeders that can closely monitor and adapt to changes in consumer preferences and market demand because of their proximity to the major trading markets.

Changing consumption patterns and the emergence of new production centers are creating new challenges for producers. Until recently, carnations were the most important cut flower traded on the world market because of their robust character, long vase life, and relatively low cost. Although carnations remain an important component of the flower trade, in the last 10 to 15 years, rising demand for roses has led to a worldwide decline in carnation production and a rapid increase in rose production. During that same period, production of carnations and roses shifted from Europe and the United States to Israel, Latin America, and Sub-Saharan Africa. In Kenya, the unit value of production under roses is expected to increase by more than 15 percent within the next year alone.

The global market is also characterized by growing competition, particularly from Latin American and Sub-Saharan African countries, such as Tanzania, Uganda, Ethiopia, and Rwanda. To some degree, these countries are benefiting from spillover in experience, skills, and resources from Kenya and other industry leaders. Competitiveness is also affected by the trade preferences that it enjoys relative to other producing countries, and by environmental and social performance demanded in the markets it caters to. Kenya is concerned about losing its status as a least developed country—a status that Uganda and Tanzania will continue to have—when the Cotonou Agreement takes effect in 2008. African producers that cater to European markets are required to adhere to higher standards of environmental performance than are Latin American producers that cater to North American markets.

But although floriculture is expanding in neighboring countries, it has not been at the expense of Kenya's flower industry. Kenya is expected to maintain competitive advantage over its neighbors because of its climatic conditions, policies, trained workforce, and support industry that has developed. It may even emerge as a service center for the region.

Production in Kenya

Export agriculture in Kenya dates back to the early days of British rule. European settlers introduced coffee-growing in the temperate Kenyan highlands at the beginning of the twentieth century and developed the first tea estates a few years later. Fruits and vegetables had been traded regionally from Kenyan coastal ports even earlier. In contrast to coffee and tea, the colonial government made little effort to develop horticultural production in Kenya, and only small-scale production and trade occurred before World War II.[4]

During World War II, the provisioning needs of Allied Forces in East Africa led to the development of the Dried Vegetable Project. The Department of Agriculture built processing factories, developed irrigation schemes, and provided inputs and technical support to more than 13,000 smallholder African farmers to produce vegetables on contract. This project was the model for later smallholder schemes and helped jump-start an export horticulture industry.

After World War II, British controls on dollar imports provided an opportunity for Kenyan farmers to move into export horticulture. The agriculture department set up research and extension programs for horticultural production and processing into canned goods for export. The government further supported the industry with price supports and other measures. By 1960, both large-scale European farmers and smallholder African farmers were supplying produce for canning.

In 1952, European farmers set up the Horticultural Cooperative Union (HCU) to facilitate the marketing and export of fresh produce. By the late 1950s, with little government involvement, HCU was exporting fresh produce on a small scale by air to the United Kingdom for the winter off-season market. Cut flowers were also being produced by European farmers and exported by air on a very small scale, while some smallholder African farmers were growing cut flowers for the local market.

Independence in 1963 brought a new government and major changes in the horticulture sector and its relationship with the government. The new government had little funding for horticultural research and extension. The Horticultural Crops Development Authority (HCDA) was established in 1967 as the implementing arm of the Ministry of Agriculture for horticulture, but it had little funding or authority to regulate the sector. Government efforts to promote horticulture centered on foreign investments in fruit and vegetable processing for export, particularly as a means to strengthen the incomes of smallholder farmers.

During this period, the government encouraged growth in tourism as a source of foreign exchange. As the tourist industry grew, so did the volume of air traffic between Europe and Kenya. Airfreight chartering facilities developed to exploit the increased capacity, although the poor storage and handling facilities at the Nairobi airport constrained growth in the export of perishable goods.

The Producers   The composition of Kenya's cut flower industry has changed dramatically during the last three decades.[5] Initial development of the sector was spurred by a single large investment by a Danish company in the late 1960s. Kenya's flower production was dominated by this operation for more than a decade. The Danish firm's success led to an influx of other foreign investors—Europeans, Asians, and Africans—as well as Kenyans. This has broadened the sector during the past 25 years, but it continues to be dominated by a small number of large operations.[6]

The initial development of floriculture in the country was rooted in the fresh produce exports industry. By the 1950s, Kenyan, European, and Asian growers were exporting fresh vegetables, and some farmers were producing cut flowers to a limited extent. The Kenyan horticulture sector expanded in the late 1960s, driven by a growing European market for out-of-season produce and for "exotic" produce for Asian and other immigrant communities. During this time, almost all fresh produce was supplied by about 150 to 200 medium- to large-scale European and Asian farmers. Much of the produce was exported by Asian traders with ties to buyers in Europe. Kenyan exports enjoyed duty-free access to the U.K. market and low tariffs on exports to the European Economic Community.

The first investments in irrigation systems occurred on large farms around Lake Naivasha outside Nairobi. Processing and marketing produce became more sophisticated, with investments in storage and transport facilities.

HCU lost its dominant role in trading due to internal conflict and the entry of new exporters. Companies that were involved in the domestic marketing of produce expanded into the export business, as did some farmers. Meanwhile, new growers, both European investors and smallholder African farmers, entered the flower sector. Growers experimented with a wider range of flowers, but production was still primarily for the local market due to the difficulties in transporting cut flowers to export markets.

The first major investment in floriculture took place at the end of this period. The Dansk Chrysanthemum Kultur (DCK) company was expanding from its European flower-production base to locations with lower labor costs and longer growing seasons. DCK received financial support from the Danish aid agency and very favorable investment terms from the Kenyan government, including a low-cost lease on land

and the right to bring in expatriate farm managers and technical expertise. DCK purchased two additional farms, at Lake Naivasha and Updown. The company experimented with the production of chrysanthemums, asparagus plumosus, and carnations at their three farms before focusing on the production of carnations at the Lake Naivasha farm.

Several developments in the 1970s facilitated the growth of the industry, but problems remained. Floriculture exports benefited from the growth of air traffic, for example, but gaining access to planting material was still a problem because of foreign exchange controls and Kenya's unwillingness to adhere to the conventions on intellectual property rights of the International Union for the Protection of New Varieties of Plants (UPOV). In other developments, DCK was purchased by local investors, and the Kenyan Horticultural Crops Development Authority (HCDA) tried unsuccessfully to develop an outgrower scheme. In 1975, the Fresh Produce Exporters Association of Kenya (FPEAK) was formed with the goal of strengthening growers' access to market information and providing technical assistance and training and lobbying for the sector.[7] FPEAK received early support from the HCDA. The industry faced a shortage of technical manpower and also labor.

In the 1980s, flower production emerged as a strong component of horticulture. The range of flowers produced expanded in response to European demand, but planting material continued to be a problem. A large local vegetable producer, Oserian, invested in floriculture. Many new small- to medium-scale flower operations were started up, some by expatriate managers and technical staff with experience working for the major flower operations. Kenyan public officials and entrepreneurs also began investing in flower farms, often with expatriate technical support or in joint ventures with foreign flower specialists. At the end of the 1980s, the industry was large enough to attract suppliers of inputs.

In the 1990s, the industry continued to attract investments. The airfreight situation improved, with the elimination of government interventions in the market. At the same time, the foreign exchange market was liberalized, and the import of inputs was streamlined.

Recent estimates are that the industry has roughly 5,000 producers, ranging from smallholder farmers to large commercial operations. However, the number of small producers, particularly outgrowers, seems to have declined considerably with the shift toward rose production. About 50 medium- to large-scale operations produce 75 percent of the total exports. The size of these operations ranges from 20 to 100 hectares, with 250 to 6,000 employees. Another 10 to 15 percent of

Table 11.2 Cut Flower Exports from Kenya, by Volume,
1995–2003
(Thousands of tons)

Flower	1995	1996	1997	1998	1999	2000	2001	2002	2003
Alstroemeria	2.7	4.2	2.7	1.9	1.7	1.1	0.7	0.6	0.8
Carnations, standard	2.3	2.5	0.7	1.0	1.1	1.4	1.2	1.1	1.2
Carnations, spray	2.2	0.5	2.1	1.8	1.9	1.2	0.4	0.2	0.1
Cut foliage	2.2	5.9	5.5	1.2	0.7	0.5	0.1	0.3	0.6
Mixed flowers	0.0	0.0	0.0	0.0	0.7	0.6	4.7	4.5	6.3
Roses	14.2	15.1	18.3	18.0	24.6	27.7	30.3	40.4	45.7
Statice limonium	4.0	4.7	3.4	3.3	2.2	1.6	1.5	1.1	0.6
Other	1.8	2.2	3.2	3.0	4.0	4.6	2.4	4.0	5.9
Total	29.4	35.2	35.9	30.2	37.0	38.8	41.4	52.1	61.0

Source: HCDA statistics, 1995–2003.

exports are produced by several dozen small- to medium-scale operations of 4 to 10 hectares, with up to 100 employees. The remaining 5 to
15 percent of exports are produced by 4,000 to 5,000 smallholder
farmers on an acre or less of land. The economies of scale are becoming increasingly important. To be viable today, operations must be as
large as 10 to 15 acres.

Product Composition   Kenya's floricultural products are primarily cut
flowers and foliage with a small amount of live plants and seeds.[8] The
composition has evolved with changing market demands and growing
producer expertise. Early development of the sector was based on the
production and export of carnations in the 1970s. As the sector grew
rapidly in the 1980s, growers diversified into roses and other flowers.
Continued rapid growth in the 1990s was primarily driven by the
increased production of roses. By the mid-1990s, roses made up about
50 percent of total flower exports. By 2000, they constituted 70 percent
of exports (table 11.2).

   Today, floriculture is dominated by the production of roses, both in
volume and in value. Since 2000, roses have constituted 70 to 75 percent of exports by volume and, depending on world prices, have contributed 70 to 90 percent of the total value of flower exports. Between
1995 and 2003 alone, Kenyan rose exports increased from 14,000 tons
to 45,000 tons. Today, roses are estimated to be grown on nearly
1,300 hectares, and in the next year alone, the area planted is expected
to increase by 200 to 300 hectares.

   The remaining 30 percent of Kenyan flower production includes a
wide range of cut flowers and foliage. In the past decade, important

flower exports have included alstroemeria, arabicum, carnation cuttings, spray and standard carnations, chrysanthemum cuttings, eryn-
gium, gypsophilla, hypericum, lisianthus, molucella, ornithogalum,
papyrus, solidaster, statice limonium, veronica, and zantedeschia/calla,
as well as cut foliage, dried flowers, and, recently, mixed flowers. While
older plant types, such as carnations, statice, and alstroemeria, continue
to play a role, producers have diversified into higher value flowers.

Production Environments   Favorable year-round growing conditions
give Kenyan growers a competitive edge over growers in regions with
cold winters, such as Europe, or hot summers, such as the Middle
East. The equatorial location provides stable and favorable year-round
growing conditions in terms of light intensity, day length, and temperature for horticulture in general and flower production in particular.
The highlands do not suffer the extreme summer heat of the lower
areas, and the differential between night and day temperatures is
essential for maintaining the best quality in many flowers, such as
roses.[9]

The main flower-growing areas are in the highlands above 1,500
meters, around Lake Naivasha, Thika, Limuru (Kiambu), Eldoret,
Athi River, Muranga, Nyeri, Kericho, and Embo. Production is clus-
tered in areas where irrigation water is available and where the necessary infrastructure exists for production (electricity and telecommunications) and for efficient transport to export markets (developed road
networks and a nearby airport).

The oldest of Kenya's flower-growing areas is around Lake
Naivasha, where lake water for irrigation provided the impetus for the
development of farms growing carnations and, more recently, greenhouse production of roses as well as open-field production of other
flowers. As the industry grew, other areas for growing were gradually
developed. Most of the newer areas are at higher altitudes than Lake
Naivasha and are dominated by the greenhouse production of high-
value flowers.

Technical requirements for production vary widely among flowers,
with the major distinction being between annual crops, such as statice
and alstroemeria, which are usually grown in open fields with a relatively low use of purchased inputs, and perennial crops, such as roses
and carnations, which are typically grown under greenhouse conditions. As noted, greenhouse production requires higher capital investment and more intensive management than open-field crops, but yields
a much higher quality product.

Even in the production of older flower types, such as carnations,
growers are increasingly moving to higher altitudes and greenhouse

production in order to produce flowers of marketable quality. Initially, carnations were produced in open fields around Lake Naivasha at 1,700 meters above sea level. Today, they are produced in greenhouses to minimize rain damage and produce more mature, higher quality stems. Growers have shifted to altitudes of 2,000 to 2,400 meters above sea level, where year-round temperatures are mild and carnations can be grown as a two-year crop and harvested year-round.[10]

The shift toward roses has also meant a move away from low-input, open-field flower production to greenhouse production at higher altitudes. As roses require mild daytime temperatures and a nighttime drop in temperature to flower well, they are best grown between 1,800 and 3,000 meters above sea level. They also require high water and fertility conditions and are easily damaged by rain and pests. Greenhouse production allows growers to more effectively manage moisture and fertility and protect against damage from weather or pests.

Harvesting, Processing, and Transporting to Markets

The importance of reliable transport and the volume of exports have led to the development of an airfreight forwarding sector specializing in transporting cut flowers to Europe.[11] The freight forwarders inspect and document flower and temperature conditions, palletize packed flowers, store them in cold-storage facilities at the airport, clear them through export customs, obtain phytosanitary certification, and load the cargo onto commercial or charter flights.

To strengthen their control over the transport process, most large flower operations, including Oserian, Sher Agencies, and Homegrown, have joint-venture arrangements with freight forwarders. In contrast, smaller exporters must rely on Kenya Airways' airfreight subsidiary, Kenya Airfreight Handling, Ltd., whose airport storage facilities are much less sophisticated, with less temperature control and greater loss of product quality.

Kenya-to-Europe routes are served by wide-body planes from about a dozen commercial airlines as well as charter companies. Freight forwarders and exporters with enough regular volume can negotiate prices for regular blocks of air cargo space. Freight forwarders typically offer airport-to-door services that include freight clearing at the destination port and transport to the receiver, which may be an import agent or an auction house. In addition to arrangements with freight forwarders, larger exporters also have their own clearance and sales agents in key European markets.

Table 11.3 Cut Flower Exports from Kenya to Different
Destinations by Volume and Value, 1998–2002
(Thousands of tons; $ thousands)

	1998		1999		2000		2002	
Destination	Volume	Value	Volume	Value	Volume	Value	Volume	Value
Netherlands	22.7	57,588	27.7	59,407	24.0	54,962	24.4	60,223
United Kingdom	6.5	14,439	5.0	18,075	5.0	22,597	7.1	25,266
Germany	3.0	7,055	1.8	4,265	2.4	5,870	2.9	7,883
Other	3.7	7,704	3.7	8,007	2.4	7,141	1.6	6,014
Total	35.8	86,786	38.3	89,754	33.8	90,570	36.1	99,386[a]

Source: UNSO/ITC Comtrade Databases.
a. Source captures only a portion of total exports.

Supply Chain Relationships

Europe today is the dominant market for Kenyan flower exports
(table 11.3). Kenya is now the second largest exporter of cut flowers
and foliage to European markets, after Colombia.[12]

In the past decade, one of the most dynamic aspects of the export and
sale of cut flowers to European markets has been supply-chain relationships. The traditional roles of various actors in the distribution and sale
channels—among them Kenyan producers and exporters, European import agents, auction houses, wholesalers, supermarket chains, and other
retailers—have blurred. In attempts to control the distribution chain
and increase competitiveness through increased vertical integration, various actors have taken on new roles and established new types of linkages, financial and otherwise, with other actors up and down the chain.

Dutch Auctions   The Dutch auctions have been at the core of the
distribution channel for Kenyan flower exporters. The auctions have
been the primary venue for the sale of flowers from producers in
Europe, Kenya, and other countries to European wholesalers, who
then sell to European retailers or reexport to other overseas markets.
Although the importance of the auctions has fallen in recent years, in
1998, two-thirds of all Kenyan flowers were still exported via the
Netherlands, and 50 to 60 percent of those were sold at auction.

The Dutch auctions, developed to serve the marketing needs of
Dutch flower and bulb producers, are cooperative ventures of Dutch
growers, although non-Dutch can buy memberships. Over time, the
auctions became the primary distribution and sales mechanism for cut
flowers sold in Europe. Today, they bring in flowers produced worldwide, many of which are reexported to overseas destinations.

Because the auctions remain at heart a marketing tool for Dutch growers, as international competition grew stiffer in the mid-1990s, those growers attempted to cut out imports through auction quotas and bans on imports in the summer, when Dutch production peaks. The restrictions provided incentives for Kenyan and other exporters to develop alternative distribution channels, including direct sales to retail buyers and even an alternative auction, the Tele Flower Auction (TFA). Ultimately, the restrictions cost the auctions both traditional suppliers and buyers, particularly large supermarket chains that developed direct relationships with growers.

Import Agents   Import agents are an important adjunct to the Dutch auctions. They provide services to facilitate the transfer of flower imports from airport to auction, providing the care necessary to ensure the quality of perishable cut flowers until they reach auction. Increasingly, agents have also offered other services for clients, including marketing information and product development and consulting services. Agents have developed financial linkages with overseas flower producers at the one end and, at the other, have begun selling directly to wholesalers and retailers, as well as through the auctions.

While many exporters work through import agents, some of the larger Kenyan operations have opened their own European sales offices to more effectively manage the supply chain from production to sale. Oserian was the first to do this, creating East African Flowers. Sher Agencies has Sher Holland. These sales offices now act as agents for multiple Kenyan producers and exporters. Oserian supplies TFA and direct sales; Sher supplies the Dutch auctions.

Wholesalers   Wholesalers take responsibility for transporting the product from auction to buyer. They are the buyers at the auctions who sell to European retail outlets, including florist shops and supermarket chains, as well as to importers for overseas markets. As exporters have begun to bypass the auction through direct sales, some wholesalers have moved upstream in the supply chain by developing direct relationships with Kenyan flower producers and opening purchasing offices in Kenya.

Retailers   Retailers represent the final market for most cut flowers. Historically, florist shops were the primary market, but today, supermarket chains overwhelmingly dominate. During the past 25 years, the rapid growth in market share of European supermarket chains has fundamentally changed their buying and marketing strategies as well as demand for imported flowers.

The two largest retail markets for Kenyan flowers, after the Netherlands, are the United Kingdom and Germany. In the United Kingdom, there have been two major shifts in the buying behavior of the super-market chains: they have eliminated intermediaries in the supply chain and developed more direct relationships with Kenyan flower exporters, and they have looked for ways to outsource more distribution and quality control (and risk) factors to importers and exporters. The result has been the proliferation of special relationships between the supermarket chains and a small number of U.K. importers and Kenyan exporters that ensure quality standards and provide some additional flexibility to the supermarkets as well as some additional security to the exporters.

Some of the largest Kenyan operations have opened up their own U.K. import offices. Companies such as Oserian and Homegrown now control the full distribution chain, from harvest, to transport, to retail market. These direct relationships between exporters and retailers have greatly increased the flow of information for large exporters.

In Germany, most Kenyan exports are handled by two large importers that service retailers in Germany and other markets in Europe and overseas. The German importers handle all transport and logistics in the supply chain, including the airfreight from Kenya. This relationship grew out of the marketing of Sulmac flowers in the 1970s and 1980s under exclusive contract to the German importer Florimex.

Economies of Scale

Because of declining returns on rose production and benefits from vertical integration, size matters. Larger producers have better access to capital for inputs and for technical and management expertise. They are better able to organize movement of perishable products to markets and to comply with codes of practice that give access to important market segments. The type of technology employed, product choice, and marketing channels are all related to firm size.[13]

The largest operations can invest in greenhouses and postharvest facilities and hire professional managers. They have relationships with freight firms to ensure timely and economical transport to markets. Their farm locations often have better access to transport, especially in the rainy season. They have linkages to European importers and agents, and sometimes even have their own European marketing offices. They have better access to timely and accurate market information than do smaller producers that depend on middlemen. They can ensure the range, quantity, quality, and reliability of supply needed to negotiate and develop relationships directly with European supermarket chains, German wholesalers, and the Dutch auctions.

They can also better afford the costs of compliance with the codes of practice that are required to meet European import demands on environmental protection and worker welfare. Large operations form the backbone of the membership of the Kenya Flower Council (KFC), which lists 36 member operations on its Web site. Compliance with the codes of practice requires management skills to understand complex requirements and make necessary management changes as well as capital to finance the changes. In the past decade, the codes of practice have become increasingly complex and expensive; they might be described as moving targets.

Medium-size firms are a mix of greenhouse and open-field operations. Those that export directly to Europe typically sell through the Dutch auctions, while others sell their product to larger exporters. Their viability depends on their ability to make investments in capital and management to meet the codes of practice and stay abreast of changing consumer preferences in varietals.

Most smallholder producers have open-field operations and limited or no irrigation facilities. Their lack of access to capital typically limits them to growing open-field plants, such as alstroemeria, and older, non–UPOV varieties. Their product is usually sold on contract to a larger producer/exporter. Lack of technical and management expertise, as well as postharvest access to cold-storage facilities and transport, make it difficult for smallholder producers to maintain the levels of product quality and consistency required to sell in European markets.

Industry Context and Policy

Context

The Kenyan floriculture sector has grown rapidly, particularly in recent years, even as the country's economy, agricultural sector, and overall exports have stagnated. A stable country, Kenya was exceptional in having fairly high levels of growth in the first two decades after gaining independence in 1963. Since the 1990s, however, the growth spurt has dwindled. Annual rates of GDP growth declined from 6.5 percent in the 1960s and 1970s to about 2.2 percent from 1990 to 2001, while the population grew at about 2.5 percent. Real GDP growth fell to 1.1 percent in 2002. Real income per head is lower, and poverty is more widespread now than in 1990 (EIU 2003).

Kenya's agricultural sector has lagged behind the general economy. The agricultural sector accounts for 24 percent of the GDP and is the largest contributor to foreign exchange. Tea, coffee, and horticultural products account for nearly 65 percent of merchandise exports. About 75 percent of the population depends on the sector for its livelihood. The agricultural sector has not had sustained growth, except for the period from 1985 to 1990, chiefly because of the inability of the government to liberalize and privatize in a timely and extensive way (IPC 2004).

The industrial sector accounts for 18 percent of the GDP, with agro-processing an important component. The services sector is the largest source of foreign exchange earnings, representing about 54 percent of the GDP. It is dominated by tourism and financial and communication services. However, its relatively high cost structure appears to have imposed a constraint on the development of other sectors of the economy that are dependent on basic services.

Exports also performed poorly between 1990 and 2000. The export of goods as a share of GDP came down dramatically from 1997 to 2000 after a modest increase from 1995 to 1996. The composition of exports changed significantly between 1990 and 2000, with the share of coffee declining dramatically, while that of tea grew significantly, and the share of horticultural products almost doubled, from 6.5 percent to 13.3 percent (IMF 2002).

Gross domestic investment fell from 20 percent in the early 1990s to about 13 percent in 2001, with the bulk of the decline in the public sector: public sector gross capital formation fell to just over 4 percent of GDP in 2001.[14] Private sector investments have not made up for the decline in public sector investments. Trade reforms carried out in the early 1990s did not strengthen private sector investment. Private sector capital formation has remained steady at around 9 percent of GDP. Kenya missed out on private financial movements in the last decade or so, largely because of poor governance, which has made it an unattractive destination for investments. General Motors' investments in Kenya in the late 1970s were the last major foreign investment in the country. The stock of FDI in Kenya is less than half of the level in Tanzania. As a recipient of FDI, between 1990 and 2001, Kenya slipped from the 90th to 118th position out of 140 countries (EIU 2003; Himbara 1994; World Bank 2004).

The role of the state in the Kenyan economy continues to be strong, and privatization has been slow to take hold. In the early 1990s, Kenya carried out structural and macroeconomic reforms, including in trade, to establish a more growth-conducive economic environment. Although there has been some transition from import-substitution to outward-oriented policies, privatization has not been adequate. Governance issues, labor unrest, power shortages, and high utility costs have affected investors' confidence. The recently elected government announced its intentions to reduce red tape and high energy and communication costs, primary concerns of the private sector. With assurances that the new government would make serious efforts to reduce corruption, bilateral donors and multinational institutions had begun to engage with the country.[15] Yet more recently, concerns have arisen again that the government is not doing enough (White and others 2004; WTO 2000).

Links and Benefits Because so few smallholder farmers are involved in the flower industry, the industry's impact on poverty is primarily from employment in medium- to large-scale flower operations and in ancillary industries. The workforce in the flower industry is composed mainly of a small number of floriculture specialists with technical or management expertise, most of whom are expatriates, and a very large number of semi-skilled farm workers, most of whom are Kenyan women. Most semi-skilled labor in the flower industry is drawn from rural areas that have some of the most severe unemployment and underemployment problems in Kenya. The flower industry employs approximately 40,000 to 50,000 workers, almost all on the two to three dozen largest operations, with another 60,000 to 70,000 workers employed in ancillary industries.

While industry wages are above Kenyan national norms, much of the semi-skilled labor is hired on a seasonal or "casual" basis. "Casual" labor comprises approximately 65 percent of the labor force, with no benefits or right to join unions. Since the late 1990s, enforcement of the rights of casual labor to permanent employment has become an issue of negotiation between worker and industry organizations. The codes of practice developed by FPEAK and KFC address some workers' rights issues, but pressure has been building for more participatory development and monitoring of codes (Barrientos and others 2001; ILRF 2003; Jaffee 1995; KFC 2004; Opondo 2001; Thoen and others 1999).

Policy

Kenya is a mixed economy. At independence, power was seized by a "conservative fraction of Kenya's rural society committed to accumulation, investment, and private property" (Bates 1989). As the strategy was to create a mixed economy, nationalization and state ownership were not widespread (Maitha 1976). Redistribution of wealth was not an important objective.

Kenya also inherited a dualistic agricultural sector with export-oriented European farms, which arose from the creation of large farms in "scheduled areas" from which Africans were excluded. The settlers were able to maintain large-scale farming because they benefited from protections that included land set-asides, favorable tax treatment, and other measures that ensured labor supply while excluding Africans from the cultivation of certain crops. Since independence, agricultural exports have not been subjected to significant levels of direct or indirect taxation (Deininger and Binswanger 1995; Senga 1976; Smith 1976; World Bank 1998).[16]

Africanization, or "Kenyanization," of the economy was a government priority after independence. Considerable efforts were made to change the dual character of the agricultural economy in which Kenyan Europeans and Asians owned all large-scale agricultural concerns and only large farms produced for the export market. The government bought a number of large farms for resettlement by African smallholders. However, Kenyanization was not particularly successful in agriculture or in other sectors (Himbara 1994).[17]

Nonmarket controls have been a significant aspect of rural life. Although private interests are protected, the government also intervenes actively in markets and maintains regulations. Most agricultural crops are produced and marketed by cooperative societies. Sweeping reforms were made in the early 1990s to encourage private sector participation in the production and marketing of crops. Most crop boards have been restructured, removing their monopoly rights in marketing and their role in pricing. One of the objectives of institutional reforms under the current rural development strategy is to simplify some 60 different statutes that govern the agricultural sector into unified legislation.

While much of Kenyan agriculture, particularly export crops, has been heavily regulated, Kenyan horticulture has had almost no government regulation. First, many of the regulations, particularly for cash crops, were inherited from preindependence days, when horticulture was practiced only on a very small scale. Second, as the horticultural sector grew after independence, the perishability of the products made it difficult for the government to intervene effectively. State efforts to provide small producers with cooperative society–based services for collecting and storing crops, and to exercise state control over marketing, were seen as inappropriate for highly perishable crops such as fruits and vegetables and, more recently, flowers. Third, many politicians and civil servants invested in floriculture enterprises, which may have had something to do with the favorable treatment accorded to the industry. Finally, floriculture has entailed large investments, and the government has been sensitive to the demands of investors that contribute to foreign exchange generation and job creation.

In recent decades, the fresh produce sector has benefited from the liberalization of Kenya's foreign exchange controls, the streamlining of import procedures, and the liberalization of the airfreight market. In the early 1990s, FPEAK lobbied effectively for the import of agricultural inputs and equipment free of duty and value-added tax. Exemptions made for the horticultural sector permit the free flow of resources, technology, and technical manpower[18] across the border, largely driven by market forces, to take advantage of Kenya's favorable agroclimatic conditions and low costs for semi-skilled labor. However, the costs of doing business in Kenya—taxes, utility, and airfreight costs—continue to be higher than in other African countries with which it competes.

The absence of a policy for horticulture is interpreted by some as neglect of an industry with substantial potential for growth. The government emphasizes that the industry is led by the private sector. A recent effort to introduce a horticultural bill that included a 3 percent tax on exports to be channeled into development activities for the sector has been withheld because of complaints from the industry through FPEAK and KFC. The Kenyan government and the industry organizations have worked closely together in trade negotiations—FPEAK represents the agricultural industry as a nonstate actor at both the World Trade Organization and the European Union–Kenya trade negotiations. They are working together now to develop alternative markets should the country lose its special trading status when the Cotonou Agreement comes into effect in 2008.

Technical Aspects of Kenyan Floriculture

Floriculture has been at the forefront of the development and commercialization of intensive agricultural technologies. Some of the earliest technologies, such as automatic irrigation, "fertigation," and tissue-culture propagation were commercialized first in floriculture.[19] Several technologies developed for floriculture have now found applications elsewhere, including soil pasteurization to minimize crop loss, mist propagation, and greenhouse and environmental controls of temperature, humidity, and light. The United States led floriculture research and development 20 years ago, developing innovations such as trickle irrigation, high-intensity discharge lighting, and mist propagation. Today, most new technology is developed in northern Europe (Etkin 1990).

This section examines how the Kenyan floriculture industry gains access to global technological developments. The means of access include institutions that facilitate transfer from other countries, local technology development, and technology dissemination through various ways, including compliance with international labor and environmental standards.

The technologies used in floriculture include the following broad categories:

- *Hardware*. Structures and equipment designed to modify the environment or control the climate during the production, processing, and transport of flowers.
- *Plant-ware*. Genetic materials that embody attributes desired in the market and that are suitable for local production. Knowledge and processes required for the production of these inputs are not included.
- *Nature-ware*. Materials and knowledge systems necessary to minimize environmental damage from the production of ornamentals, such as hydroponics, which can potentially reduce water usage; chemicals that have lower residues; or knowledge systems, such as integrated pest management, which reduces the use of chemical pesticides.
- *Software*. Knowledge systems that are essential for making the best use of material technologies and managerial capabilities. Greenhouse management is an example.

Hardware

In flower-producing areas, a key determinant of relative competitiveness in the global market is the suitability of the local climate for flower production, storage, and transport to overseas markets.[20] A major advantage of tropical flower producers over their competitors in North America, Western Europe, and Japan are climatic conditions that favor year-round flower production. Suitability of local climatic conditions to flower production is also a key determinant of relative competitiveness among the tropical production centers, as it determines the degree of scarce capital and management skills that will be required for the production of high-quality flowers. Climatic conditions affect the costs of production, postharvest storage, and transportation to market. When conditions are less than ideal, production of high-quality flowers depends on investments in hardware to modify the environment and to maintain suitable climatic conditions, all of which entail higher operating costs.

Conditions in Kenya are close to ideal for rose production, so the specifications for greenhouses used for roses are minimal. In Europe, greenhouses typically have the capacity to heat during the winter, whereas Kenyan greenhouses need only protect the plants from the elements and capture daytime heat. Most Kenyan greenhouses are simple covered structures that protect the plants against rain and wind.

Selective ventilation provides some control over the temperature. The plants are grown directly on the ground.

A few flower operations are experimenting with higher levels of environmental modification. One firm heats its greenhouses using geothermal energy, while another has greenhouses with cooling capability. Heating and cooling are expected to improve quality and quantity, with 20 percent higher yields. Greenhouses used for plant propagation are sometimes equipped for higher levels of environmental control, including heating. Among Kenyan growers, there is no consensus over the economic returns to investments in heating and cooling for flower production. Returns may depend on the market segment to which different producers cater.

There appears to be greater agreement on the potential benefits of investments in hydroponic growing systems. Some existing greenhouses are being switched to hydroponics, and all new greenhouse investments appear to include hydroponic systems.

Irrigation equipment in flower production varies from the use of simple hoses in open-field production, to drip irrigation, to highly sophisticated greenhouse irrigation and fertigation systems. Kenyan greenhouses are typically equipped with drip irrigation lines and fertigation systems, often with computerized controls.

A number of local agents supply Kenyan flower growers with imported greenhouse technologies and related inputs, primarily from Israel, France, the Netherlands, and Germany. One large firm is believed to meet as much as 80 percent of the industry's demand in Kenya. Almost all materials are imported, with only limited fabrication taking place in the country.[21] Growers are investing in new and improved technologies by upgrading as necessary, driven by market conditions, strategies, and access to resources.

Plant-ware

Flower breeders today are investing in development and marketing worldwide.[22] UPOV standards for the protection of plant breeders' rights are accepted in most countries, giving breeders the right to collect royalties from producers on every plant cultivated.[23] In Kenya, which signed the UPOV convention under pressure from the industry, breeders register their varieties with the Kenya Plant Health Inspectorate Service (KEPHIS),[24] which enforces breeders' rights. UPOV gives breeders the right to confiscate illegally grown flowers in international flower markets, so producers have a strong incentive to pay royalties if they are exporting to European or North American markets.[25]

As consumer preferences have continuously shifted during the last 15 years, and as breeders have worked to develop sturdier plant types that have longer vase lives and are more stress-resistant, breeding and selection of new varieties has driven profit margins in the rose market. The old standard of eight to ten years' life of rose plants has dropped to five, with growers replanting more frequently to respond to changing consumer preferences.

Because roses are a perennial crop with a relatively long life span, choosing which variety to cultivate may be the most important decision a rose grower makes. A poor decision is costly, as the cost of replanting roses can range from $120,000 to $160,000 per hectare.[26] The characteristics of the variety determine success in both production and marketing. In addition to incorporating varietal characteristics that meet consumer demand, rose breeders also must develop varieties specifically for different grower markets. Producers in the Netherlands, for example, are often reluctant to grow the same varieties grown by African producers, so as to avoid direct competition. Furthermore, breeders must take into account the suitability of new varieties for local growing conditions and for successful packaging and transport to market. These conditions can vary significantly from one producing country to another. Transportability, for example, may be very important for producers in southern Africa, which are farther from European markets than are East African producers.

Increasingly, northern European breeders are incorporating aspects of their plant selection and testing on-site in producing countries such as Kenya to ensure suitability for local conditions.[27] Some 15 rose breeders target producers in Kenya. A Dutch company that operates in Kenya has trial centers worldwide: in Holland for the European grower market, in Kenya for the East African market, and in Ecuador for the Latin American market. Breeders without facilities of their own contract with local producers to conduct trials.

Plant propagation has also been outsourced to local propagators in flower-producing regions such as Kenya in order to reduce labor and transport costs. Today, the larger Kenyan operations, such as Sher Agencies, Waridi, and Oserian, have in-house facilities for plant propagation. Some European propagators have also opened facilities in Kenya. Propagation services are available for smaller producers and are serving the regional market as well. Major types propagated locally include roses, chrysanthemums, carnations, and statice.

Producers negotiate with breeders to obtain planting material, which is then given to propagators for multiplication.[28] Plant propagation is done in fully controlled greenhouse environments, using grafting techniques and tissue culture. The key efficiency objectives in commercial propagation are to reduce the number of days necessary to have plants ready for planting and to improve the success rate of propagation. While mother plants may be multiplied through tissue culture, grafting is the primary means of propagation. Oserian, the world's largest producer of statice, uses tissue culture for plant multiplication and for varietal selection in order to improve uniformity and consistency in selecting for different plant characteristics. Through tissue culture, Oserian has cut its costs of importing plant materials and has created enough capacity to supply planting material on contract to other producers.[29]

Nature-ware

The rapid expansion of areas under flower production has led to serious concerns about environmental impacts, particularly to the ecosystem of Lake Naivasha, Kenya's largest freshwater lake, around which the floriculture industry is based. Concerns about worker health also have been raised, particularly with respect to the use of agricultural chemicals. The Kenyan industry has responded quickly to these criticisms, implementing technological and managerial changes and attempting to allay social concerns. In the 1990s, FPEAK lobbied the government to enact the Environmental Management and Coordination Act of 1999 and also establish the National Environmental Management Authority to implement the act.[30]

In the last decade, European buyers of goods produced overseas have increasingly insisted on common adherence to "international standards" of conduct, particularly on environmental protection and worker welfare, in part to level the playing field between developed- and developing-country producers. Codes of practice have been developed to monitor the conditions of the production and distribution of goods throughout the supply chain, from European importers and large-scale buyers, such as supermarket chains, to producer trade organizations.

In the cut flower industry, the most recognized international standard, the Milieu Project Sierteelt (MPS), was developed in 1995 for the Dutch flower auctions. In the late 1990s, Kenyan growers and exporters developed their own industry codes to try to satisfy European concerns. The two main industry trade organizations, KFC and FPEAK, have taken the lead role in developing codes for the self-regulation of their members.

KFC was formed by flower growers and exporters in 1997 largely in response to concerns to develop standards on worker welfare and environmental protection that would enable members to meet European

buyers' demands. Today, KFC has 37 members, consisting of medium-
and large-scale producers that represent about 75 percent of Kenya's
cut flower business. Members work to achieve the KFC code of practice at either the "silver" or "gold" standard by meeting requirements
on plant breeders' rights, wages, labor conditions and worker safety,
safe use and disposal of pesticides and other agrochemicals, and environmental protection. KFC also has developed projects to train
smaller producers on aspects of the code so that they meet European
buyer demands.

FPEAK was formed in 1975 to represent the interests of producers
and exporters of cut flowers and fresh vegetables and fruit by supplying market information and technical assistance and training. FPEAK
has about 80 members that grow flowers, including small- to large-
scale producers. They adhere to a voluntary code of conduct on
"responsible production, environmental protection and social accountability," which was first developed in 1996 and relaunched in
1999. Since June 2004, the code has been transformed into KenyaGAP
and is being benchmarked for equivalence with EurepGAP, an initiative of the Euro-Retailer Produce Working Group with Kenyan conditions and regulations in mind.[31]

There exist today overlapping codes of practice for worker welfare
and environmental protection. Producers seek certification from MPS,
the Flower Label Programme, and EurepGAP, depending on the market they want to access. Even if the producers have KFC certification,
they must obtain certification from other sources as well. There are
ongoing efforts to harmonize codes to fewer standards. The HCDA
has been working with KFC and FPEAK to develop a single Kenyan
national code. Meanwhile, complying with complex and often con-
flicting codes is a major barrier to participation, particularly for
smaller producers. KFC has signed an agreement with a Dutch certification body to assess compliance with international environmental
and social accountability standards. KFC estimates that the cost of
meeting new European Union requirements might increase local costs
by as much as 50 percent.

The codes of practice have been important instruments for the dissemination of improved practices and also incentives to invest in technology. The codes focus on the use of knowledge-based, rather than
chemical-based, technologies and, to some extent, require local technology development. The need for the local development of "green
technologies," such as integrated pest management (IPM), has spurred
research investments by local firms, such as Homegrown, one of the
larger growers, which has created a unit to develop IPM technologies.
Additional local research focuses on using local predators in the

natural control of pests. Research and consulting firms, such as Real IPM, have emerged to help producers shift away from the chemical control of pests and disease in order to meet certification standards.

Some donor-supported programs aim to help producers adopt more environmentally sound and worker-safe practices. The U.K. Department for International Development, in association with KFC, started a two-year project in 2001 to train smallholders to adopt the code of practices, the aims of which are to minimize pesticide use, minimize worker exposure to chemicals, and deal responsibly with the disposal of waste chemicals and organic materials. FAO has also funded research to reduce the use of methyl bromide, which is used extensively for soil fumigation.

Software

The knowledge, skills, and expertise needed in floriculture include the effective organization of labor and materials at different stages in the supply chain—production, harvesting, postharvest storage, transport. It also includes the knowledge needed to manage hardware, such as greenhouses; plant-ware, such as plant propagation; and nature-ware, such as agricultural chemicals.

Managing the structures and equipment for flower production requires technical knowledge—to maintain optimal conditions in greenhouses and managing irrigation and fertigation systems, for example. Greenhouse managers embody much of this knowledge. In Kenya, most greenhouse managers are expatriates from Europe, Israel, and India, depending on the ownership of firms, although the number of Kenyan managers is increasing.

More generally, much of the requisite knowledge comes from suppliers of imported inputs, who are thereby transferring knowledge from those at the international technical forefront. For example, breeders provide information on marketing and growing conditions, and greenhouse suppliers provide information on equipment options.

Consultants are another important means to import and share knowledge. Currently, a small group of Israeli consultants is providing support to greenhouse managers. Consultants typically service many companies, transferring what works in one to others.

Little research of relevance to flower producers takes place within the country. While the Kenyan Agricultural Research Institute has helped a local firm establish a tissue-culture laboratory, the primary emphasis of the local research stations is on aspects of horticulture, such as vegetable production, in which smallholder participation is more significant. In recent years, with the shift toward rose production, the participation of smallholder farmers in flower production has diminished substantially. Only one exporter is believed still to have substantial outgrower production for the export of summer flowers.

Jomo Kenyatta University of Agriculture and Technology has introduced a degree program in ornamental horticulture, with a component in floriculture. The first class of students graduated in 2005. The students are required to go through two eight-week internships with a floriculture business. Some producers felt that the students should be attached for longer periods to be adequately trained. Local graduates are hired by the industry. Their training, in addition to taking place in house, takes place in countries from which technologies are imported.

Perhaps the more difficult kind of knowledge for Kenyan firms to acquire is access to accurate and timely information about the major overseas markets. As previously discussed, the larger firms have developed mechanisms for maintaining effective access to market information, particularly in terms of vertical integration and long-term relationships with buyers and importers. Smaller operations face much more significant barriers. Nonetheless, any producer that has a quality product can send it to the Dutch auctions. Once a firm's reputation has been built in the auction, direct selling becomes feasible, and firms can develop long-term relationships with buyers.

What Lessons Does Kenya's Experience Offer for Other Countries?

The case of Kenyan floriculture may appear to be largely a private sector story, the moral being that achieving growth depends on leaving the private sector alone. No doubt the private sector was central to the development of Kenya's floriculture, but the circumstances are somewhat unique, and the government did make some critical interventions. The government's initial support for a large foreign investment was critical to the subsequent development of the sector into one that is largely indigenously owned by the commercial class of farmers in the country, who drew on their experiences with exports of fresh vegetables.

The producers have been able to upgrade the technologies—hardware, genetic material, and production knowledge—required to become and remain globally competitive. They have been able to access the best of technologies, because the borders are open to both material technologies and human capital. This has enabled them to move to higher value products that required more complex and capital-intensive technologies. The technologies used in the sector may not be the most advanced in the world, but they are the most appropriate—that is, in fact, what gives them the competitive advantage. The country's willingness to honor intellectual property rights has been critical to accessing high-quality genetic material, the most important input for both production and marketing. In addition to being competitive in production, the industry exercises considerable control over the supply chain—developed by virtue of the contacts and links that producers have in the importing markets. The industry has been able to introduce innovations in marketing that have threatened the dominance of Dutch auctions in flower marketing.

However, the sector has been less successful in another dimension of technological upgrading. The agro-industrial equivalent of moving from assembly of components to manufacturing of complex components requiring greater human skills is graduating to the production of the complex inputs used in the production of flowers. Many aspects of production—plant protection or weed control, for example—come embodied in inputs, thus enabling input suppliers to capture a significant portion of the value added. Kenya has been weak in gearing its research and academic institutions to become regional suppliers of technology and advanced human resources.

The country has been fortunate to have a sector that has acted collectively—in the absence of which the state would have had to provide some collective goods. An important aspect of this collective action is to have demonstrated that the sector is environmentally and socially responsible. The collective action has been forthcoming because the sector has one or more firms for which the private benefits of collective action have been significantly greater than the costs.

Endnotes

1. The sources for this section are Barrientos and others (2001), KFC (2003), HCDA (2004), IFTS (2003), Opondo (2001), and Roozendaal (1994).
2. The sources for the material in this section include ILO (2000), ITC (1997), McCulloch and Ota (2002), and ILRF (2003).
3. After harvest, cut flowers are moved as quickly as possible to packing sheds for grading and bunching. They are kept in buckets in treatment solutions of clean water and chemical preservatives to maintain freshness. Grading is mostly done by hand, but conveyor belts are used in larger operations. Flowers are then bunched and wrapped in protective corrugated cardboard or polyethylene sleeves and packed flat in cardboard boxes for transport.

4. The sources for the material in this section are Jaffee (1995, 2004), Thoen and others (1999), Dolan and Sutherland (2002), McCulloch and Ota (2002), Opondo (2001), Maharaj and Dorren (1995), Gabre-Madhin and Hagglade (2004), and HCDA (2004).

5. The primary source for the material in this section is Thoen and others (1999).

6. In 1991 the three largest companies contributed more than 80 percent of Kenya's flower exports. Foreign-owned companies had a 69 percent export share, with Kenyan European-owned companies having another 25 percent share, Kenyan African-owned companies having only a 4 percent share, and Kenyan Asian-owned companies having another 2 percent share. This situation has changed in the last decade, with dramatic growth in the sector. Kenyan Asians may now have substantially increased their share of the market.

7. FPEAK received USAID funding in the 1990s to bring smallholders into the sector.

8. The chief sources for the material in this section are HCDA (2004) and Thoen and others (1999).

9. The sources used in this section are Collinson (2001), HCDA (2004), KFC (2004), Omniflora (2004), Pertwee (2003), and Thoen and others (1999).

10. At lower elevations, summers are hot and dry, so the plants must be replanted each year. In addition, under open-field conditions, growers must contend with botrytis damage on blooms during the rainy seasons.

11. The primary sources for the material in this section are HCDA (2004) and Thoen and others (1999).

12. The sources consulted in preparing this section include Collinson (2001), Dolan and Sutherland (2002), and Thoen and others (1999).

13. The sources for the material in this section are Collinson (2001), KFC (2004), Maharaj and Dorren (1995), Roozendaal (1994), and Thoen and others (1999).

14. The government's domestic debt is about 30 percent of the GDP, and foreign debt is about 40 percent. Two-thirds of the taxes collected go to servicing this debt and paying for civil servants (White and others 2004).

15. The government enacted the Anti-Corruption and Economic Crimes Act, the Public Officer Ethics Act, and the Public Audit Bill of 2003 to systematically stem corruption.

16. One of the contributions of the World Bank to Kenyan agriculture may have been an indirect one: promoting and maintaining a competitive and liberal marketing environment so that policy did not prevent development (World Bank 1998).

17. Kenya opted for a mixed economy in which the state played an entrepreneurial role to facilitate the growth of African capitalists in the private sector. State capitalism was meant to be a transitional phase in the

establishment of a viable African commercial and industrial entrepreneurial class. However, state agencies that were expected to pass on equity to African capitalists became a financial burden. Those who benefited from state capitalism were also closely tied to the tribes in power. As groups changed power, those that had gained even a marginal foothold were quickly replaced by a new set of beneficiaries.

18. The labor ministry announced recently that it will not renew work permits for two-thirds of the 25,000-person expatriate workforce in the country. However, senior managerial and technical staff in multinational firms will be spared.

19. Fertigation systems deliver fertilizers to plants through irrigation lines and require significant investments in chemicals, delivery systems, and management.

20. The sources for this section are Collinson (2001), HCDA (2004), and Thoen and others (1999).

21. Imports of steel attract duty while imports of greenhouses do not, discouraging growth in the local fabrication of structures and equipment.

22. The primary sources for this section are Collinson (2001), *Harambee* (1997), Roozendaal (1994), and Thoen and others (1999).

23. In Kenya, royalties on roses are about $0.85 per plant, which may be paid in four annual installments of $0.22 cents.

24. Before KEPHIS was founded, breeders' rights had been enforced by the Kenya Agricultural Research Institute (KARI) earlier.

25. As a result, smallholder farmers can often afford to grow only older varieties that do not require payment of royalties, but that also fetch lower prices in the market.

26. Recent estimates for a small to medium-size firm suggest that net returns are about $35,000 per hectare, with a yield of nearly 2.2 million stems and returns per stem of $0.016.

27. One of the breeders with facilities in Kenya does varietal crossing on-site in Kenya. Half a million seeds are then sent to the Netherlands for germination. Over several seasons, through negative selection, the pool is narrowed down to perhaps 700 promising varieties, which are then tested locally. The breakeven sale to recover development costs of a variety is believed to be about five hectares.

28. Earlier, propagators paid the royalty and then collected it from producers. Now local breeders' representatives collect royalties directly from producers. The farm-to-market value chain for a medium rose indicates that farm production accounts for 29 percent of total costs, while transport and marketing account for 65 percent, and royalties for 6 percent (Global Development Solutions 2004).

29. The Kenyan Agricultural Research Institute helped the local firm Oserian set up an in-house unit in the 1990s. They developed the protocol for the multiplication of astroelmeria.

30. Sources for this section include Barrientos and others (2001), Codes of Conduct (2003), EIU (2004), KFC (2004), *New Agriculturist* (2001), Opondo (2001), Thoen and others (1999), and FAO (2002).
31. EurepGAP started in 1997 as an initiative of retailers belonging to the Euro-Retailer Produce Working Group (Eurep). It has subsequently evolved into an equal partnership of agricultural producers and their retail customers. Its mission is to develop widely accepted standards and procedures for the global certification of good agricultural practices.

References

Barrientos, Stephanie, Catherine Dolan, and Anne Tallontire. 2001. "Gender and Ethical Trade: A Mapping of the Issues in African Horticulture." Department of International Development/ESCOR Working Paper R7525. United Kingdom. [Retrieved from http://www.nri.org/NRET/genderet.pdf]

Bates, Robert H. 1989. *Beyond the Miracle of the Market: The Political Economy of Agrarian Development in Kenya*. New York: Cambridge University Press.

Collinson, Chris. 2001. "The Business Costs of Ethical Supply Chain Management: Kenya Flower Industry Case Study." Natural Resources Institute Report 2607, Project V0128. Department for International Development, United Kingdom. [Retrieved from http://www.nri.org/NRET/2607.pdf]

Deininger, Klaus, and Hans P. Binswanger. 1995. "Rent Seeking and the Development of Large-Scale Agriculture in Kenya, South Africa, and Zimbabwe." *Economic Development and Cultural Change* 43 (3): 493–522.

Dolan, Catherine S., and Kirsty Sutherland. 2002. "Gender and Employment in the Kenya Horticulture Value Chain." Globalisation and Poverty Discussion Paper 8. Institute for Development Studies, Sussex, United Kingdom.

EIU (Economist Intelligence Unit). 2003. *Kenya: Country Profile*. [Retrieved from www.eiu.com]

———. 2004. *Kenya: Country Report*. [Retrieved from www.eiu.com]

Etkin, Larry. 1990. "Floriculture in the United States: A Research Agenda." Minnesota Agriculture Experiment Station Discussion Paper. [Retrieved from www.maes.umn.edu/mnsci/floriculture.asp]

FAO (Food and Agricultural Organization of the United States). 2002. "A Thorn in Every Rose for Kenya's Flower Industry: Floriculture Presents Challenges as Well as Opportunities for Kenyan Smallholders." [Retrieved April 18 from http://www.fao.org/english/newsroom/news/2002/3789-en.html]

Gabre-Madhin, Eleni Z., and Hans de Vette. 2004. "Uganda Hortifloriculture Sector Development Study." Draft. The World Bank, Export Growth and Competitiveness Project, Washington, D.C.

Gabre-Madhin, Eleni Z., and Steven Hagglade. 2004. "Successes in African Agriculture: Results of an Expert Survey." *World Development* (32) 5: 745–66.

Global Development Solutions. 2004. *Value Chain Analysis of Selected Strategic Sectors in Kenya*. Internal World Bank report, Washington, D.C.

Harambee: Pulling Together for Kenya. 1997. Kenyan Agricultural Research Institute and the University of Missouri. [Retrieved from http://www.ssu. missouri.edu/IAP/Harambee]

HCDA (Kenyan Horticulture Crops Development Authority). 1995–2004. [Retrieved from http://www.hcda.or.ke/stats.asp.]

Himbara, David. 1994. "Failed Africanization of Commerce and Industry in Kenya." *World Development* (22): 469–82.

IFTS (International Floriculture Trade Statistics). 2003. Suffolk, United Kingsom: Pathfast Publishing. [Retrieved from www.pathfastpublishing.com]

ILO (International Labour Organisation). 2000. "The World Cut Flower Industry: Trends and Prospects." Discussion paper. Geneva, Switzerland. [Retrieved from http://www.ilo.org/public/english/dialogue/sector/papers/ ctflower/139e1.htm]

ILRF (International Labor Rights Fund). 2003. "Codes of Conduct in the Cut-Flower Industry." Working paper. Washington, D.C. [Retrieved from http://www.laborrights.org/projects/women/Flower_Paper_0903/flower_ paper_index.htm]

IMF (International Monetary Fund). 2002. "Kenya: Selected Issues and Statistical Appendix." IMF Country Report 02/84. Washington, D.C.

IPC (Investment Promotion Centre). 2004. "Key sectors: Agriculture." [Retrieved from http://www.ipckenya.org/docs/keysectmain.htm]

ITC. 1997. *Cut Flowers: A Study of Major Markets*. International Trade Centre, UNCTAD/WTO.

Jaffee, Steven. 1995. "The Many Faces of Success: The Development of Kenyan Horticultural Exports." In Steven Jaffe and John Morton, eds., *Marketing Africa's High-Value Foods: Comparative Experiences of an Emergent Private Sector*. Dubuque, Iowa: Kendall/Hunt.

———. 2004. "High-Value Horticultural Exports: Government as Facilitator." Slide presentation. Global Integration and the New Trade Agenda for Africa, The World Bank Institute Course, Washington, D.C.

KFC (Kenya Flower Council). 2003. Codes of Practice for Flower Cultivation in Developing Countries. *AccessGuide*. Rotterdam, the Netherlands: Centre for the Promotion of Imports from Developing Countries. [Retrieved from http://www.cbi.nl/accessguide/]

————. 2004. [Retrieved from http://www.kenyaflowers.co.ke/]

Maharaj, Niala, and Gaston Dorren. 1995. *The Game of the Rose: The Third World in the Global Flower Trade*. Utrecht, The Netherlands: International Books for the Institute of Development Research in Amsterdam.

Maitha, J. K. 1976. "The Kenyan Economy." In Judith Heyer, J. K. Maitha, and W. M. Senga, eds., *Agricultural Development in Kenya: An Economic Assessment*. Nairobi: Oxford University Press.

McCulloch, Neil, and Masako Ota. 2002. "Export Horticulture and Poverty in Kenya." Institute of Development Studies Working Paper 174. Brighton, England.

New Agriculturist. 2001. "Kenyans Go for the Gold Standard." [Retrieved from http://www.new-ari.co.uk/01-5/focuson/focuson2.html]

Omniflora. 2004. "The Cut Flower Production in Kenya." [Retrieved from http://www.omniflora.com/e-ken.htm]

Opondo, Mary Magdalene. 2001. "Trade Policy in the Cut Flower Industry in Kenya." Globalisation and Poverty Research Programme, Institute of Development Studies Discussion Paper. University of Nairobi, Kenya. [Retrieved from http://www.gapresearch.org/governance/HORT1.pdf]

Pathfast Publishing. 2000. *Competitiveness: Cut Flower Production: High Tropics versus Temperate Climate*. Suffolk, United Kingdom: Pathfast Publishing. [Retrieved from http://www.pathfastpublishing.com/qr27/..%5CArchive02%5CCompetitiveness%202002.htm]

Pertwee, Jeremy. 2003. *International Cut Flower Manual 2003*. Suffolk, United Kingdom: Pathfast Publishing. [Retrieved from www.pathfastpublishing.com.]

Roozendaal, G. Van. 1994. "Kenyan Cut Flower Export Blooming." *Biotechnology and Development Monitor* 20: 6–7. [Retrieved from http://www.biotech-monitor.nl/new/index.php?link=publications]

Senga, W. M. 1976. "Kenya's Agricultural Sector." In Judith Heyer, J. K. Maitha, and W. M. Senga, eds., *Agricultural Development in Kenya: An Economic Assessment*. Nairobi: Oxford University Press.

Smith, L. D. 1976. "An Overview of Agricultural Development Policy." In Judith Heyer, J. K. Maitha, and W. M. Senga, eds., *Agricultural Development in Kenya: An Economic Assessment*. Nairobi: Oxford University Press.

Thoen, Ronaldt, Steven Jaffee, Catherine Dolan, and Fatoumata Ba. 1999. *Equatorial Rose: The Kenyan-European Cut Flower Supply Chain*. Washington, D.C.: World Bank.

UNSO/ITC Comtrade Databases. http://www.intracen.org/tirc/.

White, David, William Wallis, and Nicol Degli Innocenti. 2004. "FT Kenya: Special Report." *Financial Times* (U.K.) Suppl. April 6: 1–6.

World Bank. 1998. *The World Bank and the Agricultural Sector in Kenya: An OED Review*. Report Number 18088. Operations Evaluations Department, Washington, D.C.

World Bank. 2004. *Kenya Country Brief*, an internal report. Washington, D.C. [Retrieved from http://web.worldbank.org]

WTO (World Trade Organization). 2000. *Trade Policy Review: Kenya*. Press release, January 19. [Retrieved from http://www.wto.org/english/tratop_e/tpr_e/tp124_e.htm]

Index